T0336130

Plant Evolutionary Developmental Biology
The Evolvability of the Phenotype

Compared to animals, plants have been largely neglected in evolutionary developmental biology. Mainstream research has focused on developmental genetics, while a rich body of knowledge in comparative morphology is still to be exploited. No integrated account has been available.

In this volume, Alessandro Minelli fills this gap using the same approach he gave to animals, revisiting traditional concepts and providing an articulated analysis of genetic and molecular data. Topics covered include leaf complexity and the evolution of flower organs, handedness, branching patterns, flower symmetry and synorganization, as well as less conventional topics such as fractal patterns of plant organization. Also discussed is the hitherto neglected topic of the evolvability of temporal phenotypes, including a plant's annual, biennial or perennial life cycle, flowering time, and the timing of abscission of flower organs.

This will be informative reading for anyone in the field of plant evo-devo, from students to lecturers and researchers.

Professor of Zoology at the University of Padova until 2011, Alessandro Minelli is currently serving as Specialty Chief Editor for evolutionary developmental biology for *Frontiers in Ecology and Evolutionary Biology*. He was previously Vice-President of the European Society for Evolutionary Biology. For several years his research focus was in biological systematics, but in the mid-1990s he moved his interest towards evolutionary developmental biology, the subject of his previous book *The Development of Animal Form* (Cambridge, 2003). On his retirement, he decided to study plants and write the botanical equivalent of his book on animal evo-devo.

Plant Evolutionary Developmental Biology

The Evolvability of the Phenotype

ALESSANDRO MINELLI

University of Padova

With photographs by MARIA PIA MANNUCCI

CAMBRIDGE
UNIVERSITY PRESS

CAMBRIDGE
UNIVERSITY PRESS

University Printing House, Cambridge CB2 8BS, United Kingdom

One Liberty Plaza, 20th Floor, New York, NY 10006, USA

477 Williamstown Road, Port Melbourne, VIC 3207, Australia

314–321, 3rd Floor, Plot 3, Splendor Forum, Jasola District Centre, New Delhi – 110025, India

79 Anson Road, #06–04/06, Singapore 079906

Cambridge University Press is part of the University of Cambridge.

It furthers the University's mission by disseminating knowledge in the pursuit of education, learning, and research at the highest international levels of excellence.

www.cambridge.org
Information on this title: www.cambridge.org/9781107034921
DOI: 10.1017/9781139542364

© Cambridge University Press 2018

First published 2018

A catalogue record for this publication is available from the British Library.

ISBN 978-1-107-03492-1 Hardback

Contents

Colour plates can be found between pages 244 and 245

Preface

Amongst the most characteristic features of modern science are: first,
a variety of contradictory views on the same problems, secondly,
the conflict of these with views recently accepted, and thirdly,
the extravagance of the new conceptions.

B. Kozo Poljanski, 1936, p. 479

In the late 1960s I was a student in natural sciences at the University
of Padova. I learned then from my botany professor, Carlo Cappelletti,
that the leaves of a palm growing in our botanical garden had been a
major source of inspiration for a German traveller who saw it in 1786.
Eventually, this man, whose name was Johann Wolfgang von Goethe,
encapsulated the core idea of his theory of plant morphology in the
well-known aphorism, *alles ist Blatt* (all is leaf). Goethe's palm, as it is
universally known today, is currently the oldest inhabitant of the
Padova botanical garden, where it was planted in 1586.

In my student days, I was fascinated by the modularity of the
plant body, by the geometrical and arithmetical relationships among
its parts. I dreamed of finding something comparable in animals, in
arthropods in particular, which represented anyway my favourites
among living beings. Eventually, I became a zoologist and continued
studying arthropods – although, for many years, addressing questions
other than the development and evolution of their modular,
segmented organization. It was a comparative study of segment
number in centipedes, and the vague hope of eventually explaining
the idiosyncrasies of their distribution (for example, all adult
centipedes have an odd number of leg-bearing segments) in the light of
the then recent discoveries on *Drosophila* segmentation genes, that
fuelled my quick transfer to the discipline that was soon to get its

Goethe's palm, Padova botanical garden.

name (and the corresponding nickname) of evolutionary developmental biology (or evo-devo). That was indeed the title of a slim (in its first edition) but stimulating book published by Brian K. Hall in 1992. Of course, a number of facts and ideas eventually pertaining to the new discipline had been published before, even long before, if we accept the popular idea that Etienne Geoffroy Saint-Hilaire (1772–1844) should be regarded as a precursor of evo-devo; of more recent authors, I mention here Gavin de Beer and Stephen Jay Gould, for their pages on embryos and ancestors (see Chapter 1), and Rudy Raff and Thom Kaufman, whose book on *Embryos, Genes, and Evolution* (1983) was really, for many researchers of my generation, the most important source of inspiration to decide that the study of the relationships between development and evolution was worthy of becoming one's life research field.

Since it took its first steps, the new discipline of evo-devo has manifested two characteristic traits: a strong zoological bias and the dominant impact of comparative developmental genetics. This is not to say that plant evo-devo has been totally neglected, or that evolutionary biology and comparative morphology have failed to provide their contribution. However, animals and forms of genetic control still dominate the field – witness books, articles and talks at the congresses of the two societies born since then, the European Society for Evolutionary Developmental Biology, founded in 2006, and the Pan-American Society for Evolutionary Developmental Biology, founded in 2015.

With my book on *The Development of Animal Form*, published in 2003, I tried to offer a view of animal evo-devo complementary to the (at the time, and still today, although not to the same degree) dominant molecular perspective. Already at that time it was clear to me that plants deserved a parallel effort. Thus, when eventually, in July 2011, Dominic Lewis asked me if I was ready to write another book for Cambridge University Press, I asked him, in turn, if he was ready to support the metamorphosis of an evo-devoist trained in zoology to something like a plant evo-devoist. The fact that six years later I am finally delivering this work to the publisher demonstrates two important points: first, that commissioning editors can be much more flexible, and supportive, than the academic world usually is; second, that moving from animals to plants as a professional in evo-devo is a long process that generates its first fruits only after several years. Those I am offering here are quite probably unripe, but at any rate they will fill a previously empty slot on the (book) shelves, and will hopefully stimulate new research and eventually the production of a better synthesis of the field.

Here and there, I have allowed my familiarity with animals to intrude, at a very moderate level, but I have avoided attempting the systematic comparison between animals and plants that some readers might have expected to find here. For this decision, I owe the readers some justification (provided in Section 1.2), and this will explain why I do not offer any apology for this intentional omission.

The plants this book is about are only the angiosperms, the flowering plants. The main reason for this somewhat narrow circumscription is that an adventure into further groups, both extinct and extant, would have required an additional effort that I am not ready to undertake. In the following pages, I will mention mosses, lycophytes, ferns and gymnosperms in very few places.

The way the book is articulated into chapters is easily explained, except perhaps for Chapter 9.

In the first three chapters I give a conceptual introduction to plant evo-devo, providing definitions for concepts to some extent specific to this discipline (e.g. modularity, evolvability) and redefinitions of others. A few tools are provided too, including a refreshed outline of angiosperm classification and a list of model plant species, with some comments on problems with generalizing from results obtained on them.

Chapter 4, devoted to genes and genomes in so far as they are relevant to evo-devo, includes a section on the histories of duplication of a number of important gene families and remarks on subsequent functional specializations.

Chapters 5 to 7 deal with three main architectural components of the plant body: in Chapter 5, the shoot, its meristems, its branching patterns, and the positioning and differentiation of the lateral organs; in Chapter 6, the leaves; in Chapter 7, the flowers. Only short paragraphs are devoted to roots (at the end of Chapter 5) and fruits (at the end of Chapter 7), mirroring both my meagre familiarity with these subjects and the comparatively negligible attention these plant parts have received till now in evo-devo.

General aspects of plant body architecture are discussed in Chapter 8. These include some intriguing patterns such as paramorphism (strictly similar patterning expressed along body axes of different order) and the development and evolution of unconventional structures that defy classification into one of the standard categories of plant morphology.

Chapter 9 is devoted to pheno-evo-devo, i.e. to the evolutionary developmental biology of temporal phenotypes. In my view, indeed, temporal properties such as the order of inception of floral organs or the temporal patterns of abscission of leaves or flower parts must be regarded as phenotypes generated by development and subject to evolution, on a par with morphological phenotypes.

Evolutionary trends in body complexity, convergence and parallelism, evolutionary reversal and, more controversially, saltational evolution are the subject of Chapter 10.

The very short Chapter 11 summarizes a few key ideas presented in the book and suggests a few items for a future plant evo-devo agenda.

Acknowledgements, as usual, fill the last lines of this preface, but are the first in importance. My warmest thanks are for my wife, Maria Pia Mannucci, not only for her moral support over the years (especially in the crazy months in which I was struggling to digest a huge and poorly structured amount of information extracted from the literature) but also for offering her beautiful photographs to illustrate an otherwise arid text. All illustrations in this book are by her, except for the flower of *Passiflora incarnata* (Fig. 10.3) kindly provided by our friend and excellent nature photographer Ioannis Schinezos.

Most sincere thanks to Dominic Lewis, who quietly but firmly supported this book project from its inception, and to Jenny van der Meijden, also at Cambridge University Press, who provided sensible and timely advice on how to solve the many problems of which I became aware only when I wrongly thought I was at last ready to deliver my long-overdue manuscript within a couple of weeks. Equally worthy of the warmest thanks is Hugh Brazier, my excellent copy-editor, who smoothly but firmly rescued the text from hundreds of linguistic barbarisms and other unpleasant defects.

Other people are probably not aware of their contribution to this book and must therefore not be blamed for their indirect support of a project for which I retain fully responsibility: I mean the plant scientists with whom I have enjoyed exchanges of views on matters of development and evolution, without ever mentioning this book; in strict alphabetical order, Richard Bateman, Peter Endress, Angela

Hay, Rainer Melzer, Paula Rudall, Rolf Rutishauser, Günter Theißen and Miltos Tsiantis. Their real contribution to my project is in their precious books and articles, on which, together with those of many other scientists, I have heavily (and gratefully) relied, as documented by the bulky list of references.

I Introducing Plant Evo-Devo

I.I EVO, DEVO, EVO-DEVO — A TRADING PLACE, AT LEAST

Each of them in its way, both evolutionary biology and developmental biology, deals with change in living beings, and both of them are interested in phenotypes as well as in genotypes. Nevertheless, for many decades the dialogue between these two major branches of the life sciences was very limited, and a number of authoritative scientists explicitly stressed their autonomy. To evolutionary biologists, development was a conveniently closed black box linking together genes and morphology, and this was apparently all that evolutionary biologists needed to know about development, in order to safely proceed with their study of changes of frequencies of alleles and phenotypes in natural or lab populations, on the one side, and of adaptive changes in morphology, on the other. Developmental biologists, in turn, could apparently articulate their research agendas without asking questions about the origin (whatever this word could eventually mean) of the embryos, cells and molecules forming the subject of their study.

To be honest, in the first decades following the publication of *Origin of Species* (Darwin, 1859), the embryonic development of animals had been passionately investigated in search of suggestions to be used in reconstructing ancestors or, more reasonably, as a source of data to be exploited in establishing phylogenetic relationships. In the work of Ernst Haeckel (e.g. Haeckel, 1866), this expected link between individual development and the evolutionary history of the species provided the theoretical foundation for the so-called biogenetic law (*biogenetisches Grundgesetz*), according to which ontogeny recapitulates phylogeny.

These wide-ranging, but eventually problematic, speculations were largely confined to zoology, including animal palaeontology. On the one hand, botanists were generally cool about evolution; on the other, plants have nothing equivalent to the larval stages of many animals (think of a butterfly's caterpillar, or the trochophore of a polychaete), and even the equivalence between animal embryos and plant embryos is easily a matter of dispute.

Eventually, progress in both evolutionary and developmental biology was both the cause and the effect of a shifting away from largely abstract generalizations such as Haeckel's principle of recapitulation. The concept survived, to some extent, in the pages of textbooks, but was not the subject of new, critical studies – with two major exceptions. In a couple of small but pithy books, de Beer (1930, 1940) discussed the relationships between (animal) embryos and ancestors, to conclude that the addition, in the descendant, of a novel adult at the end of a modified ontogenetic schedule in which there is a recapitulation of the adult stages of its ancestors is only one of the possible ways in which development can be modified in the course of evolution. de Beer also provided terms and definitions for a number of alternative patterns of change of developmental schedules, focusing on the timing of the ontogenetic progression towards sexual maturity, relative to somatic development. In this way, de Beer set the foundation for subsequent approaches to the evolution of development in terms of heterochrony (see Section 9.11). Renewed interest in the relationships between development and evolution exploded, however, only 3–4 decades later. With his book on *Ontogeny and Phylogeny*, Gould (1977) offered a detailed historical account of the subject, in which, elaborating on de Beer's work, he developed an articulated classification of heterochronic patterns. However, this effort was still not enough to stimulate (and legitimize) an integrated approach to developmental evolution. The turning point was provided by the advent of developmental genetics, the biological discipline which was able, at last, to open the black box by which the link between genes and phenotypes had until then been represented.

This does not mean that the relationships between genes and phenotypes are either simple or unequivocally predictable. Indeed, the complexity of the genotype → phenotype map underlies much of what is described and discussed in this book. However, with the advent of developmental genetics it was possible to perform a genetic dissection of developmental mechanisms and to compare the results between any desired pair of species – a true revolution, compared to the recent past, when experiments limited to hybridization and mutagenesis restricted the inferences about the developmental role of a specific gene to comparisons within a species or, at most, between very closely related species.

Thus, in a sense, evolutionary developmental biology entered the stage through the door of comparative developmental genetics. Without the sweeping technical progress that set in motion a revolution in this field starting from the late 1970s, we would be hardly further than this science was around the mid-twentieth century, when Huxley (1942) argued about the possible role in evolution of what he termed 'rate genes', that is, genes affecting the rate at which individual developmental processes may progress – eventually, a gene-level explanation of heterochronic patterns.

However, comparative developmental genetics is only one of the roots of evolutionary developmental biology. Nor is it just the sum of developmental genetics plus (Neodarwinian, perhaps) evolutionary biology. Evolutionary developmental biology, or evo-devo (from here on let's use the now well-established nickname), would not deserve to be described as a distinct branch of the life sciences, if its research agenda were not characterized by an original programme, based on novel concepts and addressing questions that neither of the parent disciplines could answer satisfactorily.

What then is evo-devo? Evo-devo is the study of the arrival of the fittest, complementary to the study of the survival of the fittest suggested by the traditional approach to evolution based on the Darwinian principle of natural selection. This is not to dismiss the importance of selection in evolution. The problem is that selection can only test, so to speak, the fitness of those phenotypes that happened to be produced, but it does not say if and how any hitherto unseen phenotype could eventually be produced. That is, it does not predict, from a developmental rather than an evolutionary point of view, the likely scenarios of change: in other words, what is the *evolvability* of a specific, existing (and, as a rule, positively selected) phenotype.

Development and evolution are inextricably linked together but, taken separately, neither developmental biology nor the study of evolution can provide a satisfactory account of the history of life. Despite millions of generations of purifying selection, development continues to generate, besides viable, 'hopeful' products, also innumerable 'hopeless' phenotypes, including cancer. On the other hand, we should try to explain why selection for a longer neck has failed to produce giraffes with more than the seven cervical vertebrae found in other mammals; or why no adult centipede has an even number of leg pairs (Minelli, 2009a); or why, to give at last a couple of botanical examples, no plant species is known to have bi- or tripalmate leaves, whereas other, only apparently equivalent patterns, i.e. bi- or tripinnate leaves, are found in many different species, including carrot and tomato; or why, among the huge number of mutants obtained in the model species *Arabidopsis thaliana*, not a single case of compound leaves is known in this plant, while this is the wild-type phenotype in so many of *A. thaliana*'s closest relatives (Efroni *et al.*, 2010).

Thus, let's explore here *evolvability*, 'the real focus of evo-devo', as poignantly defined by Hendrikse *et al.* (2007).

Until now, evo-devo has largely focused on animals, rather than plants. The excellent but still fragmentary evidence available to date for plants (leaves and flowers especially) is reviewed in the following chapters, alongside a large amount of facts, mainly from plant comparative morphology, which are ready for a broad, although preliminary, systematization and to be used to fuel progress in plant evo-devo.

I hope I will be able to show, in this book, how amazingly broad is the potential scope of evo-devo. Indeed, I will suggest that it extends into areas that have been hitherto largely overlooked, in both animal and plant evo-devo. This is probably due not so much to an intentionally different circumscription of the research agenda of this biological discipline, but rather to a twofold historical bias, deeply rooted in the divergent traditions of evolutionary biology and developmental biology.

As noted by Love (2010), as soon as these two disciplines try to establish a dialogue, a tension emerges as a consequence of their opposite attitudes towards variation. To the developmental biologist, variation is mainly noise, an obstacle to the use of the standard tables of ontogenetic stages painstakingly compiled for a number of model organisms, and the potential cause of difficulties in obtaining consistent results from replicated experiments. On the other hand, variation, to the extent that it has at least some level of heritability, is the core subject matter of evolutionary biology. An evo-devo approach should ideally include a population-level study of variation of some of the genes targeted by developmental genetics. This is perhaps one of the very few aspects in which plant evo-devo is possibly more advanced than its animal counterpart – but potentially useful evidence is scattered across the full range of primary and applied plant science, and to review it is beyond my current means and scope.

Additionally, as noted by Mestek Boukhibar and Barkoulas (2016), phenotypic heterogeneity is abundant even within genetically identical individuals. In particular, considerable variability in growth has been found in the leaf epidermis (Elsner *et al.*, 2012) and the shoot apical meristem (Kierzkowski *et al.*, 2012; Uyttewaal *et al.*, 2012).

A second difficulty in furthering an evo-devo approach to living systems is the widespread obsession with adaptation. Consider the traditional contrast, long championed for example by Ernst Mayr (e.g. Mayr, 1982), between functional (including developmental) biology and historical (i.e. evolutionary) biology. As a science of proximate causes, functional biology is called upon to explain how things work: specifically, in the case of developmental biology, how an organism, or a part of an organism, is built. In principle at least, this excludes any consideration of adaptation – which is exactly what evolutionary biology is about: to explain the selective advantage of the phenotypes produced by evolution compared to alternative phenotypes over which they managed to be more successful.

In practice, however, the relationships between Mayr's two biologies are not so simple – better, not so static. Eventually, all developmental systems have been produced by evolution, and all phenotypes on which natural selection operates (but also, by the way, those that evolve by genetic drift) are the products of development. Unique to evo-devo is the effort to make these links between development and evolution explicit, and to explore in which respects either area of biology needs a conceptual revisitation, precisely because of a mostly overlooked 'contamination' by the other. Winther (2015) went so far as to describe evo-devo as a 'trading zone' where a variety of disciplines, styles and paradigms are actively negotiating with one another.

In the end, this refreshed perspective can be of major consequence for both parent sciences. The advent of evo-devo has been seen by a number of biologists and philosophers of biology as one of the strongest reasons for advocating the advent of an Extended Evolutionary Synthesis, in which selection is only one of a number of forces responsible for biological evolution; the interested reader can start with the paper by Laland *et al.* (2015), or the collection of essays edited by Pigliucci and Müller (2010). However, it is from developmental biology that a major conceptual revisitation is needed.

Let's start with adaptation. In the conventional splitting of responsibilities between proximate-causes and remote-causes biology, adaptation is apparently left of the shoulder of the latter. However, many biologists are arguably reluctant to accept processes leading to non-adaptive results as legitimately belonging to developmental biology. The latter is about how a fertilized egg eventually develops into an adult animal, how a plant produces flowers, how a seed germinates, how a caterpillar metamorphoses into a butterfly. What else should belong to developmental biology? In fact, a lot of processes, including some with which the developmental biologist is obviously familiar. What would we know about normal (i.e. adaptive) development, were it not for studies on 'wrong' ontogenetic schedules that sooner or later result in hopeless phenotypes, the 'monsters' of old literature, ennobled today by the Greek name of terata, which clearly do not help them to achieve an adaptive value? The study of maladaptive development is even important per se, irrespective of its contribution to understanding normal development – when it comes to cancer, for example, or to the developmental response to the attack of pathogens. A peculiar kind of developmental process, in which plants are involved in consequence of interactions with other organisms, is represented by galls, especially those induced by insects. Irrespective of their adaptive significance for either the plant or what is usually known as their causal agent, galls should be studied as biological systems with their own, specific development, often culminating in phenotypes no less regular or recognizable than are fungal carpophores and many other biological structures.

In the pages that follow, I have tried to avoid as far as possible any consideration of the adaptive significance of plant phenotypes, e.g. leaf shape and complexity, flower symmetry, temporal patterns of abscission or sexual identity of the individual flowers within the inflorescence. Thus I invite the reader to focus on the arrival of the fittest rather than on the survival of the fittest (de Vries, 1904; Wagner, 2014), faithful to evo-devo's characteristic focus on evolvability.

Nevertheless, adaptation is always lurking around the corner, and in some instances it will necessarily resurface: see Section 1.6.

I.2   PHYLO-EVO-DEVO

To date, evo-devo has been dominated by developmental genetics: this is not necessarily a hindrance, as long as sufficient attention is paid to some fundamental aspects of evolution. Unfortunately, there is an aspect of evolution which has been often overlooked, in animal and plant evo-devo alike. This aspect is phylogeny. If we fail to follow the basic precepts of tree thinking, we can be easily led astray.

Classifying plant species based on the involvement of different genes in the control cascades responsible for perianth differentiation is not necessarily more informative than classifying the same species based on purely morphological criteria, including phyllotactic patterns or perianth complexity. To some extent, developmental genetics can contribute information to unravelling phylogenetic relationships, as morphology also can, although the single most solid line of evidence is now acknowledged to rest with comparisons of sequences of a suitably informative sample of genes. Whatever the data, however, the main point with phylogeny is that the position of a species in the tree is not dictated by its overall similarity to those on the closest branches, but based on the identification of synapomorphies, the derived features shared by the lineages emerging from the same node in the phylogenetic tree. A second point to be kept in mind is that phylogenetic relationships are, in a sense, pure topology. In other terms, no qualitative character, such as 'lower' and 'higher', should be given to two branches, based on the order in which they separate from the rest of the tree.

Nevertheless, some taxa are more conservative than others, although this does not necessarily involve all aspects of their morphology, of their genome, of their development. Moreover, the sequential arrangement of the lines on a page and of the pages in a book put a possibly unfortunate constraint on the way we can explore the phylogenetic tree in discussing an aspect of evolution, be it morphological, developmental, or other. This constraint is the linear development of the discourse, because of which examples and comparisons can only be made by cutting the tree into pieces and listing some of them, as required in any specific context, one after the other. In this way, it is seriously difficult to avoid being trapped by some form of *scala naturae* metaphor (from lower to higher), despite all the efforts of evolutionary biology to dispose of this old and still deeply entrenched frame of mind. I hope that this risk can be reduced to a minimum by making it explicit, and by listing here, in this introductory chapter of the book, the main groups of angiosperm families repeatedly mentioned throughout the book. A more detailed excerpt from currently accepted angiosperm classification is given in Table 1.1.

Table 1.1 *An outline of the classification of flowering plants according to APG IV (2016) (reproduced in modified form with permission of Oxford University Press). This excerpt is limited to the 269 families, out the 416 listed in that work, that are mentioned in this book, including the taxonomic index.*

AMBORELLALES
   Amborellaceae
NYMPHAEALES
   Hydatellaceae
   Cabombaceae
   Nymphaeaceae
AUSTROBAILEYALES
   Austrobaileyaceae
   Trimeniaceae
   Schisandraceae
MESANGIOSPERMS
MAGNOLIIDS
     CANELLALES
       Canellaceae
       Winteraceae
     PIPERALES
       Saururaceae
       Piperaceae
       Aristolochiaceae
     MAGNOLIALES
       Myristicaceae
       Magnoliaceae
       Degeneriaceae
       Himantandraceae
       Eupomatiaceae
       Annonaceae
     LAURALES
       Calycanthaceae
       Siparunaceae
       Atherospermataceae
       Hernandiaceae
       Monimiaceae
       Lauraceae

Table 1.1 (*cont.*)

CHLORANTHALES
 Chloranthaceae
MONOCOTS
 ACORALES
  Acoraceae
 ALISMATALES
  Araceae
  Tofieldiaceae
  Alismataceae
  Butomaceae
  Hydrocharitaceae
  Aponogetonaceae
  Zosteraceae
  Potamogetonaceae
  Posidoniaceae
  Ruppiaceae
  Cymodoceaceae
 PETROSAVIALES
  Petrosaviaceae
 DIOSCOREALES
  'Burmanniaceae', incl. Thismiaceae
  Dioscoreaceae
 PANDANALES
  Triuridaceae
  Velloziaceae
  Stemonaceae
  Cyclanthaceae
  Pandanaceae
 LILIALES
  Melanthiaceae
  Alstroemeriaceae
  Colchicaceae
  Smilacaceae
  Liliaceae
  Corsiaceae
 ASPARAGALES
  Orchidaceae
  Hypoxidaceae

Table 1.1 (*cont.*)

Iridaceae
Tecophilaeaceae
Asphodelaceae
Amaryllidaceae
Asparagaceae
ARECALES
Arecaceae
Dasypogonaceae
COMMELINALES
Commelinaceae
Philydraceae
Pontederiaceae
Haemodoraceae
ZINGIBERALES
Strelitziaceae
Heliconiaceae
Musaceae
Cannaceae
Marantaceae
Costaceae
Zingiberaceae
POALES
Typhaceae
Bromeliaceae
Juncaceae
Cyperaceae
Restionaceae
Flagellariaceae
Joinvilleaceae
Ecdeiocoleaceae
Poaceae
CERATOPHYLLALES
Ceratophyllaceae
EUDICOTS
RANUNCULALES
Eupteleaceae
Papaveraceae
Lardizabalaceae

Table 1.1 (*cont.*)

Menispermaceae
Berberidaceae
Ranunculaceae
PROTEALES
Sabiaceae
Nelumbonaceae
Platanaceae
Proteaceae
TROCHODENDRALES
Trochodendraceae
BUXALES
Buxaceae
CORE EUDICOTS
GUNNERALES
Myrothamnaceae
Gunneraceae
DILLENIALES
Dilleniaceae
SUPERROSIDS
SAXIFRAGALES
Paeoniaceae
Hamamelidaceae
Grossulariaceae
Saxifragaceae
Crassulaceae
Penthoraceae
Haloragaceae
Cynomoriaceae
VITALES
Vitaceae
ZYGOPHYLLALES
Krameriaceae
Zygophyllaceae
FABALES
Fabaceae
Polygalaceae
ROSALES
Rosaceae

Table 1.1 (*cont.*)

Barbeyaceae
Dirachmaceae
Rhamnaceae
Ulmaceae
Cannabaceae
Moraceae
Urticaceae
FAGALES
Fagaceae
Myricaceae
Juglandaceae
Betulaceae
CUCURBITALES
Apodanthaceae
Anisophylleaceae
Cucurbitaceae
Datiscaceae
Begoniaceae
CELASTRALES
Celastraceae
OXALIDALES
Connaraceae
Oxalidaceae
Cunoniaceae
Elaeocarpaceae
Cephalotaceae
Brunelliaceae
MALPIGHIALES
Rhizophoraceae
Ochnaceae
Clusiaceae
Calophyllaceae
Podostemaceae
Hypericaceae
Caryocaraceae
Lophopyxidaceae
Putranjivaceae
Elatinaceae

Table 1.1 (*cont.*)

Malpighiaceae
Balanopaceae
Dichapetalaceae
Chrysobalanaceae
Achariaceae
Violaceae
Passifloraceae
Lacistemataceae
Salicaceae
Rafflesiaceae
Euphorbiaceae
Linaceae
Picrodendraceae
Phyllanthaceae
GERANIALES
Geraniaceae
MYRTALES
Lythraceae
Onagraceae
Vochysiaceae
Myrtaceae
Melastomataceae
Crypteroniaceae
Alzateaceae
Penaeaceae
CROSSOSOMATALES
Aphloiaceae
Stachyuraceae
Crossosomataceae
PICRAMNIALES
Picramniaceae
HUERTEALES
Tapisciaceae
SAPINDALES
Kirkiaceae
Burseraceae
Anacardiaceae
Sapindaceae

Table 1.1 (*cont.*)

Rutaceae
Simaroubaceae
Meliaceae
MALVALES
  Cytinaceae
  Muntingiaceae
  Malvaceae
  Thymelaeaceae
  Bixaceae
  Cistaceae
  Dipterocarpaceae
BRASSICALES
  Tropaeolaceae
  Moringaceae
  Caricaceae
  Limnanthaceae
  Setchellanthaceae
  Resedaceae
  Capparaceae
  Cleomaceae
  Brassicaceae
SUPERASTERIDS
BERBERIDOPSIDALES
  Aextoxicaceae
  Berberidopsidaceae
SANTALALES
  'Olacaceae'
  Opiliaceae
  Balanophoraceae
  'Santalaceae'
  Misodendraceae
  Loranthaceae
CARYOPHYLLALES
  Tamaricaceae
  Plumbaginaceae
  Polygonaceae
  Droseraceae
  Nepenthaceae

Table 1.1 *(cont.)*

Simmondsiaceae
Caryophyllaceae
Amaranthaceae
Aizoaceae
Nyctaginaceae
Basellaceae
Portulacaceae
Cactaceae
ASTERIDS
  CORNALES
    Nyssaceae
    Hydrangeaceae
    Loasaceae
    Cornaceae
  ERICALES
    Balsaminaceae
    Marcgraviaceae
    Fouquieriaceae
    Polemoniaceae
    Lecythidaceae
    Sladeniaceae
    Sapotaceae
    Pentaphylacaceae
    Ebenaceae
    Primulaceae
    Theaceae
    Styracaceae
    Sarraceniaceae
    Actinidiaceae
    Clethraceae
    Cyrillaceae
    Ericaceae
    Mitrastemonaceae
  GARRYALES
    Garryaceae
  GENTIANALES
    Rubiaceae
    Gentianaceae

Table 1.1 (cont.)

Apocynaceae
BORAGINALES
Boraginaceae
SOLANALES
Convolvulaceae
Solanaceae
LAMIALES
Oleaceae
Tetrachondraceae
Gesneriaceae
Plantaginaceae
Scrophulariaceae
Stilbaceae
Linderniaceae
Byblidaceae
Acanthaceae
Bignoniaceae
Lentibulariaceae
Verbenaceae
Lamiaceae
Phrymaceae
Orobanchaceae
ASTERALES
Campanulaceae
Phellinaceae
Menyanthaceae
Goodeniaceae
Asteraceae
ESCALLONIALES
Escalloniaceae
DIPSACALES
Adoxaceae
Caprifoliaceae
APIALES
Araliaceae
Apiaceae

Named after the acronym of its component orders, the set of angiosperm taxa to which I will generally refer first, whenever relevant, is (i) the ANA group (Amborellales, Nymphaeales, Austrobaileyales), followed by (ii) magnoliids, of which I will more frequently mention the following families: Winteraceae, Piperaceae, Aristolochiaceae, Magnoliaceae, Degeneriaceae, Eupomatiaceae, Annonaceae, Calycanthaceae, Lauraceae and Chloranthaceae.

The next major clade is represented by (iii) the monocotyledons, or monocots, which include, among others, the Araceae, Alismataceae, Dioscoreaceae, Triuridaceae, Liliaceae, Orchidaceae, Iridaceae, Asphodelaceae, Amaryllidaceae, Asparagaceae, Arecaceae, Commelinaceae, Pontederiaceae, Musaceae, Zingiberaceae, Bromeliaceae, Juncaceae, Cyperaceae and Poaceae.

Monocots are a well-circumscribed taxon which has survived without modifications in its circumscription the momentous revolution that progressively dismantled the dicotyledons (dicots).

In the old classifications, the bulk of dicots was represented by the large clade now renamed eudicots, but also the families of the ANA group and the magnoliids, plus (iv) the Ceratophyllaceae, which are closer to the eudicots than are the monocots, but are currently recognized as the sister group to the eudicots. Within the latter, a group of orders is (sadly) known as (v) the 'lower' eudicots, to contrast it with the much more species-rich 'core' eudicots. Lower eudicots have attracted well-deserved special attention from evo-devo researchers – witness the number and importance of the studies, frequently cited in the following pages, on representatives of the Papaveraceae, Berberidaceae, Ranunculaceae and Proteaceae; other families mentioned in this book are Sabiaceae, Nelumbonaceae, Platanaceae, Trochodendraceae and Buxaceae.

Finally, the core eudicots, at the base of which (vi) a small number of families such as Gunneraceae and Dilleniaceae occupy a kind of pivotal position, are divided into two very large groups, the superrosids and the superasterids. To help readers less familiar with the current classifications of plants, I list here a selection of families belonging to each clade.

(vii) Superrosids: Saxifragaceae, Crassulaceae, Vitaceae, Fabaceae, Rosaceae, Moraceae, Urticaceae, Fagaceae, Betulaceae, Cucurbitaceae, Begoniaceae, Podostemaceae, Hypericaceae, Malpighiaceae, Violaceae, Passifloraceae, Salicaceae, Rafflesiaceae, Euphorbiaceae, Linaceae, Geraniaceae, Lythraceae, Myrtaceae, Sapindaceae, Rutaceae, Meliaceae, Malvaceae, Brassicaceae.

(viii) Superasterids: Berberidopsidaceae, Loranthaceae, Nepenthaceae, Caryophyllaceae, Amaranthaceae, Cactaceae, Hydrangeaceae, Cornaceae, Primulaceae, Theaceae, Ericaceae, Rubiaceae, Gentianaceae, Apocynaceae, Boraginaceae, Convolvulaceae, Solanaceae, Oleaceae, Gesneriaceae, Plantaginaceae, Scrophulariaceae, Acanthaceae, Bignoniaceae, Verbenaceae, Lamiaceae, Orobanchaceae, Campanulaceae, Asteraceae, Caprifoliaceae, Araliaceae, Apiaceae.

On a much broader scale, sound tree thinking will easily cancel from the evo-devo research agenda any question framed in terms of comparisons of animals versus plants, except for problems of convergent (exceptionally, parallel) evolution. The fact that convergent or parallel traits cannot contribute to reconstructing phylogenetic relationships must not deter us from addressing them in an evolutionary context (see Section 10.2). But animals and plants evolved from different, very distantly related unicellular ancestors (e.g. Cavalier-Smith *et al.*, 2014), not to mention that angiosperms (or, by the way, even embryophytes, or tracheophytes) are only a branch, although a very large and successful one, of a larger clade in which multicellularity was acquired before the emergence of land plants; therefore, all aspects of multicellular organization of angiosperms (or embryophytes, or tracheophytes) and metazoans belong to fully independent evolutionary histories.

1.3 RETHINKING DEVO IN THE LIGHT OF EVO

1.3.1 *What is Development?*

Broad and important biological notions, derived from pre-scientific concepts rooted in the human experience with a limited number of common species, mainly vertebrates and plants, are likely to retain conceptual associations that deserve critical reassessment. This is true of development (both animal, including human, and plant), which is commonly identified with the functional changes that accompany an individual's life throughout its predictable trajectory from a seed (or an egg) to an adult organism. In this traditional and naive view, there are a number of assumptions that must be revisited. These issues belong to the intellectual area where biology overlaps with the philosophy of biology and eventually lead to the contentious question, whether developmental biology can (or should) have its theory, or perhaps several theories (similar to the conceptual foundations of evolutionary biology), and, if so, what kind of theory, or theories (Minelli and Pradeu, 2014).

I am not suggesting that we have to look for a novel definition of development. I think indeed that it is quite adequate to define it as temporal change of organization along the life cycle (Minelli, 2003), although admitting that some problems remain, but alternative definitions are not necessarily better (Pradeu *et al.*, 2016). My main point here is that the notion of development implicitly taken for granted in evo-devo is, with very rare exceptions, considerably more restrictive and essentially the same as the popular concept of development as the adaptive process through which, along a series of predictable changes perhaps controlled by a genetic program, a germ (seed, egg) is turned into a reproductive mature adult plant, or animal. The most serious shortcomings of this notion are briefly discussed in the following paragraphs.

First, is development necessarily adaptive? It is certainly not so (see also Minelli, 2014) – otherwise, what could we learn about development from morphologically seriously compromised and eventually sterile mutants? Conversely, how could we describe and interpret the proliferation of cancer, or cells in culture, if not in terms of development? If we want to restrict development to adaptive processes only, how confidently would we extrapolate from the study of mutants, chimaeras etc.?

Standard development is not necessarily optimal in every respect. For example, grazed plants of *Anthyllis cytisoides* are developmentally more stable than untouched ones (Alados *et al.*, 1999). These non-adaptive aspects of development may have their counterpart in the evolution of the underlying genetic control. Townsley and

Sinha (2012) have remarked that changes in the wiring of gene regulatory networks and, in particular, of those regulating aspects of leaf development, are, in large part, non-adaptive. These authors tentatively explain this fact in the framework of the developmental system drift (DSD) paradigm (True and Haag, 2001). This suggests that non-adaptive changes in regulatory connections within a gene regulatory network accumulate over time in a stochastic way and in the absence of evolutionary pressure. This generates diversity in the regulatory wiring underlying developmental processes independent of selection. This hypothesis has been elegantly demonstrated by a comparative study of 18 494 gene transcripts in non-senescent leaves of 14 accessions representing 11 species in 6 genera of Brassicaceae, plus *Cleome pinnata* as an outgroup (Broadley *et al.*, 2008). A positive correlation was found between transcriptome divergence and phylogenetic distance between taxa, and similar results were obtained in comparisons involving pseudogenes (by definition, not subject to selection) or chloroplast genes evolving at different rates. These results are a clear suggestion that many differences in gene expression may have no adaptive significance,

A second point is that we must distinguish between development as a process (actually, a set of processes) and development as an arbitrarily delimited sequence or pattern of change. Asking whether development begins with the seed, or the egg, is a question of *developmental pattern* rather than one of *development as process*. The same applies to regeneration. We can study developmental processes involved in regeneration, and compare them to those occurring along the individual development of an uninjured individual, but it would be inadequate to regard regeneration, as such, as a developmental process. Some forms of plant regeneration are based on the activation of relatively undifferentiated cells in somatic tissues, while in others there is dedifferentiation and redifferentiation of somatic cells (Ikeuchi *et al.*, 2016). Due to their modular organization, it may be debatable if in plants the concept of regeneration can be applied with the same meaning as in animals, but this is not relevant in our context.

A third point concerns the temporal framework of individual development. Following the series of events through which a flowering and fruiting plant emerges from the seed, from its initial stages of vegetative growth to the flowering transition and beyond, is arguably the single best way to organize knowledge about plant development, but we must be aware that this optimal temporal framework applies to development as pattern, not to development as process. As a consequence, development does not necessarily begin with a seed, with the further (and more important) implication that development should not be interpreted as the deployment of the genetic information encoded in the zygotic nucleus, ready to be used to build a new individual!

If development does not necessarily begin with a seed, it also does not necessarily end with an adult plant, what is otherwise commonly taken to be the end (teleologically and temporally alike) of the individual's life. Admittedly, this adultocentric perspective is much more obsessively entrenched in respect to animal, rather than plant development (Minelli, 2003). At least, the frequent escapes of plants towards vegetative reproduction provide a sobering message in this respect.

The compatibility of one genome with two very different developmental schedules (one linked to sexual reproduction and thus going through the bottleneck of meiosis, and also encompassing embryonic development; the other bypassing meiosis, syngamy and the embryonic phase) is also a good reason to take distance from the popular metaphor of the genetic program. To be sure, there are also other reasons – think for example of phenotypic plasticity (see Section 2.5). But even within the extent to which we can acknowledge a straight control of the genotype on the production of a phenotypic trait, we must be aware of the fact that this genotype → phenotype relationship is often just a frozen accident of evolutionary history, as demonstrated by the following example. In different plants, the same gene can act as a marker of the adaxial domain of a lateral organ (that is, for example, the side of a flower which is closer to the plant's axis), or as a marker of the abaxial domain (the side of a flower which is most distant from the plant's axis). In eudicots, the bilateral symmetry of flowers such as those of *Antirrhinum* is largely controlled by the *CYCLOIDEA2* (*CYC2*) genes, which are expressed asymmetrically, usually in the adaxial part of developing flowers, except for the Asteraceae, where the same genes are instead expressed ventrally (Broholm *et al.*, 2008; Juntheikki-Palovaara *et al.*, 2014; Nicolas and Cubas, 2016). Similarly, the expression of the *YABBY* (*YAB*) genes *YAB1* and *YAB3* marks the abaxial domain (and the marginal regions) of leaf primordia in *Arabidopsis* (Sawa *et al.*, 1999a; Siegfried *et al.*, 1999); the homologue *GRAMINIFOLIA* (*GRAM*) has the same spatial pattern of expression in *Antirrhinum* (Golz *et al.*, 2004) and, more generally, the abaxial expression pattern of this gene is regarded as ancestral in angiosperms (Yamada *et al.*, 2011; Fukushima and Hasebe, 2014). However, in *Zea mays* the *YABBY* orthologue shows adaxial expression (Juarez *et al.*, 2004) and the *Oryza sativa* homologue has uniform expression throughout the primordium (Dai *et al.*, 2007).

Before leaving the issue of how to circumscribe development, a further, virtually unexplored point must be mentioned, represented by the scaffolding relationships, temporary or permanent, between the living parts of an organism (i.e. the organism in the proper sense) and dead or mineralized structures produced by it along its life cycle and representing an integral part of its developmental system (Minelli, 2016a). In plants, this applies especially to vessels, wood generally, bark and dry fruit parts, all of which, although representing a dead leftover of the metabolically active parts of the plant, are nevertheless an important scaffold to its development and are morphologically and functionally integrated with its living cells.

1.3.2 A Process View of Development

In the early years of population genetics, a key step forward was taken when a null model, the Hardy–Weinberg equilibrium, was introduced that fixed conditions for the absence of change in a population. With reference to that model, it was eventually possible to disentangle a number of factors, including mutation, natural selection and assortative mating, that cause deviation from the equilibrium and eventually, in the long term, evolution. I suggested elsewhere (Minelli, 2011) that developmental biology would benefit from similarly adopting a *principle of developmental inertia*. As noted by Griesemer (2014), it should be possible to define such a principle in adequately general terms, without explicit reference to organisms with cellular organization; nevertheless, from an operational, rather than philosophical, point of view, we may arguably be content with defining the inertial (null) dynamics of development in terms of pure growth, uniform in space and time, as an indeterminate iteration of cell proliferation.

Similar to evolutionary biology in respect to the Hardy–Weinberg principle, developmental biology is actually interested in the factors that imply deviations from the idealized null (inertial) behaviour, including positional effects, cell–cell competition and, of course, differential gene expression and its short- and long-range effects.

I.4 THE INDIVIDUAL

We can address many interesting questions of plant developmental biology without any need to make clear what we mean by 'individual plant', but this is only true of 'local' development, e.g. the differentiation of perianth organs, or the control of leaf shape. However, the problem of individuality cannot be eluded forever – but this does not mean that the problem has an obvious solution.

As argued by Pradeu (2016), the main reason behind the traditional failure is the fact that under the current concept of *biological individuality* two different and only partly overlapping notions are included: *physiological individuality* (as determined, for example, by the physical separation in respect to the other individuals or by the genetic or immunological uniqueness) and *evolutionary individuality*. In turn, the latter can be intended in two different ways, (1) by equating the individual to the interactor sensu Dawkins (1976), i.e. to each entity which may undergo natural selection, or (2) in terms of Darwinian individuals. These were defined by Godfrey-Smith (2009) as members of a Darwinian population, which is itself a population of objects characterized by variation, heritability and differential fitness. Therefore, an individual should be understood as a *reproductive unit*. A further approach to the problem, targeted in any case to a new concept of evolutionary individuality, has been proposed by Clarke (2011).

However, the problem remains, that, whatever conception of evolutionary individuals one adopts, the categories of physiological individuals and evolutionary individuals do not necessarily coincide (Pradeu, 2016).

It may thus be sensible to abandon further efforts to find a unified concept of biological individuality, to admit instead that different criteria unavoidably circumscribe different units in the living world, and to focus here, in the context of developmental biology, on the possible concepts of physiological individuality. A novel criterion of physiological individuality, based on immunology, has been defended by Pradeu (2012, 2016). Its background is that all living things have some form of immune system, including plants (e.g. Nürnberger and Brunner, 2002; Tsuda *et al.*, 2009; Dodds and Rathjen, 2010; Tsuda and Katagiri, 2010). A definition based on immunological properties would satisfy at least three criteria of individuality (Pradeu 2012): uniqueness (what makes each biological individual unique?), delineation (what are the boundaries of a biological individual?) and persistence (what ensures that something remains the 'same' biological individual despite constant change?). This approach deserves a deeper analysis than I can offer here. For most of what I will discuss in this book, it is generally sufficient to acknowledge the duality between the *genet* and the *ramet* (Harper and White, 1974), i.e. between the physiological individual based on genetic uniformity (thus including whole clones, irrespective of the possible absence of physical and physiological continuity of their parts), and the physiological individual based on physical and physiological independence.

I.5   GAMETOPHYTE VERSUS SPOROPHYTE

In all tracheophytes (vascular plants), a complex morphology is associated with the diploid, sporophytic phase of the life cycle, whereas the haploid gametophyte is much less conspicuous and of much simpler architecture than the corresponding sporophyte, even in the taxa in which it is not reduced to very few cells (or nuclei), as in the angiosperms. This association between morphological complexity and ploidy is so strict, that it may be sensible to ask whether it is causally constrained, or simply (or mainly) represents instead one of the many frozen accidents of biological history to which evo-devo is increasingly opening our eyes (for an excellent collection of examples concerning our own species, see Held, 2009).

Looking beyond the tracheophytes, but still within the embryophytes, we find that complex phenotypes can evolve also in association with haploidy, e.g. in the case of moss and liverwort gametophytes. In a very distant lineage (Heterokonta), in which multicellularity evolved independently from plants (Viridiplantae, or green plants), the brown alga *Ectocarpus* shows that the haploid gametophyte and the diploid sporophyte of the same organism can be identical, with a phenotype of some complexity. Therefore, the association between ploidy and structural complexity does not seem to be obligate.

It is therefore not so easy to explain the extremely rare occurrence of haploid sporophytes in the tracheophytes. Rarity, however, does not mean absolute lack. In the ferns, some examples of haploid individuals with morphology corresponding to the sporophyte of the species have been repeatedly found in nature, and are interpreted as produced by apogamy (asexual reproduction) from otherwise normal gametophytes (prothalli) (Sheffield and Bell, 1987).

In angiosperms, haploid individuals have been occasionally reported for a number of species, a detailed list of which was provided by Dunwell (2010). Haploid plants have been also produced in the lab, either from cultured gametophytic cells (a first example in Guha and Maheshwari, 1964) or by elimination of one of the parental genomes following an interspecific cross (early examples in Clausen and Mann, 1924; Kasha and Kao, 1970; Barclay, 1975). More recently, Ravi and Chan (2010) have shown that haploid *Arabidopsis thaliana* plants can be obtained by manipulating the centromere-specific histone CENH3: in hybrids obtained by crossing a wild type with a *cenh3*-null mutant, the chromosomes from the latter parent are eliminated. A problem with the haploid plants issued in these ways is their sterility, but they can spontaneously convert into fertile diploids through meiotic non-reduction (Ravi and Chan, 2010).

The link between the different ploidy level and the different organization of gametophyte and sporophyte is thus tenuous. This circumstance suggests two hypotheses. First, that the gametophyte-to-sporophyte transition is probably controlled by a simple machinery. Second, that whole developmental cascades, and the resulting phenotypes, can be activated in either gametophytic (haploid) or sporophytic (diploid) context following a relatively easy switch. Recent studies, most of which have been on *Aphanorrhegma patens* (= *Physcomitrella patens*), point in this direction, and some genes acting as key players in these processes have been revealed.

In this moss model species, the homeotic gene *BELL1* is a master regulator of the gametophyte-to-sporophyte transition (Horst *et al.*, 2016; Horst and Reski, 2016). In the evolution of the Viridiplantae, BELL-KNOX heterodimers became responsible for the specification of the diploid life stage already before the transition from unicellular to multicellular organization, as shown by the zygote in *Chlamydomonas*, where each of these gene families is represented by one gene only (Lee *et al.*, 2008). In the land-plant lineage, multiple duplications occurred in both *BELL* and *KNOX* genes. There are 4 *BELL* and 5 *KNOX* genes in *Aphanorrhegma*, 13 and 8 in *Arabidopsis*, 15 and 19 in *Populus*, 12 and 14 in *Oryza* (Mukherjee *et al.*, 2009; Furumizu *et al.*, 2015). This numerical increase has been accompanied by a divergent association of different paralogues with gametophyte or sporophyte development. According to Horst *et al.* (2016), this set of transcription factors is possibly the key for the divergent development of plant gametophyte versus sporophyte. In *Aphanorrhegma patens*, the *KNOX2* gene *MKN6* was identified as a repressor of apospory: targeted deletion of this gene led to the development of diploid protonemata, that is, of multicellular phenotypes typical of the gametophytic phase, but issued in this case from sporophytic cells (Sakakibara *et al.*, 2013).

The evolution of type II MADS-box transcription factors provides an example of genes with regulatory roles in the sporophyte
generation of angiosperms and other derived plants, that in basal plant
lineages instead controlled developmental processes specific to the
gametophyte (Pires and Dolan, 2012). In charophytes, a single member
of this gene family is expressed during differentiation of the reproductive cells from an also haploid parent plant (Tanabe *et al.*, 2005). In land
plants, a series of duplications gave rise to two main groups: the *MIKC^c*
and *MIKC** (Henschel *et al.*, 2002). In bryophytes (Zobell *et al.*, 2010)
and angiosperms (Kofuji *et al.*, 2003), *MIKC** genes are still expressed
in the gametophyte. In mosses and ferns, *MIKC^c* genes are expressed in
both the gametophyte and sporophyte tissues (Münster *et al.*, 1997;
Quodt *et al.*, 2007; Singer *et al.*, 2007), but in the angiosperms they are
mostly restricted to the sporophyte, where their expression qualifies
them as the most important floral homeotic genes.

I.6   TYPOLOGY VERSUS FAMILIES OF PHENOTYPES

Fifty years ago, virtually all characters used in reconstructing phylogeny
and classifying organisms were morphological. This explains why Hennig (1966) chose to call holomorphology the total set of information we
can obtain about a species. In practice, gathering it in full requires
describing a number of semaphoronts, the several distinct instantiations
through which a species' holomorphology is portioned in practice. For
many animals, this means obtaining data from both sexes for virtually
all organisms, examining different ontogenetic stages. To some extent,
Hennig's plea was simply good sense, but his remarks go far beyond that
and are relevant to what we now call evo-devo.

The main point is that what evolves is not the adult (or, by the
way, the egg or the seed, the embryo, the larva, the juvenile), but the
whole life cycle. Semaphoronts do not have an independent status of
some substance; using them as terminals in a cladistic analysis, as
in the recent examples of Lamsdell and Selden (2013) and Wolfe and
Hegna (2014), is thus unjustified, as aptly remarked by Sharma *et al.*
(2017). We should instead pay more attention to the dynamic

FIGURE 1.1 A bisexual reproductive unit with a perianth (commonly differentiated into calyx and corolla) ensheathing androecium and gynoecium is characteristic of the angiosperms, but deviations from this idealized model are frequent. In the small burnet (*Sanguisorba minor*), bisexuality is more a property of the inflorescence than of the individual, all apetalous, flowers. Within the inflorescence, the basal flowers are staminate, the intermediate often hermaphrodite, the top ones carpellate. *A black and white version of this figure will appear in some formats. For the colour version, please refer to the plate section.*

relationships between the semaphoronts of one species and replace the popular, typological descriptions of animals and plants in terms of body plans with a more flexible description in terms of clouds of interconnected phenotypes. This holds both for evolution and for development. In this way, we will appreciate the difference between limiting our study to individual phenotypes and studying instead whole families of phenotypes. Taken out of its evolutionary or developmental context, a phenotype could belong to multiple families, but variation will eventually reveal where it historically belongs and, as a consequence, what are the best questions to be asked about it. This is why, in this book, I discuss leaves and flowers, rather than *the* leaf and *the* flower (Fig. 1.1).

2 The Plant Phenospace

2.1 MORPHOLOGY, PROCESS AND TEMPORAL PHENOTYPES

To a considerable extent, evo-devo is about the so-called developmental genes (see Chapter 4), their evolution and their expression, but it is much more than that. I even dare to affirm that, at the present stage of development of the discipline, there is a strong need to focus instead on the phenotype, which is at the same time both the product of development and the direct target of selection. In other words, the observed phenotype is perhaps the fittest (but not necessarily so), and we need to investigate both its arrival and its survival.

In these first few decades of evo-devo, however, the phenotype has failed to attract attention comparable to that given to the increasingly popular developmental genes. This has caused widespread neglect of a huge amount of information, available in the botanical (and zoological) literature, that could provide valuable suggestions for the most important questions to be addressed with the help of the newly available molecular tools, and no less important suggestions about the taxa to focus on in addressing cutting-edge problems in the discipline.

But this is far from being the only reason to focus on the phenotype. In my opinion, it is time to take a broader view of what we describe as a plant's phenotype – or, if we prefer, of what we regard as the product of its development.

From time to time, a few comparative biologists have acknowledged that phenotype does not reduce to morphological characters, even if we take this notion in the widest possible sense, for example by extending it to include also the subcellular architecture (e.g. membranes, cytoskeleton). In addition to that, there is a dynamic aspect of phenotype, something we could call 'process phenotype'. Up to now, this term has only been used in a few articles on the so-called anatomical and functional 'ontologies' (e.g. Mungall *et al.*, 2010; Harris *et al.*, 2013), but even in this field other authors contrast process ontologies to phenotype ontologies (e.g. Hoehndorf *et al.*, 2010). However, even if the term was not used in that context, the concept of process phenotype was implicit in the idea, first floated in botany by Sattler (1994) and later revived by Gilbert and Bolker (2001) in a basically zoological context, that we can predicate homology of (developmental) processes in the same way as we do with morphological features.

In my view, the concept of process (as phenotype) should not necessarily be restricted to what is going on, e.g. cell proliferation, apoptosis or leaf abscission, or to gene expression, as we do when comparing transcriptomes; neither should it be considered only in terms of spatial restriction, as in the studies of floral homeotic genes, where different patterns of localization of transcripts have been used to support the theories of perianth evolution discussed in Section 7.2.3. As appreciated by a modern approach to heterochrony ('sequence heterochrony'; see Section 9.11.1), the relative time of occurrence of a developmental event, such as the inception of a petal or its abscission following the fertilization of ovules in the flower, is per se one of the many facets of the phenotype. In other words, the temporal organization of the plant (or animal) can be regarded, as such, as one of the products of development, on a par with the morphological features that are deployed along the sequence. Of course, if we accept this broader notion of development, we must also consider its implications for evo-devo. In this book, the whole of Chapter 9 is devoted to the subject of 'temporal phenotypes' (Minelli and Fusco, 2012).

2.2 MODULARITY

It is a popular notion that the organization of the plant body is modular. In principle, we can accept here Klingenberg's (2005, p. 224) definition: 'Modules are assemblages of parts that are tightly integrated internally by relatively many and strong interactions but relatively independent of one another because there are only relatively few or weak interactions between modules.'

In practice, however, delimiting modules is often contentious. The vegetative part of the plant is usually described in terms of phytomers, each of which includes a leaf and the associated node (let's refer here, for simplicity, to plants with only one leaf per node), plus an internode. However, as noted by Kaplan (2001), leaf insertion is not localized at the node and no internode 'belongs' to one leaf only. Rather, each internode is a compound structure consisting of decurrent leaf bases. This is a consequence of the fact that leaves are initiated from the periphery of the shoot apex meristem before any significant extension of the internodes. In this interpretation, in leaf abscission it is not the whole leaf that abscises, but only the upper leaf zone, i.e. the lamina and the petiole, separating from the lower leaf, i.e. the decurrent leaf base clothing the axis.

Modularity goes much beyond the simple contrast between vegetative and reproductive structures, but extends to within both domains, although with often fuzzy borders; for example, the branching structure of the inflorescences can be described as an extension of the vegetative architecture into the reproductive region of the plant.

Morphologically well-recognizable organs may well correspond to major distinct targets of selection, e.g. in terms of pollination or seed dispersal, but, as a rule, these are not coextensive with mechanistically autonomous developmental modules. The general lack of any simple correspondence between developmental modules and anatomical and functional models is arguably one of the most interesting areas of tension between developmental biology and evolutionary biology, and thus a potentially critical area for the ongoing negotiations between these biological disciplines on the common ground of evo-devo. A number of specific issues involving modularity will emerge repeatedly in the following pages, but a general issue deserves mention here. This is the fact that individual organs, including the morphologically and functionally best individualized ones, are not the product of dedicated developmental modules. A flower is not the product of a hypothetical process of anthogenesis, and even a relatively simple plant organ such as a leaf is not the outcome of an equally hypothetical phyllogenesis. The same, of course, is true of animal organs (Minelli, 2009b). Thus, the use of the term *organogenesis* – quite frequent even in the titles of books and chapters – should generally be avoided as misleading.

2.3 EVOLVABILITY

It is well known that, to Charles Darwin, the rapid emergence of the flowering plants was one of the less tractable among the big issues in the history of life; as he wrote in a letter to Joseph Dalton Hooker, dated 22 July 1879, 'The rapid development, as far as we can judge, of all the higher plants within recent geological times, is an abominable mystery.'

Some 140 years later, our appreciation of plant evolution – before the emergence of the angiosperms as well as along their explosive radiation – is much more precise than the picture Darwin could assemble even in the last years of his life, but nevertheless the whole history of the flowering plants can still be described as one of 'rapid development', compared with a number of other plant and animal clades, including taxa usually classified at the quite low rank of family, such as cockroaches (Blattidae), alderflies (Sialidae), dobsonflies (Corydalidae), whirligig beetles (Gyrinidae) and horseflies (Tabanidae), all of which were already represented on our planet when the angiosperm lineage separated from its sister group.

To help explain this puzzling macroevolutionary picture, evo-devo has much to offer, by inviting us to direct our attention towards the arrival, rather than the survival, of the fittest. Focus is thus moved from plant evolutionary history to plant evolvability.

More than other key concepts discussed in these pages (e.g. modularity), evolvability is a very contentious issue. A couple of dozen different definitions have been proposed (not all of them relevant to evo-devo), focusing variously on either the genotype or the phenotype, on the properties of the individual or those of the population, on the pattern of variation offered by the latter to natural selection or on the unequal prospects, for a population, of evolving according to this or that pathway in the multidimensional space of future possible phenotypes. Many of these concepts have been comparatively discussed by Pigliucci (2008), Brookfield (2009) and Minelli (2017). We can accept here a definition of evolvability as the capacity of a developmental system to evolve (Hendrikse et al., 2007), which depends on the potential of the developmental system to generate heritable phenotypic variation (Pigliucci, 2008) and to produce or maintain phenotypic variation over evolutionary time, enabling it to pursue diverse evolutionary trajectories (Schlichting and Murren, 2004).

Modularity and integration are two key determinants of evolvability (Hansen, 2003; Pavlicev and Hansen, 2011). Strict integration among the components of the same module, especially if well-entrenched in the plant's genetic structure in the form of extensive pleiotropy (see Section 4.6.2) or strong linkage of genes with major phenotypic effects, may limit evolvability, at least in the short term (Conner, 2012). There is widespread consensus, instead, that modularity enhances evolvability by maintaining integration among functionally related characters, while allowing at the same time for the independent evolution of other modules (Wagner, 1996; Wagner et al., 2007; Klingenberg, 2008; Hallgrímsson et al., 2009; Clune et al., 2013; Fruciano et al., 2013; Diggle, 2014).

An exemplary study of evolvability in a large plant clade is Donoghue and Ree's (2000) broad comparative analysis of flower symmetry in the asterids, where the repeated evolution of a limited number of forms of bilateral flower symmetry may reflect the constraints imposed by overall flower orientation and the underlying mechanisms of differentiation. Radial flower symmetry is generally accepted as the primitive condition in this clade, from which corollas with bilateral symmetry evolved at least eight times independently, with at least nine reversals to actinomorphy (Donoghue et al., 1998; Ree and Donoghue, 1999). A first aspect of the evolvability of flower symmetry in this clade is that transitions from a bilateral to a radial corolla are more likely to occur than changes in the opposite direction. In the two major asterid clades with zygomorphic flowers, the most common pattern is in both instances the same, a 2:3 pattern with two adaxial (upper) petals opposite to three abaxial (lower) petals. Another two patterns have evolved repeatedly: the 4:1 pattern, with four adaxial and one abaxial petal, and the 0:5 pattern, in which all five petals are shifted to the abaxial side of the flower. Other hypothetical patterns such as 3:2 or 1:4 have never evolved. According to Donoghue and Ree (2000), this canalized evolution of flower zygomorphy is largely explained by the nature of flower development in these plants. At the inception of flower organ primordia, there are five petals, one of which is in a medial position on the abaxial side of the flower. Only the three commonly observed forms of bilateral symmetry (2:3, 4:1, 0:5) are possible as long as this arrangement of primordia is maintained and the mechanism producing dorsiventral differentiation in these flowers is also retained. On the other hand, in several clades of asterids, shifts from five to four petals have occurred, sometimes within clades with zygomorphic flowers, as in the Rubiaceae; in other groups, a change from five to four has accompanied reversal from zygomorphy to actinomorphy, as in *Plantago* (Donoghue *et al.*, 1998; Endress, 1998; Reeves and Olmstead, 1998) (see Section 10.3). Much less common is a shift from a regular dorsiventral differentiation of the flower primordium to a mechanism producing asymmetric flowers, as in *Centranthus*, with differentiation of one of the two dorsal petals divergent from the other four petals (Endress, 1999).

2.4   DEVELOPMENTAL ROBUSTNESS

Developmental robustness is the ability of an organism to produce the same phenotype regardless of perturbations caused by internal or external factors. Structural conditions (from the molecular to the organismic level) and physiological mechanisms responsible for robustness have attracted increasing attention in recent years. Developmental robustness has been suggested to be an important mechanism fostering the production of morphological complexity and paving the way towards recognizable bursts of speciation (Melzer and Theißen, 2016). However, to evaluate developmental robustness in any actual case is far from easy. Interpretations are necessarily biased by the unavoidable subjectivity with which we decide which aspect(s) of a plant should be considered 'relevant' phenotypes. Let's consider a couple of examples.

The size of individual plant organs is often quite uniform within a species. This is particularly true of flower parts, e.g. the petals in *Arabidopsis*. This does not imply that these organs have the same composition in terms of cell number. It has been suggested that robustness in the size of a plant organ does not rely so much on mechanisms simply counting cells or assessing cell size, but on monitoring the organ's overall size. This may require a trade-off between cell number and cell size. Mechanisms of compensation have been suggested (reviewed in Hong *et al.*, 2016), whereby cells of larger size are produced in organs where mitotic activity has been slower, or terminated earlier, resulting in the production of organs of almost normal size. Often, the physiological limits of this kind of robustness become obvious when resource limitation associated with defoliation causes a decrease in the size of floral organs (Diggle, 2003), a change also sometimes observed in the latest flowers produced by a plant in a flowering season.

Genetic robustness is 'robustness to perturbations both in the form of new mutations and in the form of the creation of new combinations of existing alleles by recombination' (Masel and Trotter, 2010, p. 407). So, much as the term *evolvability* suggests an intrinsic ability to change, so does *robustness* suggest an intrinsic resistance to change. However, the robustness of a genetic system does not necessarily contrast with evolvability (Wagner, 2011).

If developmental robustness evolves under natural selection, there should be loci in the genome that act as suppressors of phenotypic variation. One of these putative 'canalizing genes' is *SUPERMAN*

(*SUP*). In *Arabidopsis thaliana*, an early and a late function of *SUP* can be distinguished (Breuil-Broyer *et al.*, 2016). In the first phase, this gene is involved in the specification of the boundary between male and female organs and in the control of stamen number. Later in flower development, *SUP* is involved in controlling intra-whorl organ separation within each whorl of sexual organs as well as in carpel development. It has been thus hypothesized that the developmental robustness of the flower may have been substantially improved, in angiosperm evolution, by the expression of this gene at early stages of carpel development, eventually resulting in the splitting of a primitive male–female gradient into sharply separated male and female whorls.

2.5   PHENOTYPIC PLASTICITY AND EVOLUTION

Phenotypic plasticity (reviewed, e.g., in Schlichting and Pigliucci, 1998; West-Eberhard, 2003; Fusco and Minelli, 2010) is 'a property of individual genotypes to produce different phenotypes when exposed to different environmental conditions' (Pigliucci *et al.*, 2006, p. 2363).

Production and eventual fixation of phenotypic plasticity involve a diversity of evolutionary processes collectively termed genetic accommodation (Minelli, 2015c, 2016c), whereby a phenotype first produced in direct response to an environmental condition eventually becomes genetically encoded (e.g. West-Eberhard, 2003; Schlichting and Wund, 2014). A diversity of mechanisms (genetic assimilation, the Baldwin effect, and accumulation and eventual release of cryptic variation) can in fact be involved. There is genetic assimilation (Waddington, 1953) when selection progressively reduces the genetic background for plasticity, with the result that a genetically encoded phenotype will be eventually fixed (Robinson and Dukas, 1999; Pigliucci and Murren, 2003). When plasticity enhances the survival of an individual in a new environment and selection subsequently favours overdetermination of the phenotype by accumulation of heritable variation supporting it, the process is referred to as the Baldwin effect (Baldwin, 1896). Beyond these two mechanisms, recent studies in evolutionary biology target especially the effects of accumulation and eventual release of cryptic variation; this may happen when a population has for a long time not had to contend with environments in which its phenotypic plasticity would have expressed some of the possible alternative phenotypes. Robustness allows the accumulation of this cryptic (not expressed) genetic variation (Gibson and Dworkin, 2004; Le Rouzic and Carlborg, 2008; Lahti *et al.*, 2009; Van Dyken and Wade, 2010). Eventually, environmental change can unmask it and expose it to selection, with the possible release of novel phenotypes (Barrett and Schluter, 2008; Masel and Siegal, 2009; Ehrenreich and Pfennig, 2016; Mestek Boukhibar and Barkoulas, 2016; Theißen and Melzer, 2016).

Monniaux *et al.* (2016) contrasted the robustness of floral development, allowing for a strict control of the identity and number of floral organs in each whorl, to the extensive developmental plasticity of the vegetative parts of the plant. The phenotypic expression of cryptic genetic variation is thus limited, even in fluctuating and stressful environments in which the plant's branching patterns and leaf shape and distribution are often conspicuously affected.

On the long-term evolutionary scale, the actual role of phenotypic plasticity in the production of novel phenotypes remains controversial (Schlichting and Pigliucci, 1998; Pigliucci, 2001; Wund, 2012), but there is increasing evidence, from plants and animals alike, that plasticity provides previously untested targets for natural selection, creates novel trade-offs and increases genetic variation and divergence, thus creating opportunities for diversification, speciation included (Moczek, 2010).

2.6 EVOLUTIONARY INNOVATIONS

Angiosperms evolved from ancestors that did not possess flowers; plants with sympetalous flowers evolved from plants with separate petals; and plants with syncarpous gynoecia evolved from ancestors characterized by apocarpy. A cladist would describe flowers, sympetaly and syncarpy as apomorphies, i.e. as derived states, respectively contrasting them with (primary) lack of flowers, dialypetaly (the now largely forgotten antonym to sympetaly, or gamopetaly) and apocarpy as the corresponding plesiomorphies, or primitive states. But an evolutionary biologist will probably (incautiously, perhaps!) disregard the detailed topology of the phylogenetic tree and simply use the term *innovations* for the derived features that so conspicuously characterize angiosperms, or large subclades of angiosperms, and their unquestionable success. Some purists might prefer to use the term *innovation* to denote the process by which the new feature has been introduced, and instead use *novelty* to designate the latter, but in the evo-devo literature the two terms have been applied in an unproblematic mixed way. There is no need to fastidiously fix their usage here.

Innovation is often regarded as one of the core concepts of evo-devo, along with evolvability, but this is far from being uncontroversial.

First of all, it is difficult to avoid an at least unconscious association between the notion of evolutionary innovation and the idea of progress. Who would describe the miniaturization of *Lemna* or *Wolffia*, or lack of chlorophyll in many parasitic or mycoheterotrophic plants (see Section 10.1.2), or the extreme reduction of leaves e.g. in *Cuscuta* as an innovation? The term will probably be reserved for things like the haustoria of parasitic plants. A number of clades of parasitic and hemiparasitic plants evolved independently these specialized organs that penetrate into the stem or root of the host and allow an exchange of materials between the host and the parasite through a vascular connection. If we want to characterize these organs as evolutionary innovations, we must anyway use the plural, because haustoria have different origins, from roots (*Agalinis*), from stems – either endogenously (*Cuscuta*) or exogenously (*Hydnora*) – and even from modified leaves (the rhizome scales of *Hyobanche sanguinea*). Haustoria are described as innovations, because this looks like a progressive rather than a regressive trait.

Similarly, in animals, the specialized absorbant body surface of tapeworms and their relatives is generally presented as an innovation (witness the term, neoderm, by which it is described), but in the same animals it is quite unlikely that the absence of mouth, gut and other organs will be described as innovations. Regressive traits are just that, the opposite of progressive traits; thus, by implication, they are not innovations. This biased perspective must be seriously addressed and eventually corrected.

Back to flowering plants, let's describe as innovation features such as the secondary shoot apex of the Rafflesiaceae, despite the fact that this unusual surface distinctly implies deconstruction of traditional structures and boundaries. These parasitic plants are popular for the enormous size of the flower in some species, up to more than 100 cm in *Rafflesia arnoldii*. In comparative developmental biology, however, the Rafflesiaceae are remarkable for the unusual origin of their apical surface. This morphological boundary originates secondarily by schizogeny, that is, by mutual detachment of cells along the interface between the host plant and the parasite. The same process is responsible for the opening of the radial clefts of the gynoecium, which is initially represented by a solid mass of tissue, without differentiation of carpels from the floral apex (Nikolov *et al.*, 2014).

Let's move on to a second controversial aspect of the common notion of evolutionary innovation. Even if in our view of evolution we provide a place for saltational change (see Section 10.4), it is hard to believe that innovations, as a rule, are the product of quantum change. This is certainly not the case for the flower, an innovation that eventually emerges from a number of evolutionary transitions (see Section 7.1). According to West-Eberhard (2003, p. 198), a novelty is a 'phenotypic trait that is new in composition or context of expression relative to established ancestral traits'. This definition acknowledges that novelty is never a property of a feature as a whole, because even a well-characterized feature is a composite of multiple traits, traceable to separate homologues of variously remote ancestry (Minelli and Fusco, 2005; Minelli, 2016b).

The link with the conventional notion of homology is clear in Hall's (2005, p. 549) definition of a novelty as 'a new feature in a group of organisms that is not homologous to a feature in an ancestral taxon'. But this does not help much in resolving the issue (see Brigandt and Love, 2010, 2012), because the meaning of homology is also controversial (Donoghue, 1992; Wake, 2003; Moczek, 2008; Minelli and Fusco, 2013). If homology is not all-or-nothing (Minelli, 2003; West-Eberhard, 2003), we must also accept that there is a continuum between non-novelty and novelty. Eventually, the idea that characters can 'remain themselves' and distict from the other characters throughout any possible transition in the course of evolution, except for the sudden emergence of innovations, is probably based on an idealistic interpretation of how organisms evolve (Minelli *et al.*, 2006; Minelli and Fusco, 2013).

One more question concerns the adaptive significance of innovations. According to Hallgrímsson *et al.* (2012, p. 502) an evolutionary novelty should 'involve a transition between adaptive peaks on a fitness landscape [and] a breakdown of ancestral developmental constraints, such that variation is generated in a new direction or dimension'. Innovations associated with the emergence of highly species-rich or otherwise successful clades are often described as key innovations. Endress (2001b, 2006) listed a number of features as key innovations supporting the evolutionary success of large angiosperm clades. These include the double perianth (sepals and petals) characteristic of eudicots, the syncarpy with a compitum (see Section 7.6.1.3) largely prevalent in core eudicots and monocots, the sympetaly of asterids and the fusion of stamens with petals characteristic of euasterids. Endress (2006) also remarked that key innovations may evolve slowly and also take a long time before getting fixed, an example being provided by sympetaly in the Ericales (Schönenberger *et al.*, 2005).

Evidence for a causal association between novel trait and clade success, however, is generally circumstantial, and sound causal proofs may even be beyond reach. In any case, we should always be aware of the possibility that exaptation is involved, that is, that the selective regime under which the novel trait was fixed may be other than the selective regime that currently promotes its perpetuation. Moreover, the same feature is perhaps a key innovation in some of the clades in which it occurs, but not in all. This has been suggested, for example, for the evolution of zygomorphic flowers (Endress, 2006): these are characteristic of such obviously successful clades as the Orchidaceae, the Lamiales and the Fabaceae (Faboideae), but zygomorphic flowers have also evolved in a number of other clades, most of which include only a small number of species.

In so far as this represents a way of addressing the problem of the arrival of the fittest, innovations belong to evo-devo, but the notion of 'key innovations' is mainly about adaptation, thus falling within the traditional domain of evolutionary biology. This is the case, at least, as long as we regard key innovations as advantageous phenotypes positively targeted by selection. We cannot rule out a priori, however, that different circumstances may prevail, as in the case of the nearly universal fixation of the number (seven) of cervical vertebrae in mammals: here, an adaptive significance of this number is extremely unlikely, witness its fixation in morphologically divergent lineages such as hippos and giraffes, with necks of obviously opposite proportions; the target of selection is probably something else than the number of cervical vertebrae, with which this poorly known condition, or some developmental process behind it, is nevertheless very strictly linked (Galis, 1999; Galis *et al.*, 2006). Another potential cause of fixation of what may be eventually regarded as a key innovation is a possible reduction of its further evolvability, if the apparent key innovation is a kind of 'oceanic island' lost in morphospace – that is, a condition which is not within easy reach from more common ('continental') conditions, and also from which it is difficult to move because of the lack of other 'islands' (stable phenotypes) to which it may be easy to travel. Eventually, oceanic islands may become the theatre of peculiar, highly species-rich radiations – witness the drosophilid flies on Hawaii or, indeed, the fleshy-fruited lobelioids of the same archipelago (Carlquist, 1980). These examples suggest that key adaptations could be seriously studied from an evo-devo perspective. I am not aware of targeted studies thus far, at least for plants.

Eventually, discussing innovations in terms of the adaptive advantage they may have provided to a successful clade is an interesting evolutionary question, but not one of evo-devo. Therefore, I close this section with a case story that is limited to a very small clade (just two, closely related species) and involves an apparently minor innovation, but which exemplifies several points I have discussed in these first two chapters as significant in evo-devo: the modularity of development and evolution, the not necessarily adaptive nature of development, and the importance of exaptation, eventually resulting in functional innovation. In *Erythranthe guttata* (= *Mimulus guttatus*), in the axil of each leaf there are two meristems, rather than the usual one; of these meristems, the proximal one typically remains dormant and seems not to have a function. However, in *E. gemmipara* (= *Mimulus gemmiparus*) (possibly derived from *E. guttata*), both axillary meristems are active: the distal one behaves as a typical axillary meristem, by producing either a lateral branch or a flower, while the proximal meristem becomes a vegetative propagule whose first leaves are the site of nutrient storage (Moody *et al.*, 1999).

2.7   POLYMORPHISM, SELECTION AND EVO-DEVO

In evo-devo, the focus is on the arrival of the fittest rather than on its survival. However, there are aspects of phenotypic evolution where the two aspects (evolvability and survival value) are so tightly intertwined that, for a while, evo-devo must allow selection, first gently dismissed from the stage, to enter again through the back door.

This is the case of species in which a kind of polymorphism has evolved, where the reproductive success of the individuals of one morph is strictly dependent on the complementarity between some of the latter's traits and those of another morph. In fungi or in ciliate protozoans, these morphs will generally correspond to complementary mating types, based on genetic polymorphism but indistinguishable from a morphological point of view; in animals, we would rather consider the two sexes (when distinct) between which extraordinary forms of dimorphism have evolved in many groups. In angiosperms, secondary sex characters are very elusive or non-existent: according to Lloyd and Webb (1977), the only species in which it is possible to identify with certainty the sex of an individual without considering its flowers would be *Cannabis sativa*. As a rule, at the level of physiological incompatibility, complementary individuals are not more distinguishable than are fungal or ciliate mating types; however, there are exceptions to this rule, and it is mainly in such cases that evo-devo cannot ignore functional constraints in phenotypic evolution, that is, selection.

These exceptions are the plant species with stylar polymorphisms. In these, two or more different phenotypes are present, which in terms of reproduction correspond to mutually compatible morphs, while from an evo-devo perspective they provide examples of short-range (easily evolvable) alternative phenotypes, the further evolution of which is however constrained by their tight interdependence. From a developmental perspective, stylar polymorphisms are neat examples of modularity, in which, as a rule, only the reproductive parts of the flower are involved. Most of the examples discussed below have a simple genetic basis, comparable to the genetic polymorphism on which reproductive incompatibility is based, either in morphologically non-polymorphic plants, or in other groups of organisms; however, from an evo-devo perspective, even environmentally controlled polyphenisms with functional properties like those of stylar polymorphism are equally important, the main difference with respect to genetic polymorphism being not in the ability of a species to produce different complementary phenotypes, but in the prospective fixation of the latter.

Less easy to place in this context is the occurrence of two (rarely more than two) distinct types of stamens that generally differ in their location within the same flower (Vallejo-Marín *et al.*, 2009). This condition (*heteranthery*) has been reported from more than 15 families, including Pontederiaceae, Commelinaceae, Lythraceae, Melastomataceae, Anacardiaceae, Malvaceae, Brassicaceae, Fabaceae, Malpighiaceae, Solanaceae and Scrophulariaceae, plus the Gesneriaceae, where one set of stamens produces sterile pollen (Gao *et al.*, 2006).

Four kinds of stylar polymorphism have been described: distyly, tristyly, stigma-height dimorphism and enantiostyly (reviewed in

Barrett *et al.*, 2000). The first three are style-length polymorphisms (reviewed in Barrett, 1992). By contrast, enantiostyly is an asymmetry polymorphism in which the style is deflected either to the left or right side of the flower (Jesson and Barrett, 2003).

2.7.1   Heterostyly (Distyly and Tristyly)

The populations of heterostylous plants include two (*distyly*) or, more rarely, three (*tristyly*) phenotypes. In the case of distyly, some individuals produce flowers with the stylus much longer than the stamen filaments (pin phenotype, L); the others produce flowers with stamen filaments much longer than the stylus (thrum phenotype, S).

Heterostyly has been recorded in approximately 200 genera in 28 families (Li and Johnston, 2010; Naiki, 2012), more commonly in the form of distyly (26 families, including Iridaceae, Amaryllidaceae, Saxifragaceae, Fabaceae, Oxalidaceae, Passifloraceae, Linaceae, Hypericaceae, Lythraceae, Malvaceae, Santalaceae, Plumbaginaceae, Polygonaceae, Polemoniaceae, Primulaceae, Rubiaceae, Gentianaceae, Boraginaceae, Oleaceae, Lamiaceae and Acanthaceae), more rarely as tristyly (Amaryllidaceae, Pontederiaceae, Connaraceae, Oxalidaceae, Linaceae, Lythraceae and Thymelaeaceae, all of which also contain distylous species, except for the Pontederiaceae and Thymelaeaceae). More than 100 heterostylous genera belong to the Rubiaceae. The phenomenon is also frequent in Connaraceae, Lythraceae, Primulaceae, Rubiaceae and Boraginaceae.

Heterostyly evolved independently 2–10 times in the Linaceae (McDill *et al.*, 2009), five times in the Lythraceae, four times in Pontederiaceae (Kohn *et al.*, 1996), two or three times in the Rubiaceae (Ferrero *et al.*, 2012) and at least 12 times in the Boraginaceae (Cohen, 2014), but in *Amsinckia*, a genus belonging to this family, there have also been several independent reversals to homostyly (Schoen *et al.*, 1997; Li and Johnston, 2010; Cohen, 2014).

The development of two or three distinct floral morphs is under the control of the *S* locus (Richards, 1997). In *Turnera subulata*, *Fagopyrum esculentum* and *Primula vulgaris*, all distylous, this locus behaves as diallelic (Matsui *et al.*, 2004; Li *et al.*, 2007; Labonne *et al.*, 2010). In *Primula*, long-styled individuals are homozygous recessive (*ss*), short-styled ones heterozygous (*Ss*), and the homozygous dominant genotype (*SS*) is lethal.

In *P. vulgaris*, the differences in style length and the relative position of anthers in the middle or the top of the corolla tube are mainly the effect of differential cell growth (Webster and Gilmartin, 2006).

In *Primula veris*, 73 genes showed increased expression in S-morph versus L-morph floral buds, but this does not extend, in general, to the other parts of the plant. Distyly, indeed, is a phenotypic difference restricted to the flowers, and of 113 genes showing floral morph-specific differential expression, 69 are not differentially expressed in L- and S-morph leaves of both *P. veris* and *P. vulgaris* (Nowak *et al.*, 2015).

2.7.2 Stigma-Height Dimorphism

Populations with stigma-height dimorphism include two floral morphs that differ in the heights at which stigmas are located. In the L-morph, the position of the stigma is above the anthers, in the S-morph below the anthers. This dimorphism is known for a number of species belonging to genera in which other species are distylous (e.g. *Linum, Anchusa, Lithodora, Narcissus, Primula*; reviewed in Barrett *et al.*, 2000; Ferrero *et al.*, 2009), but also, sporadically, in families with no heterostylous taxa (e.g. Asparagaceae: *Chlorogalum angustifolium*; Ericaceae: *Epacris impressa, Kalmiopsis leachiana*). Stigma-height dimorphism is common in *Narcissus*, where this condition has been recorded in a dozen species. The inheritance of style length is the same as in most distylous species, with the L-morph corresponding to the homozygous recessive genotype and the S-morph to the heterozygote (Lewis and Jones, 1992).

2.7.3 Enantiostyly

The flowers of several plants are basically zygomorphic (perianth and stamens obey mirror symmetry), but their style is deflected to the right or the left side (*enantiostyly*). In a few species, other floral whorls are also lateralized (*enantiomorphy*) (Marazzi *et al.*, 2006; Marazzi and Endress, 2008). Left- and right-deflected morphs are usually found in both enantiostylous and enantiomorphic plants, either on the same (monomorphic enantiostyly: e.g. *Cyanella*, *Senna* and *Dialium*) or on different individuals (dimorphic enantiostyly: e.g. *Wachendorfia*). In some species with monomorphic enantiostyly, the position of the two flower morphs within the inflorescence is at random, in others it is fixed, in still other species there are distinct right- and left-styled inflorescences (Barrett, 2010). In *Monochoria australasica*, the individual plant produces both left- and right-styled inflorescences, but each inflorescence contains only left-styled or right-styled flowers. In small individuals of this species, only one inflorescence (either left- or right-styled) is open on a given day, but in larger plants both left- and right-styled inflorescences may open on the same day (Jesson and Barrett, 2003). Dimorphic enantiostyly is a genetic polymorphism: left- and right-styled plants often occur in equal frequencies (Jesson and Barrett, 2002a, 2002b).

The evolution of this stylar polymorphism has been reconstructed as a two-step event, with a first transition from a straight-styled condition to monomorphic enantiostyly, occasionally followed by a transition to dimorphic enantiostyly (Barrett, 2010).

Monomorphic enantiostyly has evolved independently in at least 10 families, distributed among (eu)dicots and monocots, but dimorphic enantiostyly is limited to three monocot families (Barrett, 2002).

Other than with zygomorphy, enantiostyly is sometimes associated with a recent change in merosity (number of flower parts; see Section 7.5). In the very large genus *Solanum* (ca. 1500 species), most species have actinomorphic flowers with five sectors and with stamens of uniform size and shape. Enantiostyly, associated with unequal stamens, occurs in some *Solanum* species characterized by flowers with only four sectors, some of which, but not all, are also zygomorphic (Bohs *et al.*, 2007).

Inversostyly is another stylar polymorphism, with morphs differing for the orientation of styles, but in this case these are deflected either upwards or downwards rather than left or right; moreover, the two morphs also differ in the opposite location of two stamens. Inversostyly has been reported in *Hemimeris racemosa* (Pauw, 2005).

Two self-compatible floral morphs coexist in the populations of plants with *flexistyly* (reviewed by Barrett, 2010) ability; one of these morphs is proterogynous, that is, it functions first as a female and later as a male, and the other is proterandrous, with the male phase preceding the female one. This condition has been reported from 24 species in *Alpinia*, *Amomum* and *Etlingera*, all Zingiberaceae.

3 Tools

3.1 TERMINOLOGY

In a young scientific discipline such as evo-devo, it is vital to be flexible in articulating a research programme, but it is no less vital to define new concepts (and, if necessary, to redefine old ones) with the greatest clarity and to be careful and consistent in the usage of technical terms. However, this is not necessarily easy, especially in evo-devo, which frequently relies on concepts and terms already in use in one of the parent disciplines. Moreover, the mixed language seems to encourage general deregulation. For example, the subtle teleological wording that manifests the persistence of a misguided view of evolution (Caruso *et al.*, 2012; Rigato and Minelli, 2013) is present in the following sentence (just one example among many I could quote) from an otherwise excellent review: 'Probably because of their ability to modulate growth patterns, some TCP genes were recruited during angiosperm evolution to generate new morphological traits' (Nicolas and Cubas, 2016, p. 250). I would also strongly avoid anthropomorphic caricature, e.g. 'One of the most important developmental decisions in plants is the decision of when to flower' (Ogura and Busch, 2016, p. 106).

In other instances, a choice between alternative wordings may be less obvious, but nevertheless it may deserve a second thought, to avoid being trapped in a one-way approach to our study object. For example, it is often said (e.g. McKone and Tonkyn, 1986) that the female flower heads of *Ambrosia artemisiifolia* have one flower only. Why do we call these reproductive units inflorescences? Perhaps by comparison with the male units, or with the reproductive units otherwise universal in the family (Asteraceae). But this usage would be legitimate, strictly speaking, only if we could demonstrate that the female reproductive unit in *A. artemisiifolia* is actually derived from a conventional capitulum. Otherwise, a reproductive unit with one flower is, simply, a flower. Regretfully, many morphological terms for parts of the plant body are used, without further qualification, with two different meanings, that is, as purely descriptive labels, and in a way that implies a qualified statement of homology.

The target of my most serious objection in matters of terminology is, however, another. The word in question is one with ancient roots and with a well-known patent of nobility in the life sciences, namely *origin*. Unfortunately, the title of Charles Darwin's major work (Darwin, 1859) is much more widely known than the clear statement, on page 469 of that work, that 'no line of demarcation can be drawn between species ... and varieties'. This means that in Darwin's mind evolution was a history of transitions rather than a gallery of origins. Every step in evolution, cladogenetic and anagenetic, is indeed a mixture of conservation due to common descent combined with some modification. The more closely we are able to analyse these events, the more elusive becomes the divide between the old and the new. Similarly, in developmental biology, the better we know a process, the less we are able to fix the origin of a particular feature. In science, especially in biological disciplines with a strong historical dimension such as evolutionary biology and developmental biology, we may better ask questions in terms of transitions rather than origins (Minelli, 2015c).

3.2   MODEL SPECIES

3.2.1   Arabidopsis thaliana, *an Idiosyncratic Model Species*

The choice of the would-be model species – plants, animals and microorganisms alike – has been largely dictated by practical reasons suggesting a preference for organisms with short life cycles and high multiplication rates, and which are easy to breed in the lab under inexpensive conditions. In former times, a small number of chromosomes was an added bonus, as was a small genome size more recently. Unfortunately, subsequent comparisons with other, not necessarily model species, including close relatives of our favourite lab organisms, have shown that the latter are often very poor examples of the structural and functional properties of the whole group, e.g. bacteria, insects or flowering plants, to which we intended to extrapolate the results of the studies. More often than not, the idiosyncrasies of the model species are causally linked with some of the properties that made them so attractive. In the case of *Drosophila*, for example, the very quick embryonic development correlates with dramatic deviations from the conditions prevailing among the insects: segmentation of the egg up to the 13th run of mitoses occurs in syncytial rather than cellularized conditions, and very soon a long-germ-band embryo is formed in which the body segments appear synchronously rather than through prolonged posterior addition. The whole is obviously underpinned by spatially and temporally idiosyncratic patterns of expression of a number of key developmental genes.

What about *Arabidopsis thaliana*? First of all, its genome is far from typical (Litt, 2013), not just of flowering plants as a whole, but also of the Brassicaceae, a family of some 3700 species that radiated from their last common ancestor less than 50 million years ago (Kagale *et al.*, 2014; Hohmann *et al.*, 2015). The genome of *A. thaliana* is considerably smaller than most of the other plant genomes sequenced to date, including the genome of the closely related *A. lyrata*, from which it diverged about 10 million years ago (Beilstein *et al.*, 2010). The size of the genome of *A. thaliana* is only 125 Mb, much less than the 207 Mb of the genome of *A. lyrata*, whose size is closer to the average value in the genus; in terms of protein-coding genes, the difference is less conspicuous, but nevertheless remarkable for two species belonging to the same genus: predictions are for 32 700 protein-coding genes in *A. lyrata*, compared to the 27 000 in *A. thaliana*. In the latter species, the haploid chromosome number is reduced from eight to five, relative to other species in the same genus, and in the Brassicaceae generally (Kuittinen *et al.*, 2004; Yogeeswaran *et al.*, 2005; Lysak *et al.*, 2006; Hu *et al.*, 2011). As much as one-half of the *A. lyrata* genome appears not to be present in *A. thaliana*, and a quarter of the *A. thaliana* genome has no counterpart in *A. lyrata*, due to hundreds of thousands of deletions, plus numerous insertions and rearrangements across the genome (Hu *et al.*, 2011).

Second, using *A. thaliana* as a model for the identification of the genes involved in determining the identity of floral parts (sepals, petals, stamens, carpels) resulted in the production of the once popular ABC model, but this worked only in part in the other plants, including the second plant model species *Antirrhinum majus* (see Section 4.6.3).

Third, *A. thaliana* is a poor example of plant organization also from a morphological point of view. The lack of subtending bracts below the calyx is an idiosyncrasy of Brassicaceae, atypical for eudicots (Endress, 2006). Admittedly, rudiments of subtending bracts are present in the youngest stages of flower development; moreover, the *Arabidopsis JAGGED* (*JAG*) mutant produces subtending bracts (Dinneny *et al.*, 2004). However, the reason why subtending bracts and prophylls have been poorly studied from a developmental genetic point of view is because both organ types are lacking in *Arabidopsis*.

Fourth, despite decades of research on this model species, its flower structure is still morphologically problematic: it is still uncertain how many carpels form the *Arabidopsis* gynoecium (Vialette-Guiraud and Scutt, 2009). Several authors (e.g. Okada *et al.*, 1989) have interpreted this structure as containing two carpels, but the ovary wall contains four vascular traces, suggesting that in the wild-type *Arabidopsis* gynoecium there are four rather than two carpels (Lawrence, 1951).

3.2.2   Selecting New Model Species

Criteria for selecting model species for biological research have evolved considerably over time. Plants and animals of economic importance were long privileged, based on easier availability of financial support and better prospects of rewarding returns, not limited to scientific progress. Several plants of major agricultural or floricultural interest still feature today in the list of the most popular model species (Table 3.1). However, among those that have been added to rice, maize and tomato there are many species that were selected based on different criteria, such as short life cycle, low number of chromosomes and small genome, all traits that make their cultivation, analysis and manipulation easier, without necessarily qualifying them as ideal models of the structural and functional diversity of angiosperms.

In more recent years, phylogenetic criteria have also been taken into account. On the one hand, the list of model species has been expanded to include, for the flowering plants, representatives not only of the ANA group (*Nymphaea thermarum*) and the basal eudicots (*Aristolochia fimbriata, Aquilegia* spp., *Eschscholzia californica*), but also of other lineages of land plants: lycophytes (*Selaginella moellendorffii*) and mosses (*Aphanorrhegma patens*). On the other hand, it has been realized that detailed comparisons (genetic and developmental) between recently diverged species could dramatically improve our understanding of the evolution of morphological traits such as leaf complexity and flower architecture. This explains the presence, in a list such as Table 3.1, of three species of Brassicaceae belonging to two closely related genera (*Arabidopsis thaliana, A. lyrata* and *Cardamine hirsuta*) (Mitchell-Olds, 2001; Canales *et al.*, 2010) (Fig. 3.1), as well as the interest in studying multiple species of *Aquilegia*, rather than just a 'typical' one (Kramer, 2009a, 2009b). The latter genus is obviously attractive because of its floral innovations including petaloid sepals, nectar spurs on the second-whorl petals (Fig. 3.2) and the presence of staminodes (Litt and Kramer, 2010).

New model species mean new sources of information, but also new pitfalls to be avoided. A frequently overlooked problem is the taxonomic identity of the cultivated strains, especially when they represent multiple accessions of different geographical origin. Among the animal model organisms, the existence of multiple species among the lab strains has been recently demonstrated in the case of the leech traditionally known as *Helobdella triseriata* (Siddall and Borda, 2003;

Table 3.1 *A list of model plant species. Where the correct scientific name according to current taxonomy and the rules of the* International Code of Nomenclature for Algae, Fungi, and Plants *(McNeill et al., 2012) differs from the name in current use, the species is listed under the apparently correct name, with the name in current use added in parentheses.*

Scientific name	Vernacular name	Family	Selected references
Antirrhinum majus	Snapdragon	Plantaginaceae	Schwarz-Sommer *et al.* (2003)
Aphanorrhegma patens (= *Physcomitrella patens*)	'Moss'	Funariaceae	Nishiyama *et al.* (2003); Reski (2003); Reski and Cove (2004); Knight *et al.* (2009); Lang *et al.* (2016)
Aquilegia spp.	Columbine	Ranunculaceae	Kramer (2009a, 2009b); Kramer and Hodges (2010)
Arabidopsis lyrata	Lyrate rockcress	Brassicaceae	Mitchell-Olds (2001); Hu *et al.* (2011)
Arabidopsis thaliana	Thale cress, arabidopsis	Brassicaceae	Meinke *et al.* (1998); Arabidopsis Genome Initiative (2000); Mitchell-Olds (2001); Somerville and Meyerowitz (2002–)
Aristolochia fimbriata	White-veined hardy Dutchman's pipe	Aristolochiaceae	Bliss *et al.* (2013)
Capsella rubella	Pink shepherd's-purse	Brassicaceae	Slotte *et al.* (2013)
Cardamine hirsuta	Hairy bitter-cress	Brassicaceae	Canales *et al.* (2010); Hay *et al.* (2014)
Erythranthe guttata (= *Mimulus guttatus*)	Yellow monkeyflower	Phrymaceae	Wu *et al.* (2008)
Eschscholzia californica	Californian poppy	Papaveraceae	Carlson *et al.* (2006)

Species	Common name	Family	Reference
Gerbera hybrida	Gerbera	Asteraceae	Teeri *et al.* (2006a)
Habenaria radiata	White egret flower	Orchidaceae	
Helianthus annuus	Sunflower	Asteraceae	Gill *et al.* (2014)
Lotus corniculatus var. *japonicus* (= *L. japonicus*)	[Japanese] bird's-foot trefoil	Fabaceae	Sato *et al.* (2008)
Nymphaea thermarum	Pygmy Rwandan waterlily	Nymphaeaceae	Povilus *et al.* (2015)
Oryza sativa	Rice	Poaceae	Goff *et al.* (2002); Yu *et al.* (2002)
Petunia × atkinsiana (= *P. hybrida*)	Petunia	Solanaceae	Gerats and Vandenbussche (2005)
Phalaenopsis equestris	Horse moth orchid	Orchidaceae	Cai *et al.* (2014)
Pisum sativum	Pea	Fabaceae	Smýkal *et al.* (2012)
Populus balsamifera subsp. *trichocarpa* (= *P. trichocarpa*)	Western balsam-poplar	Salicaceae	Tuskan *et al.* (2006)
Selaginella moellendorffii	'Selaginella'	Selaginellaceae	Banks *et al.* (2011)
Setaria viridis	Green foxtail	Poaceae	Layton and Kellogg (2014)
Solanum lycopersicum (= *Lycopersicon esculentum*)	Tomato	Solanaceae	Mueller *et al.* (2005)
Zea mays	Maize	Poaceae	Strable and Scanlon (2009)

FIGURE 3.1 Brassicaceae of the genus *Cardamine* are increasingly studied as model plants, partly – although not exclusively – because of the diversity of their leaf shapes, which include pinnate and palmate. Their strict affinities to *Arabidopsis thaliana* (a plant with entire leaves) allows for interesting evolutionary comparisons. Here, large bitter-cress (*Cardamine amara*). *A black and white version of this figure will appear in some formats. For the colour version, please refer to the plate section.*

Kutschera *et al.*, 2013) and the sea squirt hitherto referred to the one species *Ciona intestinalis* (Iannelli *et al.*, 2007; Brunetti *et al.*, 2015; Gissi *et al.*, 2017). Similar problems may trouble an emerging plant model species, the grass *Setaria viridis*, whose identity is not well circumscribed in respect to other *Setaria* taxa (Layton and Kellogg, 2014).

3.3   IN SILICO MODELS

Growing plants in the lab and experimenting on them is not the only way to investigate development and the evolution of development. A different resource is offered by the computer.

Throughout the centuries, the geometric regularity of plant form has always fascinated the human mind and has inspired a diversity of efforts aimed to reduce it to simple principles. The most

FIGURE 3.2 In the 1990s, Elena Kramer and her team introduced columbines (*Aquilegia*; here a hybrid cultivar without spurs) as novel model species, the first to represent the basally branching eudicots.

obvious example is phyllotaxis (see Section 5.7), but virtually every aspect of plant morphology has been described, sooner or later, in terms of circles, of spirals, of symmetry relationships, by plant anatomists such as Nehemiah Grew (1682), by experts in fruit trees (pomologists, e.g. Manger, 1783) and even by artists and art historians (e.g. Ruskin, 1900); examples from these authors are reproduced or discussed in Minelli (2015d).

Today, the computer offers an unprecedented opportunity to simulate plant development, by using very limited amounts of factual information about real organisms. Even gene expression patterns are more likely to be predicted by the simulations than picked from the molecular lab and used to feed a computer program, although genetics must eventually be brought to a meeting point with geometry (e.g. Coen *et al.*, 2004). To be sure, the morphological agreement between the results of computer simulations and the shape and architecture of real organisms does not mean that in vivo development bears any similarity to a behaviour simulated in silico. The real value of computer simulations resides elsewhere, as demonstrated by Przemysław Prusinkiewicz with his elegant modelling of plant growth and development (e.g. Prusinkiewicz and Lindenmayer, 1990; Prusinkiewicz, 2004), and especially with his efforts to provide a general framework for the development of the inflorescence (e.g. Prusinkiewicz *et al.*, 2007).

In a time in which the metaphor of the 'genetic program for' any imaginable phenotype is still so deeply entrenched both in science and in the popular imagination, it is very noticeable that a computer scientist kept his distance from the most expected, but not necessarily best informed, approach that would have suggested treating the vegetative structure (branching pattern) of the inflorescence and the spatial distribution of flowers on it as separate features the production of which could be simulated by distinct sets of rules. In Prusinkiewicz's models, flowers and branches are not treated as completely separate sets of objects, to be combined like the balls and sticks of the traditional models of molecular structure; instead, the meristems that give rise to shoots or flowers are regarded as two extremes of a continuum. To describe variation in this continuum, Prusinkiewicz introduced a variable called vegetativeness (veg), the value of which was considered to depend on many factors, intrinsic and extrinsic alike, such as the age of the plant, the position of the meristem and its internal state, and any relevant environmental influence.

In the model, high levels of veg correspond to shoot meristem identity, and low levels to flower meristem identity. In the paper in which they laid down the fundamentals of this model, Prusinkiewicz *et al.* (2007) identified the main factors influencing veg as age, measured from the beginning of inflorescence development, and the internal state of the meristem. The plant will go on producing an indeterminate vegetative branching structure as long as veg is high and does not change with time. When veg declines in some or all meristems, the plant produces flowers. With simple additional rules, the model can produce all the main architectural types of inflorescences, i.e. panicles, cymes and racemes. If veg decreases uniformly in all meristems, all meristems will at first continue to branch but will eventually terminate in flowers, all at the same time, thus producing a panicle. Instead, if the level of veg is transiently decreased in lateral meristems, these will reach the threshold for generating flowers before the apical meristems will do the same, and a raceme will be thus generated. A third kind of inflorescence, i.e. a cyme, is formed if veg is transiently increased in the lateral meristems, with the consequence that the apical meristems will precede the lateral meristems in the production of flowers (Prusinkiewicz *et al.*, 2007; Koes, 2008; Castel *et al.*, 2010).

In this way, a model is built that is flexible enough to accommodate suggestions concerning any factor experimentally found to influence flowering transition; at the same time, the model embodies the idea that developmental choices such as shoot branching or flowering are not univocally determined by the expression of one or a few (master) genes. Still more important is the fact that veg is conceived as a continuous variable rather than two-state (e.g. shoot vs. flower) or multistate (e.g. growth without branching vs. branching vs. transition to flowering). This is the key feature of Prusinkiewicz's in silico model, in so far as it suggests that the prima facie modular architecture of the inflorescence is not necessarily generated by a sequence of morphogenetic steps involving discrete 'archetypical' units, such as internodes, prophylls, branching points and flowers. This corresponds to a view of the plant body very different from the traditional one, within which leaves, inflorescences, flowers, petals, stamens etc. are 'given' – that is, represent (1) homologues, i.e. natural units of comparison, and (2) plant organs for each of which it is sensible to ask which developmental process generates them and which are the genes involved in their specification. If the actual developmental processes in vivo have any similarity to those assumed in the simulation, their primary outcome will not be individual organs (e.g. flowers or petals) with specific qualities, but properties (e.g. 'floweriness' or 'petalness') with a specific (and perhaps temporally dynamic) spatial distribution: the spatial distribution of any of those properties and the anatomical boundaries of individual organs do not overlap and are the outcome of distinct aspects of plant development and evolution. In developmental terms, this alignment has arguably been favoured by autocatalytic enhancement of gene expression boundaries, as suggested by Theissen and Melzer (2007) (see Section 7.2.3.5); in evolutionary terms, it has probably provided a selective advantage which has favoured its eventual fixation. Nonetheless, plant organs of 'uncertain identity' are not rare and will be discussed more in detail in Chapter 8. In terms of homology, this circumstance – the not necessarily one-to-one correspondence between anatomical units (e.g. lateral organs) and specific qualities – necessitates relaxing the traditional all-or-nothing approach to homology, in favour of a combinatorial approach (Minelli and Fusco, 2013).

Those who are unwilling to pay serious attention to models of growth and development unless these are expressed in terms of gene expression should consider that in *Arabidopsis* there are indeed genes that, in a sense, affect the level of veg. These are *LEAFY* (*LFY*) and *TERMINAL FLOWER 1* (*TFL1*). In the wild type, *LFY* is expressed in the lateral floral meristems (Jack, 2004), *TFL1* in the apical meristem (Conti and Bradley, 2007). In loss-of-function *lfy* mutants, flowers are replaced by shoot-like structures, while the inflorescences produced by *tfl1* mutants are short and terminate with a flower (Jack, 2004). It can thus be concluded that *TFL1* promotes veg and *LFY* represses veg. These, and additional genetic data, were eventually incorporated in the model by Prusinkiewicz *et al.* (2007).

3.4 SERIAL HOMOLOGY

From an ontological point of view, there are three main comparative frameworks within which we may discuss homology relationships: the units to be compared (usually, body parts or their features, but also processes, as mentioned in the previous chapter) may belong (1) to different taxa, (2) to different individuals of the same species, or (3) to one individual plant.

Homologies between different species, or between different supraspecific taxa, represented in the comparative analysis either by an exemplary species or by an abstract groundplan, are the subject matter of comparative anatomy (not necessarily restricted to the adult) and phylogenetics. Their potential relevance for evo-devo and, vice versa, the potential contribution of evo-devo to homology assessment will not be discussed here, as these are general questions, not specific to angiosperms; suggestions for the interested reader include Minelli (2015a, 2015b). Homologies involving different individuals of the same species are possibly involved in studies of genetic polymorphism and environmentally induced polyphenism, but angiosperms do not offer anything comparable to the wealth of primary and secondary differences between sexes that make this subject of comparative anatomy (and developmental biology) so interesting in zoology. Far more important, in botany, are the comparisons between different parts of the same individual plant. We open in this way a short discussion on serial homology.

By addressing a question of serial homology, we implicitly assume that we know where the series begins and where it ends, the comparison being thus restricted to an unequivocally delimited set. This is not necessarily clear in plants, witness Goethe's idealization, that in a plant *alles ist Blatt*, all is leaf. Serial homology in plants is indeed more intriguing when comparisons challenge conventional organ identity, such as leaf vs. bract vs. sepal vs. petal, or fertile stamen vs. staminode vs. petal (see Section 7.2.3.2). The main proposals thus far advanced to overcome the strictures of traditional morphology are the dissection of homology into a positional component versus one of special quality (Albert *et al.*, 1998; Kirchoff, 2000; Baum and Donoghue, 2002; Hawkins, 2002; Ronse De Craene, 2003; Jaramillo and Kramer, 2007; Rasmussen *et al.*, 2009) and the continuum or Fuzzy Arberian Morphology discussed in Section 8.5.1.

More than in other aspects of comparative biology, the study of serial homology requires a factorial approach (Minelli and Fusco, 2013; Minelli, 2016b), through the dissection of the phenotypes under comparison and, whenever possible, of the underlying developmental processes, in basic unit characters, in order to single out those that are actually relevant to the issue. The developmental processes by which position is specified are not necessarily coextensive with a set of serially homologous organs, but may extend beyond it, or, on the contrary, fail to cover the positional specification of all members in the series.

Moreover, an organ's special quality also results from a number of partly sequential, partly overlapping developmental processes, many of which are not specific for that organ, and the notion of organogenesis is reduced to an umbrella term for a temporal window of development, defined by the final product rather than by the underlying dynamics.

Interesting comparisons are those that involve a series of elements separated either by a temporal divide, or by a structural divide, or both. The temporal divide refers to the transitions from a developmental phase to the next (see Section 9.3). Uncontroversial examples of heterogeneous series of organs among which there is homology (but always a qualified homology of which we need to specify the terms) are embryonic versus post-embryonic leaves, i.e. cotyledons versus 'true' leaves; juvenile (e.g. rosette) versus adult (e.g. cauline) leaves. Comparisons become perhaps more interesting when these involve elements separated by a structural, rather than temporal, divide, as between leaf and stipule (sometimes a puzzling relationship: see Section 8.5.3), or between major and minor leaves produced by the same anisophyllous branch, as in *Theligonum cynocrambe* and *Anisophyllea disticha*; see also Section 8.2.3). The most attractive (and most popular) set of plant organs, the homology of which has long attracted most interest, are the parts of the flower. As we will see in Chapters 4 and 7, problems of homology between floral parts are particularly intricate because of three circumstances: (1) the frequent uncoupling of positional homology versus homology based on special quality, (2) the widespread occurrence of multiple floral organs deriving from common primordia (but eventually diverging in the differentiated stage, e.g. sepals and petals in *Bletia*, petals and stamens in the Primulaceae, and even sepals and stamens in *Astrantia major*), and (3) the co-occurrence, in the same floral organ, of parts that would qualify as typical of different kinds of floral organs, e.g. sepal and petal, or petal and stamen.

Progress in evo-devo suggests novel conceptual tools for dealing with comparisons of this kind. Decoupling positional homology from special-quality homology represented a fundamental step towards developing a factorial (combinatorial) approach to the homology of plant organs, but a further step is arguably needed, as discussed below in Section 7.2.3.2.

4 Genes and Genomes

The C-value, i.e. the amount of DNA contained in a haploid nucleus, expressed in picograms, varies enormously in the angiosperms, from 0.06 in *Genlisea margaretae* and *G. aurea* (Greilhuber *et al.*, 2006) to 152.20 in *Paris japonica* (Pellicer *et al.*, 2010). Values higher than 40 have been found nearly exclusively in monocots (102 species belonging to the Alstroemeriaceae, Melanthiaceae, Liliaceae, Asparagaceae, Amaryllidaceae and Commelinaceae), otherwise only in four species of *Viscum* and in *Hepatica nobilis* (Bennett and Leitch, 2012). At the opposite end of the range, very small amounts of nuclear DNA are found in some species of *Utricularia* (a close relative of *Genlisea*), with 1 C = 0.10 (Greilhuber *et al.*, 2006), and in *Fragaria viridis*, with 1 C = 0.14 (Antonius and Ahokas, 1996).

The mean haploid genome size of *A. thaliana* has been determined as 0.21 pg (Schmuths *et al.*, 2004), but this model species has a high level of endopolyploidy affecting nearly all parts of the plant: cotyledons, roots, lower leaf stalks, lower leaves, upper stem, upper leaves, flower stalks, sepals and petals (Barow and Meister, 2003).

4.2 DUPLICATIONS OF GENOMES AND GENES

The genomes of flowering plants are typically large and contain many genes, the majority of which evolved during the past 250–300 million years through extensive duplication events (Lynch and Conery, 2000).

All extant angiosperms are palaeopolyploids containing the remnants of at least two whole-genome duplications (Jiao *et al.*, 2011). One of these events occurred before the radiation of the seed plants, the other at the base of the angiosperms. Furthermore, a hexaploidy event predates the emergence of the core eudicots (Jaillon *et al.*, 2007; Jiao *et al.*, 2012; Vekemans *et al.*, 2012a). Additional, more recent duplications affected the genome of different plant lineages. At least 50 independent ancient whole-genome duplications are distributed across flowering plant phylogeny. The genomes of some extant species carry the remains of up to six successive genome duplications (Van de Peer *et al.*, 2009; Wendel, 2015). It has been suggested (Paterson *et al.*, 2004; Soltis *et al.*, 2009) that the diversification of some of the most speciose groups of angiosperms, e.g. Fabaceae, Poaceae and Asteraceae, has been in some way fostered by polyploidization events at the root of each of these clades. In the case of the Asteraceae, the rapid diversification of the many tribes may have been a consequence of whole-genome duplication events that occurred during the last 40 million years (Kim *et al.*, 2005a; Barker *et al.*, 2008).

In *Arabidopsis* there is evidence of three duplication events, known as γ (prior to the monocot–eudicot split), β (before the diversification of the eudicots but after the monocot–eudicot divergence) and α (after the divergence of the *Arabidopsis* lineage from other eurosids) (Bowers *et al.*, 2003).

Gene duplications in selected gene families involved in the control of major developmental events such as specification and patterning of flower organs deserve closer attention.

PISTILLATA (PI)

A duplication that occurred approximately 260 million years ago (Kim *et al.*, 2004), i.e. before the emergence of the flowering plants, generated the paralogous *PI/GLO* and paleo*AP3/DEF* lineages.

In *Arabidopsis*, *PI* is required to specify both petal and stamen identities (Goto and Meyerowitz, 1994); this is also true of *GLOBOSA* (*GLO*), its *Antirrhinum* homologue (Tröbner *et al.*, 1992).

The genome of *Aquilegia vulgaris* contains only one *PI* gene, which appears to have a role in specifying both petals and stamens (Kramer *et al.*, 2007). In many other plants, duplicate *PI* genes are known. Two *PI* lineage genes are present, for example, in *Papaver somniferum*, following a recent gene duplication event that occurred during the evolution of the Papaveraceae. Of these *PI* genes, only one is involved in both petal and stamen specification (Drea *et al.*, 2007). Two *PI* paralogues (*GLO1* and *GLO2*, resulting from a recent duplication event and both involved in petal and stamen development) have been found also in petunia and tomato (Vandenbussche *et al.*, 2004; Geuten and Irish, 2010).

In the Ericales, a widespread phenotypic correlate of the presence of duplicated *PI* genes, due to an ancient duplication event at the base of the order, is the fusion of stamens to form a ring- or tube-like structure. For example, within the Styracaceae, two *PI* copies are present in *Halesia*, with fused stamens, but only one *PI* copy is found in *Styrax*, with unfused stamens. Similarly, within the Actinidiaceae, two *PI* copies are present in *Saurauia kegeliana*, a species with fused stamens, but only one in *Actinidia chinensis*, with unfused stamens. In other families of the Ericales, *Impatiens glandulifera* (Balsaminaceae) and *Gustavia brasiliensis* (Lecythidaceae), both with fused stamens, have two copies of *PI*, due to a more recent duplication event. Exceptions to this correlation between gene copy number and androecial phenotype are found, however, in the Primulaceae: *Jacquinia auriantaca* and *Clavija latifolia* possess one *PI* copy only, but a staminal tube is present in both species (Viaene *et al.*, 2009).

APETALA3/DEFICIENS (AP3/DEF)

AP3 is another gene required for specifying both petal and stamen identities in *Arabidopsis* (Bowman *et al.*, 1989; Jack *et al.*, 1992). Similarly, in *Asparagus officinalis* the paleo*AP3* gene is expressed in the inner whorl of tepals and in the stamens (Park *et al.*, 2003).

In the Ranunculaceae, the *AP3* lineage has undergone at least two duplication events, giving rise to three paralogous lineages (*AP3-1*, *AP3-2* and *AP3-3*), which are found throughout the family (Kramer *et al.*, 2003). *AP3-3* orthologues are expressed in petal primordia and developing petals, and in species lacking petals there is no

AP3-3 expression. Loss of *AP3-3* expression has occurred several times independently, by different mechanisms, involving either deletions or insertions in the coding, promoter or intronic regions (Zhang *et al.*, 2013a). In *Aquilegia*, the *AP3-1* paralogue is broadly expressed in the primordia of petals, stamens and staminodes, but limited to the early stages, while later, at the time of initiation of carpels, its expression is restricted to the staminodes. In the same plant, the expression of the *AP3-2* paralogue begins somewhat later and initially extends over stamens and staminodes, then it disappears from the staminodes but still later it shows up in the petals. The *AP3-3* paralogue is only expressed in petals (Drea *et al.*, 2007; Rasmussen *et al.*, 2009). Divergent functional specialization (*subfunctionalization*) of the *AP3* paralogues is thus combined, in *Aquilegia*, with the involvement of a paralogue in a new function (*neofunctionalization*) (Kramer *et al.*, 2007).

During floral ontogeny, the expression of the paleo*AP3* genes in the petals of the Papaveraceae becomes restricted to the margins and tips of these organs (Kramer and Irish, 1999), in contrast to the sustained expression of their homologues throughout the whole organ, required for petal development in *Arabidopsis* and *Antirrhinum* (Kramer and Irish, 1999; Drea *et al.*, 2007). In *Papaver somniferum*, *AP3-1* is required for petal identity, whereas *AP3-2* controls stamen identity (Drea *et al.*, 2007).

In grasses (maize and rice), the expression of paleo*AP3* genes is required for the development of the stamens and of the ensheating elements known as the lodicules (Ambrose *et al.*, 2000; Nagasawa *et al.*, 2003). Parallelism between the roles of paleo*AP3* genes in monocots and eu*AP3* genes in eudicots suggests homology between lodicules and petals (Whipple *et al.*, 2004) (see Section 7.2.3.8).

In orchids, four paralogous lineages of *AP3* homologues are involved in the evolution of the strongly differentiated petaloid organs. The combined expression patterns of these paralogues are responsible for the strong differentiation between first and second perianth whorls and also between lateral petals and labellum within the second-whorl organs (Mondragón-Palomino and Theißen, 2008; Mondragón-Palomino *et al.*, 2009) (see Section 7.4.7).

At the base of the eudicots, a duplication in the paleo*AP3/DEF* lineage produced two lineages known as eu*AP3* and *TOMATO MADS BOX GENE6* (*TM6*) (Kramer *et al.*, 1998, 2006; Kramer and Irish, 1999; Causier *et al.*, 2010b). Members of the eu*AP3* paralogous lineage are only found in the so-called higher eudicots (Irish, 2006).

TM6 was lost in *Arabidopsis* and *Antirrhinum*, but is still present in tomato and tobacco; *AP3* was lost instead in papaya (Causier *et al.*, 2010b).

In petunia and tomato, *AP3* controls both petal and stamen development, while *TM6* is involved only in the development of stamens (de Martino *et al.*, 2006; Rijpkema *et al.*, 2006). This divergent expression pattern of the *DEF/TM6* paralogue correlates with the loss of a promoter element.

AGAMOUS/PLENA (AG/PLE)

The eu*AG* lineage and the *PLE* lineage arose about 120 million years ago in consequence of a duplication event (Davies *et al.*, 1999; Kramer *et al.*, 2004; Causier *et al.*, 2005).

The non-orthologous genes *AG* and *PLE* are responsible for specifying the C function in the flowers of *Arabidopsis* and *Antirrhinum*, respectively (Davies *et al.*, 1999; Kramer *et al.*, 2004; Causier *et al.*, 2005). In *Arabidopsis*, *AG* controls the determinacy of the floral meristem and the identity of stamens and carpels, and it has also some function in carpel and ovule development (Bowman *et al.*, 1989). In *Antirrhinum*, *PLE* has functions similar to *AG* in *Arabidopsis*, whereas the *Antirrhinum* orthologue of *AG*, *FARINELLI* (*FAR*), is involved in controlling stamen identity and, later, in fruit maturation (Davies *et al.*, 1999). A more recent duplication of *PLE* in the Brassicaceae has given rise to *SHATTERPROOF1* and *SHATTER-PROOF2* (*SHP1/2*). These paralogues play a novel role in *Arabidopsis* fruit development, an example of neofunctionalization (Liljegren *et al.*, 2000). In the phylogenetic lineages to which the two model species belong, the duplicated copies of these genes have thus undergone independent subfunctionalization (Causier *et al.*, 2005; Airoldi *et al.*, 2010).

The divergence of functions between the different *AG* paralogues present in rice and maize is very similar in these two grass species, suggesting that subfunctionalization began before the divergence of their lineages (Mena *et al.*, 1996; Yamaguchi *et al.*, 2006).

Another example of subfunctionalization is provided by the two *AG* genes of *Thalictrum thalictroides*, of which *AG1* exhibits typical C-class function, whereas *AG2* is not expressed in stamens, but only in carpels and ovules (Galimba and Di Stilio, 2015).

APETALA1/FRUITFULL (AP1/FUL)

AP1 is one of the main promoters of floral meristem identity: the flowers of loss-of-function mutants produce bract-like organs instead of sepals and lack petals. In the axils of the first-whorl substitutes of sepals, new floral meristems are produced that reiterate this pattern: the final phenotype is thus a 'branched flower' (Irish and Sussex, 1990; Bowman *et al.*, 1993).

In the *AP1* lineage, a major gene duplication event occurred at the base of the core eudicots, leading to the eu*AP1*, eu*FUL* and *FUL*-like gene clades (Litt and Irish, 2003).

In *Arabidopsis*, *AP1* (which belongs to the eu*AP1* clade) is required for both petal and sepal development, but it has an additional role in meristem specification (Irish and Sussex, 1990; Bowman *et al.*, 1993). However, the orthologues of this gene in other angiosperms do not seem to have any role in specifying the identity of flower organs, and only affect floral meristem identity. This was first discovered for the *Antirrhinum* orthologue of *AP1*, *SQUAMOSA* (*SQUA*), and later confirmed, for example, for the *AP1* orthologues in *Pisum sativum* (Berbel *et al.*, 2001; Taylor *et al.*, 2002) and tomato (Vrebalov *et al.*, 2002).

FUL has an additional function in promoting floral meristem identity, but also in fruit development (Gu *et al.*, 1998; Ferrándiz *et al.*, 2000); in other plants (*Petunia, Betula*), a role for the eu*FUL* gene in the control of floral induction has been suggested (Immink *et al.*, 1999; Elo *et al.*, 2007).

Two duplications in the *AP1/FUL* lineage have been demonstrated for the Poaceae, in which three gene clades are found. *FUL1*, *FUL2* and *FUL3* have broad expression domains that are not confined to the floral organs, including the accompanying glumes, but extend to roots, leaves and stems. The many roles hypothesized for these genes include floral induction and specification of floral meristem identity, followed by a later role in the differentiation of the spikelets (Trevaskis *et al.*, 2003; Preston and Kellogg, 2006, 2007, 2008). A wheat *AP1* paralogue is involved in the flowering transition dependent on vernalization (Danyluk *et al.*, 2003; Murai *et al.*, 2003; Trevaskis *et al.*, 2003; Shitsukawa *et al.*, 2007).

SEPALLATA (SEP)

In this large gene family, two major clades were the product of a duplication at the base of the angiosperms. These clades are known as *SEP3* and *LOFSEP* or *AGL2/3/4*, respectively. Both clades have undergone numerous additional duplications, often accompanied by clade-specific losses of gene copies. Within the *LOFSEP* clade, a duplication is correlated with the emergence of the core eudicots (Malcomber and Kellogg, 2005; Zahn *et al.*, 2005).

Four *SEP* genes are found in *Arabidopsis* (Pelaz *et al.*, 2000; Ditta *et al.*, 2004). Floral organ identities are completely lost in the quadruple loss-of-function mutant, suggesting a role of these genes in the specification of all floral organ types (Ditta *et al.*, 2004). Four *SEP* genes have also been reported in tomato (Hileman *et al.*, 2006) and at least six in petunia (Immink *et al.*, 2003).

In *Gerbera*, the *SEP1* orthologue is involved not only in controlling the architecture of the capitulum (Teeri *et al.*, 2006b), but also in fixing the determinate character of its meristem (Uimari *et al.*, 2004).

Following a monocot-specific gene duplication, in the Poaceae, the members of the *SEP* subfamily have undergone significant subfunctionalization (Vialette-Guiraud and Scutt, 2009); some of these genes control the development of the grass-specific spikelet meristems (Malcomber and Kellogg, 2004; Cui *et al.*, 2010; Gao *et al.*, 2010; Kobayashi *et al.*, 2010).

CINCINNATA (CIN) and *CYCLOIDEA (CYC)*

Members of the diverse *TEOSINTE BRANCHED1/CYCLOIDEA/PROLIFERATING CELL FACTOR (TCP)* family of genes coding for transcription factors are the leaf development gene *CIN* (Howarth and Donoghue, 2006) and the *CYC* lineage. Within the latter, the *CYC1*, *CYC2* and *CYC3* clades originated from two major duplication events just before the radiation of the core eudicots (Howarth and Donoghue, 2006). Of these paralogues, *CYC1* has the most prominent role in the control of the production of lateral branches (Nicolas and Cubas, 2016). The *CYC/TB1* clade underwent at least two duplications also in the evolutionary ancestry of the Poaceae, giving rise to three lineages (Mondragón-Palomino and Trontin, 2011), one of which includes all the genes known to control lateral shoot development in this family.

Multiple copies of *CYC2* genes, issued from further independent duplications, are found in most members of the rosid and asterid clades with zygomorphic flowers (Howarth and Donoghue, 2006). In these plants, a pair of *CYC2* clade genes is usually involved in establishing floral zygomorphy (see Section 7.4.6). Recent duplications of *CYC* genes have also been identified within a single genus, e.g. in *Hiptage* (Zhang *et al.*, 2016).

YABBY (YAB)

The *YAB* gene family originated approximately with the emergence of the seed plants (Floyd and Bowman, 2007). The expression of these genes is largely confined to developing leaves, including cotyledons, and floral organs (Sawa *et al.*, 1999a; Siegfried *et al.*, 1999; Watanabe and Okada, 2003; Toriba *et al.*, 2007; Floyd and Bowman, 2010).

Five *YAB* gene groups were originated by a series of duplications. In some plants, e.g. in *Arabidopsis*, the expression of two of these gene groups – *CRABS CLAW* (*CRC*) and *INNER NO OUTER* (*INO*) – is restricted to the floral organs, whereas the members of the *FILAMENTOUS FLOWER* (*FIL*), *YAB2* and *YAB5* gene groups are also expressed in leaves (Bartholmes *et al.*, 2012).

CUP-SHAPED COTYLEDON (CUC)

The *CUC* genes are expressed in the shoot apical meristem during advanced embryonic stages and are found again at the boundary between the apical meristem (a region with indeterminate growth) and the lateral determinate organs (Aida *et al.*, 1999; Ishida *et al.*, 2000; Takada *et al.*, 2001; Vroemen *et al.*, 2003; Hibara *et al.*, 2006).

The *CUC* gene family comprises two groups, issued from a duplication predating the last common ancestor of the extant angiosperms. A first group, the *NAM*-like *CUC* genes, includes genes regulated by the microRNAs miR164 (see next section) such as *CUP* of *Antirrhinum*, *NAM* of *Petunia*, and *CUC1* and *CUC2* of *Arabidopsis* (Vialette-Guiraud *et al.*, 2011). The genes of the second group, which have lost regulation by miR164, include *CUC3* of *Arabidopsis* and its orthologues.

4.3 MIRNAS

An important contribution to the regulation of gene expression is due to small polynucleotides known as the microRNAs, or miRNAs. These are single-strand RNA molecules of approximately 21 nucleotides that hybridize to complementary sites in target genes. miRNAs are also known in animals, with a diversity of molecules that roughly increases in direct relationship with the animal's body complexity: Erwin *et al.* (2011) postulated, indeed, that the expansion of gene regulation by miRNAs has been causally involved in the evolution of complex body plans and, conversely, that evolutionary simplification is accompanied by loss of miRNA families.

There is a difference between animal and plant miRNAs in the specificity of their action: animal miRNAs bind to multiple, partially complementary sites in untranslated regions of their messenger RNA (mRNA) targets, whereas plant miRNAs generally bind only to one, strictly complementary site within the coding sequence of their target mRNAs (Vazquez, 2009).

The most important targets of plant miRNAs are transcription factors that have, in turn, a critical role in regulating the expression of other genes. In this way, miRNAs become themselves part of important regulation pathways. Processes in which plant miRNAs are involved include those that affect leaf morphogenesis, flower morphology, organ polarity and number, phase transitions (see Section 9.3.3.1), root initiation and vascular development (Vazquez, 2009). Only a couple of examples are given here, but miRNAs will be mentioned again in other sections.

In *Arabidopsis thaliana*, the *miR164* family is represented by three genes, one of which (*miR164a*) controls the depth of leaf sinuses (Nikovics *et al.*, 2006) and, together with a second copy (*miR164b*), has a role in controlling lateral root branching (Guo *et al.*, 2005). The third gene of this family (*miR164c*) is involved instead in regulating petal number (Baker *et al.*, 2005). Another family (*miR165/166*) has roles in meristem maintenance, in vasculature patterning and in the specification of leaf polarity (reviewed in Chen, 2012).

In *Cabomba aquatica*, the *miR164a* homologue is more highly expressed in submerged than in floating leaves, while the expression of the *miR164b* homologue is limited to the submerged leaves (Jasinski *et al.*, 2010). This suggests an involvement of these microRNAs in the mechanisms controlling the dissection of the submerged leaves in this water plant.

Comparative studies on miRNAs are still limited to a small number of species. Sequence and expression data from more numerous taxa well distributed along the branches of the phylogenetic tree may prove useful, as suggested by the following. Similar to animals, new regulatory functions controlled by miRNAs have continuously emerged in plant evolution, but many important functions in plant development of these RNAs turn out to be conserved between *Arabidopsis*, rice, ferns and mosses (Willmann and Poethig 2007;

Vazquez, 2009). Of the 69 miRNA families shared between
A. thaliana and *A. lyrata*, 22 are also present in maize and rice
(Fahlgren *et al.*, 2010).

4.4 MITOCHONDRIAL AND PLASTID GENOME

With a size ranging from 66 to over 2400 kb (Fauron *et al.*, 2004;
Skippington *et al.*, 2015), plant mitochondrial genomes are much
larger than those of animals (14–18 kb) or fungi (70–100 kb). The
smallest among the plant mitochondrial genomes has been found
in the hemiparasitic mistletoe *Viscum scurruloideum*. This unusual
size reduction is the result of the loss of a number of genes, including
the sequences that encode for the respiratory complex I (NADH
dehydrogenase), which in all other plants thus far examined (more
than 300 species) is encoded by nine mitochondrial genes plus a
number of nuclear genes (Skippington *et al.*, 2015).

Sequence evolution is generally slow in plant mitochondrial
genomes, but exceptions are known. In *Plantago*, for example, substitution rates at some sites, but not others, are exceedingly fast and
surpass by an order of magnitude the fastest-evolving mitochondrial
genome of animals. These unusually high substitution rates are
apparently limited, in *Plantago*, to the mitochondrial genome and
do not extend to the chloroplast and nuclear genes studied by Cho
et al. (2004).

Compared to the mitochondrial genomes of plants, plastid
genomes are much more conservative: those of the 139 seed plant
species with completely sequenced and annotated plant genomes listed
by Daniell *et al.* (2016) range from 107 kb to 218 kb. Most of the lowest
values are from gymnosperms, including the 107 kb of *Cathaya argyro-
phylla*; the smallest angiosperm plastid genome in the list (114.8 kb)
belongs to *Secale cereale*, the largest to *Pelargonium* spp. Chloroplast
genome size varies independent of nuclear genome size. In this general
context, unique is the plastid genome of the parasitic *Cytinus hypocistis*, slightly larger than an animal mitochondrial genome, 19.4 kb
in size and reduced to 23 genes. All coding regions of the plastid

genome of *Cytinus* are characterized by very high substitution rates compared with those of non-parasitic members of the Malvales (Roquet *et al.*, 2016).

4.5 HORIZONTAL GENE TRANSFER

Horizontal gene transfer (Table 4.1) is much more frequent in mitochondrial than in nuclear and plastid genomes, and is particularly common in parasitic plants (Davis and Xi, 2015). For example, in the Rafflesiaceae, Xi *et al.* (2013) found that 24–41% of mitochondrial genes have likely undergone horizontal gene transfer. A large-scale transcriptome analysis of *Rafflesia cantleyi* identified at least 31 genes obtained from its hosts (Xi *et al.*, 2012). Horizontal gene transfer in both directions has been identified in the mitochondria of most lineages of parasitic plants (Davis and Xi, 2015). Even in a hemiparasitic and morphologically 'conventional' plant such as *Striga asiatica*, a complete sequence analysis of the mitochondrial genome revealed at least three protein-coding genes deriving from its monocot hosts (Yoshida *et al.*, 2016).

An exceptional condition has been found in the enormous mitochondrial genome of *Amborella trichopoda* (3.9 Mb), which contains six genome equivalents of foreign mitochondrial DNA, acquired in part from other angiosperms, in part from green algae and mosses. This includes the acquisition of entire mitochondrial genomes from three green algae and one moss species (Rice *et al.*, 2013). Most *Amborella* transgenes are pseudogenes, in contrast to the transferred genes in the Rafflesiaceae, which have retained their gene structures and probably also their functions (Rice *et al.*, 2013; Xi *et al.*, 2013), while possibly contributing to the evolution of the parasite.

In parasitic plants, reciprocal influences between the host's and the parasite's metabolism may be important even in the absence of gene transfer. Through symplastic connections with its hosts, *Cuscuta pentagona* takes up host mRNAs. Kim *et al.* (2014) found that mRNAs move in either direction. In experiments where *Arabidopsis* plants were used as hosts to *Cuscuta*, nearly half of the expressed transcriptome of *Arabidopsis* was identified in *Cuscuta*.

Table 4.1 *Foreign genes in plant mitochondrial genomes, derived from horizontal gene transfer, often exchanged with **parasitic plants** (in **bold**), largely after Richardson and Palmer (2007).*

Recipient	Donor	Gene(s)	Current state	Reference
Actinidia	Monocot	*rps2*	Replacement of a previously lost nuclear gene	Bergthorsson *et al.* (2003)
Amborella	Moss	*cox2, nad2, nad3, nad4, nad5, nad6, nad7*	Duplication	Bergthorsson *et al.* (2004)
Amborella	Angiosperms	*atp1, atp4, atp6, atp8, atp9, ccmB, ccmC, ccmFN1, cox2* (2 copies), *cox3, nad1, nad2, nad4, nad5, nad7, rpl16, rps19, sdh4*	Duplication	Bergthorsson *et al.* (2003, 2004)
Apodanthaceae	Fabales	*atp1*	?	Nickrent *et al.* (2004)
Betulaceae	?	*rps11*	Replacement of a previously lost nuclear gene	Bergthorsson *et al.* (2003)
Bruinsmia	Cyrillaceae	*atp1*	?	Schönenberger *et al.* (2005)
Caprifoliaceae	Ranunculales	*rps11*	Replacement of a previously lost nuclear gene	Bergthorsson *et al.* (2003)
Phaseolus	Angiosperm	*cp pvs-trnA*	Novel gene	Woloszynska *et al.* (2004)
Plantago	**Orobanchaceae, Convolvulaceae**	*atp1*	Duplication	Mower *et al.* (2004)
Rafflesiaceae	Vitaceae	*nad1B-C*	?	Davis and Wurdack (2004)
Sanguinaria	Monocot	*39 rps11*	Chimerism	Bergthorsson *et al.* (2003)
Ternstroemia	Ericaceae	*atp1*	?	Schönenberger *et al.* (2005)

4.6 COMPARATIVE DEVELOPMENTAL GENETICS

4.6.1 Developmental Genes?

Are there true 'developmental genes'? Yes, but a qualified yes (Minelli, 2003). Indeed, the patterns of expression of many of the genes listed in Table 4.2 are strictly limited to and correlated with specific times and events in development. Mutations in these genes may critically and conspicuously alter the normal course of development, sometimes with dramatic effects on the resulting phenotype. However, we must refrain from taking any of them as 'the gene for' the shape of a leaf or the symmetry of a flower. Very aptly, Cronk's useful book, in which plant architecture is described from the perspective of the genetic control of development, is entitled *The Molecular Organography of Plants* (2009): here, genes perform as actors in a play of which the resulting phenotype is nevertheless the real focus.

Comparisons with animals, where in any case the concept of developmental genes must also be treated with the utmost caution (Minelli, 2003), are not easy. First, in angiosperms, zygotic genome activation occurs already in the zygote (Lau *et al.*, 2012), and thus early plant embryogenesis is mostly under zygotic control (Nodine and Bartel, 2012): this is very different from early embryogenesis in animals, which is to a substantial extent under the control of maternal genes. Second, and more important, animal developmental biology focuses, to a very large extent, on embryos, on those early developmental steps through which the developing animal emerges, in the vast majority of cases, with distinct architectural traits, even in the numerous instances in which these define a larval body fated to undergo extensive rearrangement at metamorphosis. In plants, I would dare to say – contrary to traditional and current usage – that no embryo exists, strictly comparable to animals, because most of plant morphogenesis will start later, often much later, with the sorting out, as founders of lateral organs, of groups of cells that will later become the playground for the expression of a number of 'developmental genes'.

Table 4.2 *Short-form and full names for a selection of genes or gene families with major developmental effects.*

Short name	Full name
AG	*AGAMOUS*
AGL	*AGAMOUS-LIKE*
AN	*ANGUSTIFOLIA*
ANT	*AINTEGUMENTA*
AP	*APETALA*
ARF	*AUXIN RESPONSE FACTOR*
ARGOS	*AUXIN-REGULATED GENE INVOLVED IN ORGAN SIZE*
ARP	*ACIDIC REPEAT PROTEIN*
AS	*ASYMMETRIC LEAVES*
BOP	*BLADE-ON-PETIOLE*
BEL	*BELL*
BP	*BREVIPEDICELLUS* (= *KNAT1*)
BRC	*BRANCHED*
CAL	*CAULIFLOWER*
CEN	*CENTRORADIALIS*
CHSU	*CHORIZOPETALA SUZANNE*
CIN	*CINCINNATA*
CLV	*CLAVATA*
CRC	*CRABS CLAW*
CRL	*CROWN-ROOTLESS*
CUC	*CUP-SHAPED COTYLEDON*
CYC	*CYCLOIDEA*
CYL	*CYCLOIDEA-LIKE*
DEF	*DEFICIENS*
DICH	*DICHOTOMA*
DIV	*DIVARICATA*
DL	*DROOPING LEAF*
ELF	*EARLY-FLOWERING*
ETT	*ETTIN*
EVG	*EVERGREEN*
FAR	*FARINELLI*
FBP	*FLORAL BINDING PROTEIN*
FD	*FLOWERING LOCUS D*

Table 4.2 (*cont.*)

Short name	Full name
FDH	*FIDDLEHEAD*
FIL	*FILAMENTOUS FLOWER*
FIM	*FIMBRIATA*
FLC	*FLOWERING LOCUS C*
FLO	*FLORICAULA*
FRI	*FRIGIDA*
FT	*FLOWERING LOCUS T*
FUL	*FRUITFULL*
GLO	*GLOBOSA*
GRAM	*GRAMINIFOLIA*
HSF	*HAIRY-SHEATH-FRAYED*
HST	*HASTY*
HTH	*HOTHEAD*
IDS	*INDETERMINATE SPIKELET*
INCO	*INCOMPOSITA*
INO	*INNER NO OUTER*
JAG	*JAGGED*
KAN	*KANADI*
KNOX	*KNOTTED-LIKE HOMEOBOX*
LAS	*LATERAL SUPPRESSOR*
LD	*LUMINIDEPENDENS*
LEP	*LEAFY PETIOLE*
LFY	*LEAFY*
LMI	*LATE MERISTEM IDENTITY*
LOB	*LATERAL ORGAN BOUNDARIES*
LOF	*LATERAL ORGAN FUSION*
LUG	*LEUNIG*
MADS-BOX genes	Family named after the four founder members, *MCM1, AG, DEF, SRF*
MAW	*MAEWEST*
MPF	*MATURATION PROMOTING FACTOR2*
NAM	*NO APICAL MERISTEM*
NLY	*NEEDLY*
NUB	*NUBBIN*
OBO	*ORGAN BOUNDARY*

Table 4.2 (*cont.*)

Short name	Full name
PAN	*PERIANTHIA*
PEBP	*PHOSPHATODYLETHANOLAMINE-BINDING PROTEIN*
PHAN	*PHANTASTICA*
PHB	*PHABULOSA*
PHD	*PLANT HOMEODOMAIN*
PHV	*PHAVOLUTA*
PI	*PISTILLATA*
PIN	*PIN-FORMED*
PLA	*PLASTOCHRON*
PLE	*PLENA*
PSD	*PAUSED*
PTL	*PETAL LOSS*
RAD	*RADIALIS*
RANACYL	*RANUNCULACEAE CYCLOIDEA-LIKE*
RAY	*RAY*
RCO	*REDUCED COMPLEXITY*
REP	*RETARDED PALEA*
REV	*REVOLUTA*
ROT	*ROTUNDIFOLIA*
RS	*ROUGH SHEATH*
SEP	*SEPALLATA*
SEP1 = AGL2	
SEP2 = AGL4	
SEP3 =AGL9	
SEP4 =AGL3	
SHP	*SHATTERPROOF*
SHP1 = AGL1	
SPL	*SQUAMOSA PROMOTER BINDING PROTEIN-LIKE*
SQT	*SQUINT*
SQUA	*SQUAMOSA*
STK = AGL11	*SEEDSTICK*
STM	*SHOOTMERISTEMLESS*
STP	*STAMINA PISTILLOIDA*
SUP	*SUPERMAN*

Table 4.2 (*cont.*)

Short name	Full name
TB	*TEOSINTE BRANCHED*
TCP	*TEOSINTE BRANCHED1/CYCLOIDEA/ PROLIFERATING CELL FACTOR*
TFL	*TERMINAL FLOWER*
TL	*TENDRIL-LESS*
TM6	*TOMATO MADS BOX GENE6*
TOR	*TORTIFOLIA*
TS	*TASSELSEED*
UFO	*UNUSUAL FLORAL ORGANS*
ULT	*ULTRAPETALA*
UNI	*UNIFOLIATA*
WOX	*WUSCHEL-RELATED HOMEOBOX*
WUS	*WUSCHEL*
YAB	*YABBY*
ZFL	*ZEA LEAFY*

Rules for writing the names of genes (in full and in short form), their mutants and the proteins they encode are not uniform in botany. For example, the recommended format is (i) *GENE* (*GE*) [wild type], *mutant*, PROTEIN for *Arabidopsis* (The Arabidopsis Information Resource, www.arabidopsis.org); (ii) *gene* for *Zea mays* (A Standard for Maize Genetics Nomenclature, www.maizegdb.org/nomenclature); (iii) *GENE* for *Oryza sativa* (McCouch, 2008); (iv) *Gene, mutant* for *Triticum* (McIntosh, 1988); (v) *Gene* for *Solanum lycopersicum* (Tomato Genetics Resource Center, http://tgrc.ucdavis.edu). Increasingly frequent is the use of plant-name acronyms before the name of the gene, e.g. *GhCYC2* for the *CYC2* gene of *Gerbera hybrida*. In this book, gene names are given in uppercase italics, as in this table; the proteins they encode are also uppercase but not italicized; the few loss-of-function mutants cited in the text are given in lowercase italics. Plant-species acronyms have been used very sparingly.

Shall we therefore say that the development of a leaf, or a flower, begins at the time of expression of the 'boundary genes' (see Section 5.6.1) through which an organ primordium is isolated from the shoot apical meristem? Again, yes and no. It would be uselessly fastidious, however, to try to split this hair, and the argument is better brought to an end here by saying, with Oyama (2000), that a gene 'initiates' a sequence of events only if our investigation starts at that point (Oyama, 2000).

One more question, however, should be asked, to better clarify the developmental role of genes in the broader context of our evo-devo perspective: the question is whether it is genes that have generated new forms or if, instead, already existing forms have been attractors for the expression of genes that have eventually acquired, with time, a major function in the perpetuation of those phenotypes. The latter option, although perhaps counterintuitive and hardly reconcilable with the common view of DNA as the blueprint for the phenotype, is arguably closer to evolutionary history, as a rule at least. Budd (1999) expressed it by saying that, in a sense, it is form that captured genes more than genes that created form. The concept has been further defended (e.g. by Jablonka, 2006; Schwander and Leimar, 2011; Uller and Helanterä, 2011) under the possibly preferable terms of 'genes as followers', as opposed to 'genes as leaders' of phenotypic evolution.

This said, we must acknowledge the important role in morphogenesis of a number of transcription factors, many of which are mentioned in this book. These proteins are commonly described as responsible for 'master regulatory switches' in gene regulatory networks (e.g. Niklas, 2016). Whatever the meaning we are ready to assign to this qualification, the multiple contributions of these genes to the evolution of development cannot be overlooked. This may happen through gene duplications followed by subfunctionalization or neofunctionalization, by changes in their spatial expression patterns, by changes in the upstream regulatory genes by which their expression is modulated, or by modifications of their nucleotidic sequence, especially if these affect their DNA binding domains, with obvious consequences for the expression of the downstream genes or gene networks regulated by them.

Compared to animals, plants have a greater diversity of transcription factors; this suggests a more important role for regulation at the level of transcription (Mitsuda and Ohme-Takagi, 2009). More than 2000 transcription factors belonging to 21 gene families have been found in *Arabidopsis* (Perez-Rodriguez *et al.*, 2010).

Of these, the *TCP* family is plant-specific. *TCP* genes control leaf and flower size and shape as well as the suppression of shoot branching. These genes fall into two classes, class I and II. Within class II, two lineages found in angiosperms, *CIN* and *CYC/TB1* genes (Martín-Trillo and Cubas, 2010; Nicolas and Cubas, 2016), are repeatedly mentioned in this book.

A large family of transcription factors is represented by the homeobox genes, characterized by a highly conserved sequence of 180 nucleotides encoding a DNA binding domain of 60 amino acids (the homeodomain). Isolated for the first time from *Drosophila melanogaster*, homeobox genes were found to be involved in many aspects of development (Gehring *et al.*, 1994; Bürglin, 2005). Homeobox genes are known from all major eukaryotic lineages (Derelle *et al.*, 2007). Those recorded from plants can be classified into 14 classes (Mukherjee *et al.*, 2009; Mukherjee and Brocchieri, 2010), six of which deserve a mention here (many of these will be discussed in other chapters): *HD-ZIP III, WUSCHEL-RELATED HOMEOBOX (WOX), KNOTTED1-LIKE HOMEOBOX (KNOX1), BELL (BEL), PLANT HOMEODOMAIN (PHD)*, and *LUMINIDEPENDENS (LD)*. Among these classes only *BEL* and *KNOX* are found in red algae, whereas unicellular green algae possess also members of the *WOX* class. The whole set of 14 classes are found only in land plants, including mosses and vascular plants. In different land plants, including mosses and flowering plants, multiple gene copies are observed for most classes, suggesting that these had already differentiated in the common ancestor of mosses and vascular plants.

HD-ZIP III genes work as master regulators of embryonic apical fate (Smith and Long, 2010). Specifically, their products are involved in specifying adaxial fate in lateral organs (Ochando *et al.*, 2006). *KNOX1* genes are involved in establishing and maintaining the shoot apical meristem (Hake *et al.*, 2004; Ikezaki *et al.*, 2010). *BEL* genes are expressed in leaves and ovules during development (Cole *et al.*, 2006). Maize *PHD* genes are mainly expressed in meristematic cells (Über-lacker *et al.*, 1996). The floral repressor *FLC* encodes an *LD* class transcription factor. *WOX* genes are expressed in the boundary between the adaxial and abaxial sides of the leaf primordia.

4.6.2   Pleiotropy

The popular but misleading metaphor of the genes as the repositories of the 'program for' the construction of the organism may suggest that, as soon as we know the complete gene sequence of a plant or animal species, we also know how to make one or many of its sort. However, this is far from true. The path leading from genotype to phenotype (the genotype → phenotype map; cf. Alberch, 1991; Pigliucci, 2001; West-Eberhard, 2003; Draghi and Wagner, 2008) is complex and not neces-sarily predictable: the expression of variation depends on the structure of the underlying gene networks and the structure and robustness of the developing system (Wagner and Altenberg, 1996; Kirschner and Gerhart, 1998; Wagner, 2005; Hansen, 2006; Kaufmann *et al.*, 2010).

One of the few possible generalizations about the genotype → phenotype map is that it is dominated by pleiotropy, i.e. by the multiple phenotypic effects of mutations affecting a single locus. To some extent, what we call pleiotropy is just a consequence of how we describe the phenotype, that is, how we partition it into a list of characters. However, the pleiotropic effects of any given gene are far from random and often cluster around recognizable 'hubs' manifesting the modularity of development (Wagner and Altenberg, 1996). Genes acting as upstream regulators of developmental or cellular gene expression, usually as highly connected nodes in a network, are more likely to have conspicuous pleiotropic effects (de Bruijn *et al.*, 2012).

We can expect that pleiotropy will have an effect on evolvability, but the nature, the extent and even the polarity (positive or negative correlation) of this link is far from obvious. On the one hand, a higher degree of independence in respect to other characters is likely to increase the freedom of a character to evolve (Hendrikse *et al.*, 2007), but an increase in the number of correlated characters may translate into a growing number of constraints, i.e. a kind of antagonistic effect of pleiotropy on evolvability (Kirkpatrick, 2009; Walsh and Blows, 2009). The latter view was anticipated by Fisher's (1930) demonstration that the probability of a random mutation being favourable decreases steeply with the number of traits affected by it (Minelli, 2017).

Selected examples of major pleiotropic effects of plant genes with important developmental functions are presented in the following paragraphs.

Let's begin with *LFY*. Moyroud *et al.* (2010) proposed that this gene had an ancestral role in regulating cell division, and that in seed plants it acquired novel functions in meristem growth and specification of floral identity. These two functions are obvious in some angiosperm species, such as maize or pea; in other taxa the first might be cryptic, as in *Arabidopsis*, or the second might be reduced, as in rice, where *LFY* is otherwise required for the ramification of the inflorescence meristem (Rao *et al.*, 2008). Additionally, *LFY* genes promote indeterminate growth in the compound leaves of some legumes (see Section 6.4.2). According to Moyroud *et al.* (2010), this could represent a novel phenotypic outcome directly derived from an ancient role of *LFY* that in other plant species is less obvious or results in different phenotypes. What prima facie would seem different, independent roles of *LFY* – in the development of compound leaves, inflorescences in grasses, or axillary meristems in *Arabidopsis* – are probably only multiple expressions of this gene's capacity to stimulate meristematic growth.

The *MADS*-box genes code for a large family of transcription factors named from the four founder members (*MCM1, AG, DEFA, SRF*) (Schwarz-Sommer *et al.*, 1990) and are best known for their multiple functions in the specification of the identity of floral organs (see Section 4.6.3), but some of them have less conventional functions, either in the flower or outside it. The members of the *STMADS11* subfamily evolved novel functions in the transition from the vegetative to the reproductive phase, as well as in flower and fruit development. In *Betula pendula*, for example, the *MADS4* homologue is involved in the initiation of inflorescence development as well as in the transition from the vegetative to the reproductive phase of development (Elo *et al.*, 2007). In peach (*Prunus persica*), six genes of this subfamily are associated with a temporal phenotype (see Chapter 9), specifically with floral bud dormancy (Jiménez *et al.*, 2009; Li *et al.*, 2009).

In tomato, the *AP1* orthologue *MC* controls the indeterminate character of the inflorescence meristem as well as sepal size (Vrebalov *et al.*, 2002), but has also a function in the development of the abscission zone of the pedicel, with obvious consequences for seed dispersal (Nakano *et al.*, 2012).

In *Arabidopsis*, the *FUL* gene controls the development of the carpel, as shown by the deformed siliques, overfilled with seeds, representing the phenotype of its loss-of-function mutants (Gu *et al.*, 1998); this gene has also a role in the specification of meristem identity, as in its mutants the floral meristems may be replaced by leafy shoots (Ferrándiz *et al.*, 2000; Airoldi and Davies, 2012). A role in the regulation of the floral transition is also played in grasses by two *FUL* paralogues, *FUL1* and *FUL2* (Preston and Kellogg, 2007), and, in *Arabidopsis*, by the related gene *AGL17* (Han *et al.*, 2008), otherwise known for its potential role in root morphogenesis (Burgeff *et al.*, 2002).

CYC2 genes, as mentioned, are best known for their control of flower symmetry, but in some species the same genes control floral organ number or promote stamen abortion. This is the case of the two *CYC2* paralogues in *Antirrhinum*, *CYC* (Luo *et al.*, 1996) and *DICHOTOMA* (*DICH*; Luo *et al.*, 1999). The double loss-of-function mutant has radially symmetrical flowers, with six (instead of five) sepals, petals and stamens. In maize, the related *TB1* gene is exemplary for its pleiotropy: at lower nodes it prevents axillary bud outgrowth, in the female inflorescence it promotes the shortening of the internodes; additionally, it probably also promotes stamen abortion in the female florets (Doebley *et al.*, 1997; Wang *et al.*, 1999; Hubbard *et al.*, 2002).

Many genes involved in carpel development have additional effects on the development of other plant organs. For example, *CRC* is also important in *Arabidopsis* for nectary development, a function conserved across the Brassicales but also found in other orders, including the closely related Malvales and the more distantly related Solanales (Lee *et al.*, 2005). The unrelated gene *SUPPRESSOR OF OVEREXPRESSION OF CO1 (SOC1)*, as well as the *AG* clade genes *SHP1*, *SHP2* and *STK*, all involved in the reproductive transition and in carpel development, have also been found to play a role in the periodic spacing of lateral roots (Moreno-Risueno *et al.*, 2010).

In petunia (*Petunia* × *atkinsiana* [= *P. hybrida*]), *MAEWEST* (*MAW*) and *CHORIZOPETALA SUZANNE* (*CHSU*) are required for petal and carpel fusion, and also for the lateral outgrowth of the leaf blade (Vandenbussche *et al.*, 2009).

4.6.3 Floral Homeotic Genes

The most popular success of the early years of plant molecular developmental genetics was undoubtedly the explanation of the control of organ flower identity through the so-called ABC model (e.g. Bowman *et al.*, 1989; Carpenter and Coen, 1990; Schwarz-Sommer *et al.*, 1990; Coen and Meyerowitz, 1991). According to this model, the specification of each flower organ as a sepal, petal, stamen or carpel would depend on the local expression of a different subset of three homeotic genes, called A, B and C. A sepal would form where only A is expressed, a petal instead where A and B genes are expressed together, a stamen where B (but not A) is expressed together with C, and a carpel, finally, would form where only C is expressed. Genes with A, B or C function were soon identified in the two main model species. Class A genes would thus be represented by *APETALA1* (*AP1*) and *APETALA2* (*AP2*) in *Arabidopsis*, by *SQUAMOSA* (*SQUA*) in *Antirrhinum*; class B genes by *APETALA3* and *PISTILLATA* (*PI*) in *Arabidopsis*, by *DEFICIENS* (*DEF*) and *GLOBOSA* (*GLO*) in *Antirrhinum*; class C genes by *AGAMOUS* (*AG*) in *Arabidopsis*, by *PLENA* (*PLE*) in *Antirrhinum*.

During the last three decades, the picture has been extensively clarified and modified. It soon became clear that the model, in its attractive simplicity, could not be generalized to all angiosperms without important qualifications. A first important point was that, even between two core eudicots such as *Arabidopsis* and *Antirrhinum*, the (in part, putative) A, B and C functions were not always on the shoulders of the same genes (see Section 7.2.3.4). In particular, no gene with a function comparable to the putative A function in *Arabidopsis* was identified in other plants.

On the other hand, mainly based on studies in *Petunia*, it was suggested that an additional class of genes was involved in the specification of flower organs: the ABC model was thus expanded to an ABCD model, by the recognition of class D genes required for ovule formation (Colombo *et al.*, 1995; Angenent and Colombo, 1996). Representatives of the *FBP7/11* gene clade (D-lineage) appear widely conserved across the angiosperms (Kramer *et al.*, 2004), and most of the identified orthologues, including those of grasses (Dreni *et al.*, 2007), exhibit ovule-specific expression. However, the phenotypic effects of these genes are extensively pleiotropic (e.g. in *Aristolochia*; Suárez-Baron *et al.*, 2017). Despite the confirmed role in the production of ovules and thus of seed (e.g. in the grapevine; Ocarez and Mejía, 2016), it may be questioned if this gene deserves to be included in the standard combinatorial set of homeotic genes involved in the specification of organ flower identities.

Further studies suggested that the A-, B-, and C-function genes need other cofactors to produce floral organs (Pelaz *et al.*, 2000; Honma and Goto, 2001). According to a revised interpretation of the model, known as the quartet model (Theißen and Saedler, 2001), floral organ identity would be regulated by multimeric protein complexes, with a further component (the E function) providing proteins that act as mediators for linking the proteins encoded by the other homeotic floral genes (Melzer *et al.*, 2009): hence an ABCDE model was suggested (Immink *et al.*, 2009; Rijpkema *et al.*, 2010). The proteins encoded by these genes form dimers readily and probably bind to their consensus binding site in DNA as protein quartets (Theißen and Saedler, 2001; Melzer *et al.*, 2009). In this interpretation, the classical ABC code for floral organ determination in *Arabidopsis* becomes AA-EE (or AE-AE) for sepals, AE-BB for petals, BB-CE for stamens, CC-EE (or CE-CE) for carpels, and finally CE-DE for ovules (Immink *et al.*, 2010). In most flowering plants there is no A function, and sepals would be encoded by EE-EE (Broholm *et al.*, 2014).

In *Arabidopsis* the E function is encoded by four genes belonging to the *SEP* subfamily (Pelaz *et al.*, 2000): in the *sep1 sep2 sep3* triple mutant, sepals replace the floral organs of all other whorls (Pelaz *et al.*, 2000) and the addition of the *sep4* mutation results in the conversion of all floral organs into leaves (Ditta *et al.*, 2004). In other plant species such as petunia (Vandenbussche *et al.*, 2003; Rijpkema *et al.*, 2009), rice (Ohmori *et al.*, 2009; Cui *et al.*, 2010; Gao *et al.*, 2010; Li *et al.*, 2011) and maize (Thompson *et al.*, 2009), genes of the closely related *AGL6* subfamily are also responsible for the E function.

The molecular identification of the genes responsible for the different functions is thus revised. The floral homeotic genes respectively responsible for the A, B, C+D and E functions fall into separate clades: *SQUA* for class A genes, *DEF* or *GLO* for class B, *AG* for class C+D and *AGL2* or *AGL6* for class E (Theißen *et al.*, 1996; Rijpkema *et al.*, 2009; Melzer *et al.*, 2010).

A different version of the combinatorial model of flower organ identity is known as the BC model (Schwarz-Sommer *et al.*, 1990; Litt, 2007), more recently reformulated as the (A)BC model (Causier *et al.*, 2010a). According to the latter, both *AP1*-like and *SEP* genes would be responsible for the (A) function, but this would be denied a major role in the specification of perianth organs. The main roles of (A)-class genes would instead be the establishment, starting from the floral meristem, of a floral context where sepals only would form, as the default floral organ identity, unless otherwise specified by the expression of B- and C-class genes. Thus, in the strictest sense, only the B and C functions would be required for the specification of floral organ identity; the 'molecular code' would in fact reduce to B for petals, B+C for stamens and C for carpels (summarized in Ambrose and Ferrándiz, 2013). The evolution of a perianth composed of organs of different identity (sepals and petals), as eventually fixed in the core eudicots (see Section 7.2.3.3), would have been dependent on a whole-genome duplication that gave rise to the eu*AP1* clade (Litt and Irish, 2003).

5   Shoot and Root: Meristems and Branching

5.1   THE CELLULAR BASE OF PLANT DEVELOPMENT

During embryogenesis, only the primary meristems (the shoot and root apical meristems) are specified. The other meristems, which are established post-embryonically, are designated as secondary meristems (Greb and Lohmann, 2016). In later stages of development, stem cells can also be produced from differentiated tissue (Terpstra and Heidstra, 2009).

Throughout its activity (that is, in the case of the shoot and root apical meristems, for the whole of a plant's life) a meristem continues to split off cell groups as primordia of new organs, but the number of stem cells of which it is composed remains remarkably stable (Laufs *et al.*, 1998; Lyndon, 1998; Carles and Fletcher, 2003).

In the case of the shoot apical meristem, this number ranges from ca. 50 cells in *Arabidopsis thaliana* to over 1000 in *Glebionis segetum* (= *Chrysanthemum segetum*), and the basal diameter of the apical dome varies from 50 to 3000 µm (Mauseth, 1991). Root apical meristems are much larger, with 125 000–250 000 cells filling ca. 2 mm of apical tissue (Francis, 2008).

The size of the primordium does not necessarily affect the resulting phenotype. However – to take this time an example from the gymnosperms – in *Pinus ponderosa*, the number of cotyledons (from 6 to 16) is directly correlated with the size of the germ *at the time of cotyledon inception* and thus, indirectly, with the size of the seed (Buchholz, 1946).

Leaf initials may differ in number between species (e.g. ca. 100 cells at primordium initiation in *Arabidopsis thaliana*, ca. 150 in tobacco) (Hudson and Jeffree, 2001); leaflets of compound leaves derive from only 1–4 founder cells, at least in *Cardamine* (Barkoulas *et al.*, 2008).

The relationships between the number of cells initially recruited to the leaf primordium and the final size of the leaf are poorly known. A positive correlation is suggested by the *Arabidopsis* mutant *struwwelpeter* (*swp*), in which a reduced primordium gives rise to a smaller leaf containing fewer cells than in a wild-type plant's leaf (Autran *et al.*, 2002). In most cases, there is instead a direct correlation between the number of cells eventually formed at the end of the proliferative phase and the size of the lateral organ (Meyerowitz, 1997). Therefore, the final leaf size is largely determined by the duration of the period during which cells divide (Gonzalez *et al.*, 2012).

In plants, cell lineage seems to play no role in determining cell fate (Clark, 2001; Pallakies and Simon, 2010). This has been confirmed through the study of chimaeras, where different cell lineages are often visible even to the unaided eye: clonally distinct cell lineages are formed in the shoot meristems, but the cells derived from these lineages do not have fixed developmental fates; their patterns of division and differentiation will be based instead on positional information (Szymkowiak and Sussex, 1996).

In most plants, initiation of leaves requires a transition of cells from a pluripotent fate to a determinate fate. We will see later the role of genes in this transition and in the production of a boundary separating the indeterminate shoot apical meristem from the determinate lateral (leaf) meristem.

Lateral roots in *Arabidopsis* derive from a small pool of founder cells. The first division of these cells is asymmetric and precisely regulated, and determines the formation of a layered structure, but the pattern of subsequent cell divisions is not stereotypic, although characterized by a regular switch in division plane orientation. von Wangenheim *et al.* (2016) suggest that lateral root morphogenesis is based on a limited set of rules that determine cell growth and division orientation, and propose that self-organizing, non-deterministic modes of development account for the robustness of organ morphogenesis. This is in stark contrast to early embryogenesis, where divisions are extremely regular (Scheres *et al.*, 1994; Yoshida *et al.*, 2014; Gooh *et al.*, 2015).

5.2 GROWTH

5.2.1 Arrest Front and Organ Growth

Differences in organ size are due mainly to changes in cell number caused by alterations in the duration of cell proliferation.

In the leaves of *Arabidopsis*, the transition from cell division to cell expansion starts from the distal part of the leaf and ends in the pedicel (Andriankaja *et al.*, 2012).

The transition from cell proliferation to cell expansion is often described (e.g. White, 2006) as a dynamic 'arrest front', but more recent studies (Ichihashi *et al.*, 2011) indicate the presence of a proliferative region at the base of the leaf, which remains of quite constant size for some days before rapidly disappearing.

A number of genes influence organ size through a positive or negative effect on cell proliferation or cell expansion (Egea-Cortines and Weiss, 2013). In growing organs, auxin fosters the expression of *AUXIN-REGULATED GENE INVOLVED IN ORGAN SIZE (ARGOS)* (Hu *et al.*, 2003; Feng *et al.*, 2011). ARGOS, in turn, promotes the expression of *AINTEGUMENTA (ANT)*; the latter is required to maintain the expression of the cyclin CYCD3;1, whose loss would cause premature termination of cell division in leaves (Mizukami and Fischer, 2000; Dewitte *et al.*, 2007). Other pathways affecting organ size have been reviewed by Powell and Lenhard (2012).

5.2.2 Relative Growth: Resupination

In several plants, the final position of flowers and leaves is the result of a repositioning caused by torsion of the peduncle (resupination) or, less commonly, by twisted growth of the plant organ itself.

Resupinate leaves are known in a number of monocots, such as *Allium ursinum*, the grass genus *Pharus*, and especially in *Alstroemeria*.

In *A. pulchella* and *A. ligtu*, the angle of torsion is 0° for leaf 1, 90° for leaves 2–4 and 180° for all remaining leaves (Czapek, 1898). Resupination of flowers by torsion of the pedicel is conspicuous in *Lobelia* and in many orchids, but in other orchids and in *Medicago* the inverted position of the flower results from torsion of the ovary (Rudall and Bateman, 2004). Torsion in the middle of the flower is known in *Lonicera*. In *Narcissus*, bending may involve only the pedicel or extend to the lowermost part of the flower. Further examples of resupination are also known from Papaveraceae, Violaceae, Balsaminaceae, Acanthaceae, Lamiaceae, Zingiberaceae and other families (Goebel, 1920; Donoghue and Ree, 2000).

5.2.3 Mechanics

Tissue mechanics may have an important role in plant morphogenesis (Hamant *et al.*, 2008; Boudaoud, 2010; Mirabet *et al.*, 2011) at the level of cell growth, as a consequence of the orientation of the cellulose fibrils that make up the cell wall, but also at the organ level, especially in the apical meristem, the central zone of which is stiffer than the fast-growing periphery: this difference in mechanical properties apparently limits growth in the centre while allowing organ bulging to the flanks (Kierzkowski *et al.*, 2012; Maizel, 2016). The initiation of an organ from the shoot apical meristem requires a highly localized surface deformation, which depends on the demethylesterification of cell-wall pectins (Peaucelle *et al.*, 2008).

5.3 GROWTH VERSUS DIFFERENTIATION

Roughly speaking, the plant phenotype is the result of two main sets of processes, those responsible for growth (cell division and elongation) and those responsible for differentiation and patterning. A contentious issue is still, however, the extent to which cell-level processes eventually contribute to plant shape. According to one view (the so-called *organismal theory*), the effects of cell-division patterns on plant morphology are negligible, the latter being instead the result of processes occurring at organismic level (Kaplan and Hagemann,

1991; Kaplan, 1992). A different view (the so-called *cell theory*) posits instead that cell behaviour is the major factor in determining developmental processes (Evered and Marsh, 1989; reviewed in Tsukaya, 2003). Cell behaviour includes both the rate and the orientation of cell division. In *Antirrhinum majus*, the orientation of cell divisions, rather than their rate of division, determines the differential growth of petals (Rolland-Lagan *et al.*, 2003).

There are reasons to say that neither theory explains plant morphogenesis in a satisfactory way, especially because in different contexts (species, organ, or developmental phase) a morphogenetic process may depend more on cell autonomous processes or on regionally (or globally) controlled processes.

While acknowledging that the plant's overall architecture is determined by patterning genes, Meijer and Murray (2001) stressed the interconnections between cell division and morphogenesis and pointed to the control of the cell-division cycle as a pivotal aspect of the elaboration and execution of developmental programs. Since the first experiments with plant tissue culture in vitro (Gautheret, 1934; for a history of this topic, see Thorpe, 2007), it is widely known that cell division without patterning produces disorganized callus tissue. This is clearly different from the development of a plant from a zygote, or even from an asexual propagule, where a multicellular organism is produced through coordinated cell divisions (Steeves and Sussex, 1989; Lyndon, 1998). Meijer and Murray (2001) argued for an integrated view of cell division in development.

The disruption of normal cell division has little consequence for the form of organs, e.g. leaves, but affects their size. In experiments with *Nicotiana*, the inhibition of cell division resulted in leaves smaller than normal, which contained fewer but larger cells (Hemerly *et al.*, 1995). This demonstrates that the slower rate of cell division did not affect the expression of leaf patterning genes. But organ size is also regulated. Extensive work by Hirokazu Tsukaya and his group (e.g. Tsukaya, 2002, 2006, 2008; Ferjani *et al.*, 2007; Kawade *et al.*, 2010) has provided solid evidence in support of the existence of a phenomenon of 'compensation', involving a non-cell autonomous signal which is triggered when cell number falls below a critical threshold (Fujikura *et al.*, 2009) and a leaf of normal size is nevertheless produced, due to larger cell size.

5.4   THE SHOOT APICAL MERISTEM

In the first land plants, apical growth was limited to the gametophyte (Graham *et al.*, 2000), similar to extant liverworts and hornworts, where the sporophyte has no apical meristem (Cooke *et al.*, 2003).

5.4.1   Regulation and Maintenance of the Shoot Apical Meristem

The homeobox transcription factor *WUS* is essential for the maintenance of the shoot apical meristem as a locally restricted population of stem cells (Laux *et al.*, 1996). *WUS* expression is regulated by the *CLAVATA* (*CLV*) genes (Schoof *et al.*, 2000).

In grasses the function of *WUS* seems to have diversified, irrespective of the number of paralogues present in the different species, e.g. one in rice, two in maize. These genes are expressed in the periphery of the lateral meristem and in leaf primordia, but only transiently in the shoot apical meristem. During the reproductive phase, however, the same genes are expressed in the central region of the shoot apical meristem, together with *CLV1* orthologues (Nardmann and Werr, 2006). In contrast to the small cluster of cells that would contribute to form a new leaf in *Arabidopsis*, in grasses the leaf founding cells are recruited from the entire circumference of the shoot apical meristem (Nardmann and Werr, 2006; Miwa *et al.*, 2009).

In many angiosperms, the boundary separating the indeterminate shoot apex and a lateral organ with determinate growth, e.g. a leaf, is largely set through the KNOXARP (KNOX and ARP) regulatory module (Waites *et al.*, 1998; Timmermans *et al.*, 1999; Tsiantis *et al.*, 1999; Byrne *et al.*, 2000; Reiser *et al.*, 2000; Semiarti *et al.*, 2001). The *ARP* and class I *KNOX* genes are expressed in mutually exclusive domains of the developing shoot meristem (Carraro *et al.*, 2006; Scofield and Murray, 2006).

Class I KNOX proteins are required to avoid cell differentiation in the shoot apical meristem and to maintain it in its indeterminate state. Thus, KNOX1 proteins are usually expressed in vegetative and floral meristems and are excluded from leaf primordia and floral organs (Vollbrecht *et al.*, 1991; Long *et al.*, 1996; Reiser *et al.*, 2000).

The ARP proteins are named after ASYMMETRIC LEAVES 1 (AS1) from *Arabidopsis* (Byrne *et al.*, 2000), ROUGH SHEATH2 (RS2) from maize (Tsiantis *et al.*, 1999) and PHANTASTICA (PHAN) from *Antirrhinum majus* (Waites and Hudson, 1995). These proteins are expressed specifically in leaf initials, complementary to the expression of the *KNOX* genes in the shoot apical meristem (Waites *et al.*, 1998; Tsiantis *et al.*, 1999; Byrne *et al.*, 2000).

The KNOX family includes two classes of proteins (KNOX1 and KNOX2), members of both of which have been found in all embryophyte groups. Class I KNOX proteins are involved in the regulation and maintenance of the shoot apical meristem (e.g. Reiser *et al.*, 2000; Hake *et al.*, 2004), whereas the possible functions of *KNOX2* genes in the angiosperms are poorly known; in the moss *Aphanorrhegma patens*, as mentioned in Section 1.5, the *KNOX2* gene *MKN6* has been identified as a repressor of apospory. Four *KNOX1* genes – *SHOOTMERISTEMLESS* (*STM*), *BREVIPEDICELLUS* (*BP*), *KNAT2* and *KNAT6* – are found in *Arabidopsis* (Floyd and Bowman, 2007); nine genes are known from maize (Reiser *et al.*, 2000), five from rice (Sentoku *et al.*, 1999), five also from tobacco, four from tomato (Reiser *et al.*, 2000).

Throughout the shoot apical meristem, expression of *KNOX1* genes is required to maintain its meristematic identity, for instance, *KN1* in *Zea mays* and *STM* in *Arabidopsis* are expressed everywhere in the meristem, except for the sites of leaf initiation (Long *et al.*, 1996; Byrne, 2012). Lateral organs are initiated from the flanks of the shoot apical meristem at regularly spaced sites (see Section 5.7 on phyllotaxis), at sites where *KNOX1* genes are not expressed and where auxin maxima are established thanks to the auxin efflux carrier

PIN-FORMED1 (PIN1) (Long *et al.*, 1996; Reinhardt *et al.*, 2000, 2003; Benková *et al.*, 2003; Bainbridge *et al.*, 2008).

5.4.2 Terminal Leaves

In most plants, the shoot apical meristem maintains its indeterminate character beyond the inception of the last leaf. There are exceptions, however, in which the meristem eventually gets exhausted and the shoot ends in a terminal leaf, as in the grass *Gigantochloa* (Arber, 1928), in the seedlings of other monocots (Troll, 1937) and in some Podostemaceae, e.g. *Mourera fluviatilis*. In the latter, the terminal flower-subtending leaf of the inflorescence is initiated first; all other bracts are incepted, in two rows, in basipetal sequence (Rutishauser and Grubert, 1999). This sequence of basipetal inception of lateral appendages is found nowhere in shoots, whereas it is quite common for the leaflets of compound (imparipinnate) leaves in many angiosperms (Rutishauser and Sattler, 1997; Gleissberg and Kadereit, 1999); this is one of the many aspects in which plants of the Podostemaceae depart from the typical rules of construction of the other flowering plants and often show a kind of 'identity crisis' of their parts (Rutishauser *et al.*, 2008); see Section 8.5.6.

5.5 BRANCHING AND AXILLARY MERISTEMS

Branches are generated from axillary meristems at the base of leaves. Axillary meristems are set aside early, analogous to the germ line of many metazoans. Axillary buds can remain dormant for long periods before elongating to give rise to new branches.

During the vegetative phase of development, the formation of axillary meristems is regulated by the *LATERAL SUPPRESSOR* genes – *Ls* in tomato (Schumacher *et al.*, 1999) and *LAS* in *Arabidopsis* (Greb *et al.*, 2003). Their action terminates with the transition to the reproductive phase of development, so axillary shoots can be formed at this stage also in the mutants, and inflorescence branching is not affected.

In ferns and mosses, the apical meristem represented by a single cell can directly split and give rise to two or more new axes, producing a dichotomy or a polytomy. This terminal branching is not known in gymnosperms and occurs quite rarely in angiosperms, owing to the multicellular organization of the apical meristem and its patterns of cell division. Examples of terminal branching are found in *Mammillaria* (= *Dolicothele*), *Asclepias* and a few monocots: some Arecaceae (*Chamaedorea*, *Eugeissona*, *Hyphaene*, *Nypa*, *Oncosperma*), *Flagellaria* and *Strelitzia* (Barthélémy and Caraglio, 2007). Dichotomous branching also occurs at root apices (Peterson, 1992). Dichotomously branched leaf blades (also found in juvenile leaves of ferns) are rare, an example being *Drosera binata* (Troll, 1937–1943).

5.6 LATERAL ORGANS

5.6.1 *Boundaries*

Boundaries between different cell types, especially between the stem cells of meristems and the cells of a lateral primordium, are established by narrow stretches of cells with particular shapes and reduced growth. Boundary cells isolate the primordium as a group of cells with determinate fate and also serve as organizing centres for the patterning of adjacent tissues (Dahmann *et al.*, 2011; Žádníkova and Simon, 2014; Huang and Irish, 2016).

The first boundaries established in the embryo separate cotyledons from the shoot meristem. Later, during the whole vegetative growth phase, boundaries are established at the time a cluster of founder cells segregates from the shoot apical meristem: this helps not only to separate the lateral organs from the shoot apical meristem, but also to ensure separation of organs from one another (Aida and Tasaka, 2006; Rast and Simon, 2008). In boundary cells, mitotic activity is very slow, compared to both the neighbouring meristem and the lateral organ primordium (Žádníkova and Simon, 2014).

Several genes are involved in boundary formation. In the shoot apical meristem of *Arabidopsis*, these include *CUC*, *LATERAL ORGAN BOUNDARIES* (*LOB*), *BLADE-ON-PETIOLE* (*BOP*), *AS* and *LATERAL ORGAN FUSION* (*LOF*) (Arnaud and Laufs, 2013).

Pivotal in this process, and also better known, is the role of the three transcription factors CUC1, CUC2, CUC3 (Aida *et al.*, 1997; Takada *et al.*, 2001; Vroemen *et al.*, 2003), which regulate *TCP* and *KNOX* genes to control lateral organ outgrowth. The activity of the *CUC* genes is controlled, in turn, by miR164. The additional role of *CUC* genes in controlling the formation of serrations, lobes and the leaflets of compound leaves (Blein *et al.*, 2008; Berger *et al.*, 2009) is discussed in Section 6.4.

Brassinosteroids may also play an important regulatory role in boundary formation (Bell *et al.*, 2012; Gendron *et al.*, 2012). These phytohormones stimulate growth by promoting cell division and elongation; their activity is low in the boundary regions where, as mentioned above, mitotic activity is slow. The expression of boundary genes (*CUC1*, *CUC2*, *CUC3* and *LOF1*) is inhibited by brassinosteroids, and thus their expression in the boundary region is possibly dependent on the local reduction in brassinosteroid signalling. High brassinosteroid levels cause defects in organ separation (Žádníkova and Simon, 2014).

In addition to the boundaries between the apical meristem and the lateral organ primordia, other boundaries are also specified in plants. In flowers, boundaries between sepals are maintained by locally expressed *CUC* genes and *PETAL LOSS* (*PTL*). Specifically, *CUC* genes prevent the boundary between contiguous sepals from differentiating into sepal tissue, while *PTL* controls cell proliferation in the boundary regions (Griffith *et al.*, 1999).

In leaves, the expansion of the lamina depends on maintaining the boundary between the adaxial (external, lower) and abaxial (internal, upper) side of the primordium. In *Arabidopsis*, this boundary is marked by the expression of *WOX3* and *WOX1* (Nakata *et al.*, 2012).

5.6.2   Establishing the Axes of Lateral Organs

The enormous variation in morphology among the lateral organs of angiosperms results largely from differential growth along three axes: adaxial–abaxial, proximal–distal and left–right.

Mutants in which both the adaxial–abaxial axis and the proximal–distal axis of the leaf are simultaneously affected are known both in *Antirrhinum* (Waites and Hudson, 1995, 2001) and in *Arabidopsis* (Eshed *et al.*, 1999; Long and Barton, 2000). More intriguing are the multiple effects of other genes, also studied in *Antirrhinum*, e.g. *DICHOTOMA*, which affects the adaxial–abaxial axis of the flower and the left–right axis of individual floral organs (Luo *et al.*, 1999), and *HANDLEBARS*, which affects the adaxial–abaxial axis of leaves and the left–right axis of floral organs (Waites and Hudson, 2001).

Adaxial identity is determined by the expression of several members of the III HD-ZIP class and by the ARP transcription factor PHANTASTICA (PHAN). The latter was characterized by Waites and Hudson (1995), based on a mutation in *Antirrhinum* in which older leaves developed ectopic axes of growth on the upper surface of leaves (to be compared with the petals of *Menyanthes trifoliata* described in Section 7.2.3.6). *PHAN* orthologues with similar ectopic features were found in tomato (Kim *et al.*, 2003) and tobacco (McHale and Koning, 2004). In the absence of the dorsalizing factor encoded by the *PHAN*, a needle-like (unifacial) leaf is formed, covered with an abaxial epidermis. However, this role of the PHAN transcription factor in promoting adaxial character, as found in *Antirrhinum majus* (Waites and Hudson, 1995; Waites *et al.*, 1998; Byrne *et al.*, 2000), does not extend to its orthologous genes *RS2* in maize (Timmermans *et al.*, 1999) and *AS1* in *Arabidopsis* (Tsiantis *et al.*, 1999). This suggests a diversity of regulatory mechanisms responsible for adaxial–abaxial patterning (Fukushima and Hasebe, 2014). These include the III HD-ZIP class transcription factors PHABULOSA (PHB), PHAVOLUTA (PHV) and REVOLUTA (REV) (McConnell and Barton, 1998; McConnell *et al.*, 2001; Otsuga *et al.*, 2001; Emery *et al.*, 2003; Prigge *et al.*, 2005; Byrne, 2006). These factors repress the adaxial expression of *AUXIN RESPONSE FACTOR3/ETTIN* (*ARF3/ETT*) and *ARF4*. The precise expression pattern of these genes in the adaxialization pathway is controlled by microRNAs (miR165 and miR166 in *Arabidopsis*:

Reinhart *et al.*, 2002; Rhoades *et al.*, 2002; Emery *et al.*, 2003; Tang *et al.*, 2003; Kidner and Timmermans, 2010).

Abaxial cell identity is determined by the combined action of members of the *KANADI* (*KAN*) (Eshed *et al.*, 2001, 2004; Kerstetter *et al.*, 2001; Emery *et al.*, 2003) and *YABBY* (*YAB*) families (Sawa *et al.*, 1999a, 1999b; Siegfried *et al.*, 1999; Stahle *et al.*, 2009; Sarojam *et al.*, 2010).

5.7 PHYLLOTAXIS

Phyllotaxis is the spatial arrangement of lateral organs, either leaves or flower parts. With some exceptions, this arrangement is regular and mostly fixed for each plant species, apart from the changes that often accompany the transition from leaf to floral phyllotaxis (see Section 7.3). Some phyllotactic patterns are very easy to observe: the whorled one, with three or more lateral organs emerging from the same node, regularly spaced around the circumference; the decussate one, which can be described as a whorled pattern with two lateral organs per node, which are thus emerging from opposite positions (Fig. 5.1). In whorled phyllotaxis, the lateral organs of each pair (or whorl set) occupy the bisectors of the angles formed by two contiguous leaves of the previous (or next) whorl; in the case of decussate phyllotaxis, this translates into having all the leaf pairs located in either of two perpendicular planes crossing along the stem or branch axis. Regularity is less obvious when only one lateral organ is bound to a node, except for the case in which all lateral organs are found in the same plane and simply alternate from one direction to the opposite one, i.e. when the divergence angle between them is 180°. When the divergence angle between a leaf and the next one is smaller than 180°, we can imagine an acropetal spiral ordinately joining all points of emergence of the lateral organs. Large amounts of ink have been spent in describing spiral phyllotaxis, explaining it in terms that range from number theory to geometry, from physics to cell biology; this literature is not reviewed here, except to mention that virtually all phyllotactic patterns follow Hofmeister's rule (named after

FIGURE 5.1 Valerian (*Valeriana officinalis*), a plant with decussate leaves.

Hofmeister, 1868), i.e. the new lateral organ tends to form as far away as possible from previously initiated organs (Braybrook and Kuhlemeier, 2010). A possible explanation for Hofmeister's rule is that each primordium is the source of an inhibitor that decreases in strength with time and/or distance. The next new organ will arise at the site of lowest inhibition within the differentiation zone around the apical meristem. It is currently acknowledged that phyllotactic patterning is based on a positive feedback loop between auxin and its transporter, PIN1 (see, e.g., Kuhlemeier and Reinhardt, 2001; Reinhardt *et al.*, 2003; Smith *et al.*, 2006).

Fascinating is the old remark, that the possible divergence angles (expressed as fractions of 360°) observed in different plants with spiral phyllotaxis follow (more or less precisely) the following series:

$$\frac{1}{2}, \frac{1}{3}, \frac{2}{5}, \frac{3}{8}, \frac{5}{13}, \frac{8}{21} \cdots$$

where the numbers 1, 2, 3, 5, 8, 13, 21 ... form a set known as the Fibonacci sequence.

In fact, the arrangement of leaves in a number of plants does not fit precisely into any of these basic patterns. The main deviation is the overall spiral twist of what would otherwise be a distichous arrangement, as found, for example, in *Ischnosiphon* spp. and *Phenakospermum guyanense* (spirodistichous pattern; Bell, 2008).

A related condition is helical anisoclady: in some plants with whorled (including decussate) phyllotaxis with two axillary buds per whorl, these buds are of unequal size and the larger buds of the different whorls, in the sequence, are arranged along a helix. Examples of helical anisoclady in plants with decussate phyllotaxis are not rare among the Caryophyllaceae (Loiseau, 1969; Rutishauser, 1981, 1998; see also Fig. 5.2); among those with whorls of four or more elements, this pattern occurs e.g. in *Galium* (see Section 8.5.3) and in *Limnophila* (Rutishauser, 1999).

In some plants with whorled phyllotaxis, the number of leaves per whorl is not fixed. In the aquatic plant *Ceratophyllum submersum*, there are between 7 and 10 leaves per whorl, occasionally up to 12; the number increases slowly during a first phase of growth, but decreases when the shoot is close to stopping growth. Other examples (*Utricularia purpurea*, *Limnophila* spp.) were discussed by Rutishauser (1999).

Another deviation from the standard arrangement of lateral organs in plants with whorled phyllotaxis is the occurrence of superposed whorls, where the organs are arranged in the same radii in all whorls, rather than alternating between a set of radii and the bisectors of the angles formed by them; in this case, Hofmeister's rule is violated (Rutishauser, 1999). Examples of intermediates between alternating and superposed whorls have been described in the Rubiaceae

FIGURE 5.2 Chickweed
(*Stellaria media*), an example
of helical anisoclady.

Rubieae with tetramerous leaf whorls that show different degrees of spiral twisting: for example, in *Cruciata glabra* and *Asperula cynanchica* subsequent whorls are nearly superposed, whereas *Galium rubioides* shows nearly alternating whorls (Fukuda, 1988).

Experimentally disrupted phyllotactic patterns may recover quickly, with the further growth of the plant following the arrangement characteristic of the species. However, predictable developmental switches from one pattern to another occur frequently during the life of a single plant. For example, in (eu)dicots, the decussate position of cotyledons is often followed by spiral phyllotaxis. In the lianoid Vitaceae, phyllotaxis is spiral during the juvenile stage, but changes to distichous after about node 6; this is the time the plant forms the first tendrils (Gerrath *et al.*, 2001). Tendrils and other lateral organs also follow phyllotactic patterns in the same way as do typical leaves.

In *Arabidopsis thaliana*, the embryonic leaves and the two first vegetative leaves show a decussate pattern before switching to a spiral phyllotaxis for later vegetative leaves and flowers, and finally to a whorled pattern for floral organs (Palauqui and Laufs, 2011). Also in *Arabidopsis*, meristem size contributes to the robustness of phyllotaxis (Landrein *et al.*, 2015).

In many representatives of the Poaceae, the apical meristem changes from the distichous phyllotaxis of the leaf-bearing shoot to the spiral phyllotaxis of the inflorescence (Kellogg *et al.*, 2013), but some grasses do not undergo this phyllotactic change. In the Pooideae (e.g. *Hordeum, Avena, Stipa, Poa*), the inflorescence is distichous like the arrangement of the leaves. No shift in phyllotaxis is also observed in some Panicoideae and in the bamboo *Fargesia*. In all these cases, the distichous distribution of the primary branches of the inflorescence is apparently derived (Malcomber *et al.*, 2006). If so, it represents an example of secondary reduction in developmental complexity, through the retention in the flowering phase of a structural trait that was originally abandoned at the transition from the vegetative to the reproductive phase.

5.8   HANDEDNESS

In plants, left–right asymmetry may occur at different levels, from the shape of the leaf lamina to the winding direction of climbing plants and the arrangement of petals in some patterns of aestivation (see Section 7.4.2). In some instances, handedness (Endress, 1999; Hashimoto, 2002) is under genetic control, in others it is a stochastic effect. Twining plants generally show fixed handedness, consistently either right- or left-handed, suggesting genetic control: hop (*Humulus lupulus*) and honeysuckle (*Lonicera* spp.) form left-handed helices, *Convolvulus* and many others, right-handed ones. This should be compared with non-enantiomorphic asymmetry, the condition found in some plants (e.g. *Lathyrus* and *Vigna*) with very elaborate asymmetric flowers, of which only one of the two potential mirror-image morphs exists in nature (Endress, 2012).

Some specific patterns of handedness have a taxonomically restricted distribution. Flowers with contorted petals of which both morphs (the sinistrorse and the dextrorse) occur on the same individual are largely restricted to rosids, among the eudicots, and are not known in asterids. On the other hand, the occurrence of only one morph in an individual (sometimes, in the whole species, or even in a larger taxon) is characteristic of asterids, but is rare in rosids, except for the Myrtales (Endress, 1999).

Enantiomorphic mutants have been described for *Arabidopsis*, which cause either dextrorse or sinistrorse twist of petioles, by chiral cell expansion. There is strong evidence that cytoskeleton microtubules are involved in regulating this anisotropic expansion of axial and lateral organs (Furutani *et al.*, 2000).

Spiral phyllotaxis is not controlled genetically. At the level of the inflorescence, a consistently left- or right-handed helical arrangement of flowers is found in *Spiranthes australis*: the lack of genetic control is demonstrated by the fact that all individual plants, either right- or left-handed, produce, if self-pollinated, an equal number of right- and left-handed offspring (Callos and Medford, 1994). In

Arabidopsis, the phyllotactic handedness is stochastically determined during embryonic development by slight deviations from mirror symmetry between the two cotyledons (Woodrick *et al.*, 2000).

5.9 FUSION OF PARTS

Structures deriving by synorganization of two or more lateral organs are very frequent, obvious examples being sympetalous corollas and syncarpic gynoecia. The morphological diversity of these structures and their occurrence in different taxonomic groups, as well as the evolvability and actual evolution of synorganization, are discussed in Section 7.6. Here, only cellular and genetic aspects of the phenomenon are briefly discussed.

In several instances, the organs involved in synorganization retain anatomical independence, and acquire only a limited degree of coherence, e.g. through interposition of secretions or by interaction of trichomes; more commonly, however, the individual organs are in actual anatomical continuity. Based on the developmental processes that eventually culminate in synorganization, two conditions are distinguished, referred to as congenital and postgenital fusion (Endress, 2006). In postgenital fusion, the organs subsequently involved in synorganization are free at inception, but the epidermis of one eventually fuses with the contiguous epidermis of the other organ. Congenital fusion is the production of a synorganized structure from a single meristem corresponding to the otherwise distinct meristems of the organs involved.

Synorganization without fusion is observed in *Geranium robertianum*. Apart from the syncarpous gynoecium, all other synorganized parts of the flower are simply held together by architectural modifications: this occurs among the five sepals, as well as between sepals and petals and between stamens and carpels (Endress, 2010d).

The distinction between congenital and postgenital fusion is clear-cut; however, the current terminology is somehow confusing, because it fails to separate evolutionary change from differences in developmental processes. In developmental terms, a fusion of previously separate organs occurs in the cases of synorganization described as deriving from postgenital fusion, but no fusion at all is observed in the other cases. Rather than a 'congenital fusion', it is arguably better to speak of 'congenital lack of independence', or some equivalent term. It is only from an evolutionary perspective that a fusion has occurred, as an apomorphy (perhaps also an adaptively important innovation) of a small or large clade. In the following, I will still use 'congenital fusion', but the reader should always be aware of its possible ambiguity.

Postgenital coherence may occur at different structural levels: (1) at the subcellular level, either by secretion or by interdigitation of cuticular projections (e.g. between the sepals of the Cephalotaceae); (2) at the cellular level, by interplay of hairs (e.g. between the sepals of the Campanulaceae) or by interdigitation of epidermal cells (e.g. between the tepals of the Proteaceae); or (3) at the supracellular level, by hooking together of organs, e.g. between the petals of the Oxalidaceae.

Several genes play a role in the interaction between organ surfaces; among these are *FIDDLEHEAD* (*FDH*) (Lolle *et al.*, 1992; Yephremov *et al.*, 1999; Pruitt *et al.*, 2000), *LEUNIG* (*LUG*) (Chen *et al.*, 2000, 2001), *ANT* (Liu *et al.*, 2000) and *HOTHEAD* (*HTH*) (Krolikowski *et al.*, 2003).

5.10   INFLORESCENCES

5.10.1   The Architecture of Inflorescences

The inflorescence of a plant is the system of branches that bears the flowers.

Endress (2010b) suggested a simple classification of inflorescences based on three criteria: the different patterns of branching, the relative degree of elongation of axes of different orders, and the extent to which the basic branching patterns are iterated.

Branching patterns are often stable in larger clades (Endress, 2010b); individual genera seldom include species with racemes and species with cymes (Prusinkiewicz *et al.*, 2007). However, indeterminate and determinate inflorescences may occur in the same species: in *Impatiens balsamina*, both indeterminate and determinate inflorescence varieties are found (Benlloch *et al.*, 2007).

Following Endress (2010b), we can distinguish two basic branching patterns that occur in inflorescences. In the racemose pattern, the first-order axis has a variable number of lateral branches, but there are no higher-order branches. The first-order axis can be closed (terminated by a flower) or open (without terminal flower). In the first case, the inflorescence is determinate, in the second, indeterminate. In the cymose pattern, the first-order axis never has more than two second-order axes, but these can similarly branch one or more times; the first-order axis is commonly terminated by a flower, but not always.

Thus, in a raceme, the number of branching orders is fixed (not more than two), while the number of second-order branches is variable. On the contrary, in a cyme, the number of lateral branches of each axis is fixed, while the number of branching orders is variable. Among the cymes, two main morphologies can be distinguished: the dichasium, in which all branching axes have two lateral branches, and the monochasium, in which all axes have only one lateral branch. There are also cymes including both dichasial and monochasial parts, the former being generally expressed basally, followed by monochasial apical parts, and inflorescences which start with racemose branching, followed by a cymose pattern. The latter inflorescences are called thyrsoids if they have a terminal flower, thyrses if they do not. The basic module of the complex inflorescence of *Euphorbia*, the cyathium (Fig. 5.3), is a thyrsoid; in turn, numerous cyathia are commonly arranged in a thyrsoid (Endress, 2010b).

5.10.2 Gene Expression, Inflorescences and Flower Identity

Excellent reviews of the genetic mechanism of specification of floral meristems in *Arabidopsis* are Jack (2004), Vijayraghavan *et al.* (2005) and Blázquez *et al.* (2006).

FIGURE 5.3 Part of the inflorescence of a Mediterranean spurge (*Euphorbia characias*). The individual cyathia, the characteristic pseudanthia forming *Euphorbia* inflorescences, are, in turn, strongly synorganized inflorescences. A cyathium is formed by four or, more commonly, five bracts including five groups of male flowers plus one female flower, together with nectar glands of characteristic shape. Each male flower is reduced to an anther, and the female flower is represented by the gynoecium only.

In *Arabidopsis, LFY* – a floral meristem identity gene implicated in both flowers and inflorescences – is strongly expressed in the whole floral meristem from the earliest stages of development (Weigel *et al.*, 1992), but is also expressed in leaf primordia; the activation of *AP1* is required to eventually trigger flowering (Hempel *et al.*, 1997; Parcy, 2005; Benlloch *et al.*, 2007).

In *Antirrhinum*, the functional equivalent to *Arabidopsis LFY* is *FLORICAULA* (*FLO*), while *SQUA* has the same function as the *Arabidopsis AP1* in determining flower meristem identity. Mutations in these genes cause conversion of flowers into shoots, but the resulting phenotypes are different in the two model species (Benlloch *et al.*, 2007). In the loss-of-function *squa* mutant, the floral meristems are replaced by vegetative shoots from which flowers are rarely produced (Huijser *et al.*, 1992); the *ap1* mutant, instead, produces branched flowers without petals, but bearing normal stamens and carpels.

The expression of *AP1* and *LFY* is largely restricted to lateral (floral) meristems; in the apical (inflorescence) meristem their expression is prevented by *TFL1* in *Arabidopsis* and by *CENTRORADIALIS* (*CEN*) in *Antirrhinum* (Bradley *et al.*, 1996, 1997).

TFL1 specifies shoot identity and also acts as a repressor of flowering, as demonstrated by the early flowering of loss-of-function mutants (Shannon and Meeks-Wagner, 1991; Schultz and Haughn, 1993). The expression pattern of *TFL1* is complementary to that of *LFY* and *AP1*: this gene is strongly expressed in the centre of the meristems of the main and lateral shoot inflorescences, but not in the floral meristems (Benlloch *et al.*, 2007).

CEN is only expressed in the inflorescence meristem, while the domain of expression of *TFL1* extends to both vegetative and inflorescence shoot meristems. Mutations in *TFL1* or *CEN* change the inflorescence from indeterminate to determinate, causing the conversion of the shoot apical meristem into a terminal flower (Benlloch *et al.*, 2007). These genes integrate signals from multiple pathways involved in the transition to flowering (Coen *et al.*, 1990; Weigel *et al.*, 1992; Ma, 1998; Blázquez and Weigel, 2000; Zhao *et al.*, 2001; Parcy, 2005).

A gene involved, in *Arabidopsis*, in the control of the transition between the specification of flower meristem and the activation of the flower homeotic genes is *UNUSUAL FLORAL ORGANS* (*UFO*), which is expressed in a region-specific pattern in both shoot and flower meristems (Parcy *et al.*, 1998). In association with *LFY* and other genes, *UFO* appears to activate expression of the B-function gene *AP3* and the C-function gene *AG*. It has been suggested that the temporal order in which all these genes are activated may recapitulate the order in which their expression became involved in flower specification and patterning along the evolution of the angiosperms; if so, the highly conserved geometry of angiosperm flowers may still preserve a trace of the co-option of a meristem patterning system already operating before the emergence of the flowering plants (Parcy *et al.*, 1998; Rudall and Bateman, 2010).

In addition to the homologues of the floral meristem identity genes of *Arabidopsis*, meristem identity genes unique to grass species have been isolated. In rice, the *LFY* homologue *RFL* also controls inflorescence structure, but in a different way than in (eu)dicots. The complex architecture of the rice panicle inflorescence, with primary branches and secondary branches bearing the spikelets, results from the specification of meristem identities that are absent in model eudicot inflorescences. *RFL* is expressed in the early inflorescence meristem and also controls branching at the whole-plant level (Rao *et al.*, 2008). In maize, *LFY* genes control the branching of the male inflorescence (the tassel) (Souer *et al.*, 2008), similar to their role in rice. In maize, floral meristem initiation is controlled by an *AP2*-like gene, *ids1*, and by the related gene *sid1*. In grasses, the *AP2* genes might replace *LFY* to function in floral meristem identity (Chuck *et al.*, 2008).

In the lateral, secondary inflorescences of legume model species, homologues of *LFY* and *AP1* (Hecht *et al.*, 2005; Domoney *et al.*, 2006) perform functions similar to those of *LFY* and *AP1* in the inflorescence of *Arabidopsis*. UNIFOLIATA (*UNI*), the *LFY* homologue in legumes, is expressed in floral meristems; its mutations cause an inflorescence to develop instead of a flower (Hofer *et al.*, 1997). In *Pisum* the two functions of the *Arabidopsis TFL1* gene, in controlling flowering time and in fixing the determinate character of the meristem, are controlled by two different *TFL1/CEN* paralogues: mutations in *TFL1c* cause early flowering but do not affect meristem determinacy; the opposite is the effect of mutations in *TFL1a* (Foucher *et al.*, 2003).

In inflorescences of the cyme type, the fact that the lateral and eventually the apical meristems produce flowers sequentially rather than all at the same time requires more complex regulation. This has been documented in the inflorescences of some representatives of the Solanaceae (petunia and tomato), in which the *LFY* homologues *ALF* and *FA* are expressed in a different and wider pattern than *LFY* in *Arabidopsis* (Souer *et al.*, 1998; Molinero-Rosales *et al.*, 1999). These genes, already expressed during the vegetative phase, are first expressed, in the inflorescence, in the apical meristem and show up in lateral meristems after some delay (Rijpkema *et al.*, 2010). In addition, a regulator that seems specific for cymes has been discovered in *Petunia*. This factor, *EVERGREEN* (*EVG*), stimulates proliferation in the meristems of the lateral inflorescences and contributes substantially to the diversification of inflorescence architecture (Rebocho *et al.*, 2008).

As mentioned above, most floral meristems are determinate. In *Arabidopsis* this pattern of growth of floral meristems is controlled by the expression of the *AGAMOUS* (*AG*) transcription factor (Barton, 2010).

5.10.3 Genes Controlling Floral Identity in Capitula

The capitulum is a complex inflorescence formed by small flowers or florets tightly packed together; it is characteristic of the Asteraceae and the Caprifoliaceae Dipsacoideae (formerly a separate family, Dipsacaceae). Capitula are either homogamous (all florets in a capitulum are either male or female, or hermaphrodite) or heterogamous (florets with different sexual conditions coexist in the same capitulum); see Section 8.3. In gross morphology, florets are either similar throughout the capitulum, or differentiated into two distinct sets, the disc and the ray. In heterogamous capitula, the outer (ray) florets are usually zygomorphic, with elongated petals, and are either female or sterile. The inner (disc) florets are typically small, actinomorphic and hermaphrodite.

In Asteraceae (reviewed in Broholm *et al.*, 2014) and Dipsacoideae (Carlson *et al.*, 2011), the specification of ray versus disc floral
morphologies and thus the complexity of the capitulum correlates
with characteristic expression patterns of the *CYC2* genes, which
have undergone a number of duplication events along the history of
these clades.

In *Helianthus*, *Senecio* and *Gerbera*, *CYC* homologues are
expressed preferentially, or uniquely, in the outer part of the capitulum,
where ray flowers will differentiate (Broholm *et al.*, 2008; Kim *et al.*,
2008; Chapman *et al.*, 2012). Different *CYC* paralogues are responsible
for differentiation of ray flowers in different representatives of the
Asteraceae (Chapman *et al.*, 2012); this suggests that ray flowers have
evolved multiple times in the family (Panero and Funk, 2008).

In *Senecio vulgaris*, the production of ray florets is controlled by
the *RAY* locus, which corresponds to two linked *CYC2* genes; both
of them are upregulated in ray flower primordia exclusively (Kim
et al., 2008).

In *Gerbera hybrida*, at least three *CYC2* genes are involved in
the regulation of ray floret development, with likely redundant functions. Upregulation of these genes (*GhCYC2*, *GhCYC3* and *GhCYC4*)
during the early stages of development promotes cell proliferation in
the abaxial side of the corolla, resulting in the production of a ligulate
flower, and suppresses the development of the stamens (Broholm
et al., 2008; Juntheikki-Palovaara *et al.*, 2014). *GhCYC2* is expressed
both in disc and in ray flowers, but the different (and insufficiently
known) context causes this expression to have opposite effects:
increased petal size in disc florets, reduced petal size in ray florets
(Broholm *et al.*, 2008).

Phenotypically unusual mutants, some of which have been
found to occur in natural populations, help us to understand the
involvement of *CYC2* genes in controlling the morphology of florets,
and thus the evolution of the capitulum of the Asteraceae. In
Helianthus annuus, a naturally occurring mutant (*tubular ray flower*,
tub) has radially symmetrical ray florets; a similar phenotype in

FIGURE 5.4 In the Asteraceae with radiate capitula (here, *Cosmos* sp.), the marginal (ray) florets are zygomorphic. In most instances, the adaxial side of these zygomorphic florets is very reduced and the corolla takes the form of a ligule corresponding to its abaxial side; however, mutants are known, such as the one in the photo, in which the imbalance between adaxial versus abaxial sides is very reduced and the florets approximate radial symmetry. *For the colour version, please see the book's front cover.*

another representative of the Asteraceae is shown in Fig. 5.4. This is a consequence of the lack of the normal *CYCc* gene expression; in the *tub* mutant, this gene bears a transposon insertion which results in the production of a premature stop codon (Fambrini *et al.*, 2011; Chapman *et al.*, 2012). In another sunflower mutant (*double-flowered, dbl*), in which *CYC2c* is expressed ectopically throughout the capitulum, the disc florets are converted into ray-like florets (Chapman *et al.*, 2012).

In the capitula of *Knautia macedonica* (Caprifoliaceae Dipsacoideae), morphological differences between internal and external florets are due to the differential expression of six copies of *CYC*-like genes, with a general trend towards a reduction in adaxial expression and an increase in abaxial expression in internal florets compared to external florets (Berger *et al.*, 2016).

5.11 PROGRAMMED CELL DEATH AND ABSCISSION

Abscission, the process by which parts of the plant are shed (Roberts et al., 2002), takes place primarily at predictable sites (abscission zones), usually located at the base of a leaf, a flower, a fruit or a seed (Roberts and González-Carranza, 2013).

Abscission of plant parts is an active physiological process due to dissolution of the cell walls (van Doorn and Stead, 1997).

The cell layers where abscission takes place are anatomically distinct long before organ separation begins (Roberts and González-Carranza, 2013), but the abscission zone does not develop further before abscission-promoting signals are registered by its cells (Nakano et al., 2012). Cell division in future abscission zones was observed in the catkin stalks of Salix and Castanea, but not in the flowers of Begonia and Phaseolus vulgaris, nor in the pedicel of several Solanaceae, including tobacco and tomato (van Doorn and Stead, 1997).

5.12 ROOTS

The development and evolution of roots has not attracted attention in the same way as leaves and flowers, and even large-scale questions of homology and phylogeny are still unsettled. This obvious neglect is in part explained by the fact that roots are among the simplest of the plant organs.

Roots are present in nearly all vascular plants, exceptions being Psilotum among the ferns and, among the angiosperms, the strongly miniaturized duckweeds Wolffia and Wolffiella and some epiphytic bromeliads, most conspicuously Tillandsia species.

Among the vascular plants, the earliest evidence for the presence of roots is in some representatives of the lycophyte clade of the Early Devonian, but on the phyletic line that includes the angiosperms the first fossil roots are no older than the Middle Devonian (Gensel et al., 2001; Raven and Edwards, 2001). Similar to leaves (see Section 6.1), roots have evolved at least twice in land plants. Moreover, within the euphyllophytes, differences in the anatomy of embryonic roots suggest that roots also evolved independently in ferns and in seed plants (Raven and Edwards, 2001).

In most flowering plants (with the major exception of monocots; see below), the primary root deriving from the elongation of the embryonic radicle grows into the plant's main subterranean axis, which may or may not develop lateral roots of one order only, or secondarily branching into lateral roots of higher order (Petricka *et al.*, 2012). In *Arabidopsis*, lateral roots arise along the primary root at predictable sites (Dubrovsky *et al.*, 2006). The temporal and spatial distribution of lateral roots along the primary root axis is mediated by genes with oscillating expression. Two distinct sets of genes (one of 2084 genes, the other of 1409 genes) were identified, oscillating synchronously among each set and in antiphase with those of the other set (Moreno-Risueno *et al.*, 2010). An unexpected discovery among these genes was the overrepresentation of those coding for transcription factors such as *SHP1*, *SHP2*, and *SEEDSTICK* (*STK*), all of which are otherwise involved in carpel and ovule development and in seed dispersal (Dinneny and Yanofsky, 2005), and also of *AGL20*, which regulates the transition to the plant's reproductive stage (Yant *et al.*, 2009).

The primary root is frequently reduced in water plants, e.g. *Utricularia*, *Ceratophyllum*, *Trapa*, *Nymphaea lutea* and the Podostemaceae (Arber, 1920), but this trend has systematic character in monocots, where the root system derived from the primary root is short-lived and is soon replaced by adventitious roots (nodal roots or crown roots) developing from the shoot. In rice, two mutants, *crown-rootless4* (*crl4*) (Kitomi *et al.*, 2008) and *OsGnom1* (Liu *et al.*, 2009b) have been isolated, both of which lack crown roots and show a reduced number of lateral roots (Coudert *et al.*, 2010).

Adventitious roots develop also in many eudicots, in addition to the persisting primary root. These plants, e.g. *Humulus lupulus*, *Fragaria* spp., *Rubus* spp. and *Saintpaulia* spp., are popular for their capacity to propagate vegetatively from rhizomes, stolons, stems or leaves, respectively (Bellini *et al.*, 2014).

6   Leaves

6.1   THE EVOLUTION OF LEAVES

Goethe's (1790) dictum that everything in a plant is leaf (*alles ist Blatt*) is often mentioned as emblematic of what we now call plant evo-devo, provided of course that we translate this sentence from the old wording of idealistic morphology into the modern language of evolutionary biology. There are problems, however, because this perspective takes one of the plant's parts, the leaf, as given, that is, as the safe starting point out of which all other parts of the plant eventually derive. Retelling this story in terms of changing patterns of gene expression does not remove the inadequacy of taking the leaf as the default state of all plant organs. This view has indeed two weak points. First, are we sure that sporophylls, that is, the lateral appendages of the shoot, bearing the plant's reproductive organs, are modified leaves rather than vice versa? Second, if the leaf is a plant's lateral organ, this means that the shoot came first, followed by the leaf. This would challenge the leaf's primacy as the archetypical module of plant construction. Additionally, this opens new questions – that is, are shoot and leaf necessarily distinct, and, if so, how did this morphological dichotomy arise?

These questions cannot be satisfactorily addressed unless we broaden the phylogenetic perspective beyond the limits I have chosen for this book, because shoots and leaves are not limited to the flowering plants. Even restricting our scope to vascular plants, we find that the 'leaves' of the earliest (Upper Devonian) fossil seed plants were highly ramified, and that many of them lacked a lamina, while others developed a flat surface only at the end of several orders of dichotomous branching. These lateral organs were thus slightly modified copies of the main stem (see Section 8.4), with which they shared two characteristic traits: apical growth and extensive, mainly dichotomous branching (Floyd and Bowman, 2010). Both traits were subsequently lost and occasionally regained, even in recent evolutionary times: witness the apical growth of the leaves of some flowering plants (see Section 8.5.2).

Two kinds of leaves are distinguished: the microphylls of lycophytes (e.g. *Lycopodium, Selaginella*) and the megaphylls of the euphyllophytes, including the flowering plants (Ambrose and Ferrándiz, 2013). Microphylls are generally small, have a single unbranched vascular strand and lack a leaf gap, the characteristic break in the stem's vascular tissue found in megaphylls above the point of attachment of the leaf vascular trace. In addition, megaphylls are generally large and have complex venation. However, there are exceptions to these diagnostic criteria, especially among the microphylls of fossil lycophytes.

Microphylls and megaphylls are considered non-homologous. Moreover, palaeobotanists estimate that megaphylls evolved independently between two and nine times (Tomescu, 2009).

An underlying semantic difficulty casts a shadow over all these comparisons: no clear, indisputable definition of 'leaf' seems to be available. This is widely acknowledged by plant morphologists and developmental biologists. Here is an example showing the difficulty of attempting a simple definition: 'At least two developmental characters define most if not all leaves: namely (1) leaves arise from a relatively large number of progenitor "founder cells" recruited from the periphery of the shoot apical meristem; (2) leaves are dorsiventrally asymmetric (bifacial) at their inception' (Scanlon, 2000, p. 31). But unifacial, rather than bifacial, leaves are not rare, especially in some monocot families, as discussed in Section 6.3.1.

6.2   LEAF MORPHOGENESIS AND ITS GENETIC CONTROL

Efroni *et al.* (2008) distinguished a phase of primary leaf morphogenesis, characterized by lateral and distal expansion of a flat lamina mainly due to intense mitotic activity, and a phase of secondary leaf morphogenesis, characterized by organ enlargement due to extensive cell expansion and histogenesis often associated with multiple endocycles, and limited mitotic divisions (cf. Donnelly et al., 1999); but this formal distinction is not accepted by others, e.g. Tsukaya (2013).

When the leaf lamina of Arabidopsis thaliana or Nicotiana reaches about one-tenth of its final size, mitotic divisions gradually cease, beginning with the distal tip (Poethig and Sussex, 1985; Donnelly et al., 1999; Nath et al., 2003), then cell expansion and differentiation progresses towards the leaf base. This is reflected in temporally dynamic changes involving more than half of the transcriptome of the growing blades (Nath et al., 2003; Schmid et al., 2005; Efroni et al., 2008).

In compound leaves, initiation and growth of the leaflets is also frequently basipetal, but in some species it is acropetal (Steeves and Sussex, 1989).

Specialized parts of the leaf emerge as evolutionary novelties in several angiosperm clades. An obvious example is the grass family. Here, in addition to the characteristic proximal sheath surrounding a segment of the shoot above the node, two specialized regions, the ligule and the auricle, differentiate at the boundary between blade and sheath. The leaf of grasses is regarded as homologous to the conventional angiosperm leaf; however, the ligule and the auricle have no clear homologue in the other (especially eudicot) groups. The production of the ligule is indeed regulated by genes (Langham et al., 2004; Lee et al., 2007) co-opted from roles in patterning of floral parts, such as anthers, and in the regulation of floral transition. At least in maize, this double role has been facilitated by gene duplication (Townsley and Sinha, 2012).

The genetic control of leaf morphogenesis involves the acquisition and maintenance of the adaxial–abaxial polarity and the establishment of specializations along the mediolateral and proximodistal axes (Tsukaya, 2006; Efroni et al., 2010; Nakata et al., 2012).

Proximal–distal differentiation involves the so-called *ACIDIC REPEAT PROTEIN* (*ARP*) genes, represented by *AS1* and *AS2* in *Arabidopsis*, *RS2* in maize and *PHAN* in *Antirrhinum* (Kidner and Timmermans, 2010). Expression of these genes is found at the site of leaf initiation and is maintained in young leaf primordia; eventually, they will promote distal identity (Waites *et al.*, 1998; Tsiantis *et al.*, 1999; Byrne *et al.*, 2000; Katayama *et al.*, 2010).

Many leaves are articulated into modules whose development is to some extent regulated independently from the others. Activation of the *LEAFY PETIOLE* (*LEP*) gene converts the proximal part of the leaf from petiole into leaf blade (van der Graaff *et al.*, 2000).

Regulation of adaxial–abaxial patterning is not necessarily uniform along the proximodistal axis of the leaf. This explains the usual differentiation between petiole and lamina, but more complex elaborations are possible. It has been suggested that the pitcher, tendril and basal laminar portion of the leaves of the pitcher plant *Nepenthes* correspond to the lamina, petiole and leaf base of conventional bifacial leaves, respectively (Kaplan, 1973), but no experimental evidence is available to date. A detailed and comparative study of leaf morphogenesis in *Nepenthes* should be a priority in plant developmental genetics (Tsukaya, 2014).

Genes responsible for the curvature of the leaf lamina have been studied in *Antirrhinum majus*. In this species, the leaves of the *cin* mutant are crinkly rather than flat. According to Nath *et al.* (2003), leaf cells in the mutant are less sensitive to a signal for mitotic arrest; as a consequence, the cell-cycle arrest front moves from the leaf tip towards the base more slowly and along a different path than in the wild type, causing extra mitoses and thus excess growth around the margin.

6.3 BIFACIAL, UNIFACIAL AND PELTATE LEAVES

In most leaves with bifacial laminae, simple and compound alike, the upper (adaxial) surface is different from the lower (abaxial) surface. Specification of adaxial and abaxial cells is required for the formation of a leaf blade (Waites and Hudson, 1995; McConnell *et al.*, 2001).

The transcription factor coded by the gene *PHAN* is involved in maintaining the leaf adaxial domain in the leaf primordium (Waites *et al.*, 1998; Sun *et al.*, 2002), but also in controlling the architecture of compound leaves: restriction of the adaxial domain consequent to a reduced expression of this gene causes pinnate compound leaves to change into palmate compound leaves; this has been observed in a number of plants in which compound leaves have evolved independently, such as *Pachira aquatica*, *Akebia pentaphylla*, *Oxalis regnellii* and *Schefflera actinophylla* (Kim *et al.*, 2003). In the case of pinnate leaves, *PHAN* is expressed along the entire adaxial region of the leaf primordium; this is reflected, in the mature leaf, in the presence of a distinct adaxial domain in the petiole and the rachis (Champagne and Sinha, 2004).

6.3.1   Bifacial and Unifacial Leaves

In most angiosperms, the leaf lamina is bifacial. Its expansion occurs along the border between the adaxial and abaxial domains (Waites and Hudson, 1995). In *Arabidopsis*, expression of *FIL/YAB1* and *YAB3* marks the abaxial domain and the marginal regions of leaf primordia (Sawa *et al.*, 1999a; Siegfried *et al.*, 1999).

Monocot leaves tend to show less differentiation between adaxial and abaxial tissues than those of the other angiosperms (Hudson and Jeffree, 2001).

In the cases in which either adaxial or abaxial identity is not specified in the leaf primordia, a stick-like or cylindrical leaf will eventually result, as in most species of *Juncus* and in many *Allium* species, e.g. *Allium schoenoprasum*. Some unifacial leaves, however, are flat, but in such cases, e.g. in *Iris*, the leaves are expanded perpendicular to the direction in which the lamina of a conventional bifacial leaf would be expanded (Tsukaya, 2014). Yamaguchi *et al.*'s (2010) comparison of leaf development in two *Juncus* species sheds light on the role of genes in regulating the differentiation between adaxial and abaxial domains in these unconventional leaves. Both species have unifacial leaves, but those of *J. prismatocarpus* are ensiform, while those of *J. wallichianus* are cylindrical (terete), as typical for the genus. The authors found that the development of the flat leaf lamina of *J. prismatocarpus* was in part sustained by enhanced expression of the *DROOPING LEAF* (*DL*) gene in the midrib, and interpreted this finding as indicating that the flat structure of these unifacial leaves is a consequence of extensive thickening of a lamina that cannot expand laterally, thus causing the polarity of the flat plane in unifacial leaves to be perpendicular to that in bifacial leaves.

In ensiform leaves, such as those of *Iris* species, the two flat surfaces of the blade are symmetrical and their equivalence is mirrored, in the leaf anatomy, in the symmetrical orientation of the vasculature (Arber, 1921). In these leaves, the *ARF3/ETT* orthologue is expressed throughout the entire surface of the flattened blade (Yamaguchi *et al.*, 2010).

Terete-like and ensiform-like leaves are found also in a few eudicots, but the actual polarity of their surfaces has not been determined with reasonable certainty. A blade homologue is not necessarily present; for example, the phyllodes of some *Acacia* species are composed mainly of a rachis–petiole complex (Arber, 1918, 1921; Kaplan, 1970). The primordia of the eventually terete leaves of the succulent *Kleinia rowleyana* (= *Senecio rowleyanus*) start differentiating a flat lamina with conventional dorsiventral (abaxial–adaxial) polarity, but in consequence of sustained adaxial growth, they eventually develop into terete-like organs (Hillson, 1979). The growth of the eventually ensiform leaves of another succulent, *Senecio crassissimus*, starts in a similar way. In this case, the abaxial surface expands in the mediolateral direction, whereas the adaxial surface does not (Ozerova and Timonin, 2009).

6.3.2 Peltate Leaves

Peltate leaves are those in which the unifacial petiole is attached to the abaxial side of the lamina, often, as in *Nelumbo* (Fig. 6.1), close to the geometrical centre of a circular lamina. Examples of peltate leaves are known in at least 40 families (Gleissberg *et al.*, 2005). In very early stages, their primordia are indistinguishable from those of

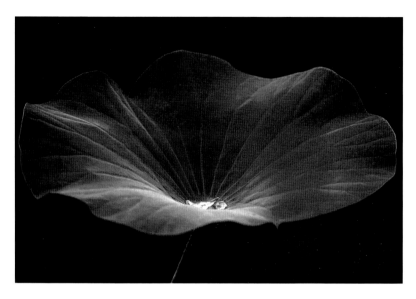

FIGURE 6.1 The sacred lotus (*Nelumbo nucifera*) is one of the most popular examples of plants with peltate leaves. *A black and white version of this figure will appear in some formats. For the colour version, please refer to the plate section.*

conventional bifacial leaves (Hagemann and Gleissberg, 1996). Their growth starts deviating from the usual path when the characteristic 'cross zone' is formed, which corresponds to the basal portion of the growing lamina, where leaf margins are congenitally fused (Fukushima and Hasebe, 2014). Heart-shaped non-peltate leaves such as those of *Farfugium japonicum* (Fig. 6.2) may suggest a transition from conventional to peltate morphology.

In the leaf primordia of the peltate-leaved species *Tropaeolum majus*, the expression domain of the *FIL* homologue is initially restricted to the abaxial side, similar to the localization of the *FIL/YAB1* transcripts in the conventional bifacial leaves of *Arabidopsis*. However, in *T. majus*, the expression domain of the *FIL* homologue expands later into the adaxial side of the petiole (Gleissberg *et al.*, 2005). This has been interpreted as the generation, in the cross zone, of a new boundary between the adaxial and abaxial surfaces under the influence of a modified expression pattern of the *FIL* homologue (Fukushima and Hasebe, 2014).

FIGURE 6.2 Leaf development in *Farfugium japonicum* may deserve a comparative study with respect to peltate leaves.

Plant morphologists distinguish two types of peltate leaves: in hypopeltate leaves, the adaxial surface is larger than the abaxial one, causing the lamina to assume an umbrella-like shape; in epipeltate leaves the opposite is true, as in *Nelumbo*. When the curvature due to the differential growth of the two surfaces is high, the lamina takes on a sac-like shape (epiascidiate or hypoascidiate). Well-known epiascidiate leaves are those of several carnivorous plants (Nepenthaceae, Sarraceniaceae, Cephalotaceae). Hypopeltate or hypoascidiate leaves are found in a number of genera, e.g. among the Melastomataceae, Ericaceae, Apocynaceae and Asteraceae.

Mutants with epipeltate leaves are known in plants with conventional bifacial leaves, e.g. the *phan* mutant of *Antirrhinum* (Waites and Hudson, 1995). Epipeltate leaves are produced by loss-of-function mutants of the adaxial determinants *AS1*, *AS2* and *HD-ZIP III* genes (Sun *et al.*, 2002; Xu *et al.*, 2002, 2003; Prigge *et al.*, 2005). Correspondingly, hypopeltate leaves are produced by gain-of-function

FIGURE 6.3 Pitcher leaf of
Sarracenia sp.

mutants of *HD-ZIP III* genes, determining adaxialization of the basal
part of the leaf (McConnell and Barton, 1998; McConnell *et al.*, 2001;
Zhong and Ye, 2004).

In the carnivorous *Sarracenia purpurea* (Fukushima *et al.*, 2015),
standard expression of adaxial and abaxial marker genes is followed by
peculiar cell division patterns of adaxial tissues, causing the lamina to
take the characteristic shape of the epiascidiate pitcher leaf (Fig. 6.3).

6.4 LEAF MARGINS AND LEAF COMPLEXITY

6.4.1 *From Leaf Serration to Leaf Dissection*

As mentioned in the previous chapter, the formation of auxin
maxima marking the sites of leaf initiation in the peripheral zone
of the shoot apical meristem is accompanied by downregulation of

the expression of *KNOX1* class genes, e.g. *STM*, in the incipient leaf primordia (Long *et al.*, 1996). This expression is reactivated in the leaf primordia of many plants, and this correlates with the development of marginal indentation, production of lobes or eventual emergence of compound leaves (Bharathan *et al.*, 2002; Hay and Tsiantis, 2006). The correlation between these different forms of leaf complexity and the expression of *KNOX1* genes along the leaf margin is better appreciated in a comparison between closely related species, one of which has simple leaves and the other has compound leaves.

I also mentioned that expression of *CUC2* and *PIN1* is required to release the development of lateral organs from the shoot apical meristem. With the exceptions mentioned below, reactivated expression of the same genes along the margin of a leaf primordium, together with its antagonist *miR164A*, determines the production of serrations (Nikovics *et al.*, 2006), lobes or leaflets. In contrast to a previous interpretation, *CUC2* is now known to promote outgrowth of teeth rather than suppression of growth at the sinuses (Kawamura *et al.*, 2010).

As in the case of the lateral meristems, 'convergence points' of PIN1 accumulation are responsible for auxin activity maxima along the margins of the leaf; besides driving the formation of the major leaf veins (Scarpella *et al.*, 2006), these convergence points also define the positions where marginal serrations will develop (Hay *et al.*, 2006; Scarpella *et al.*, 2006), similar to the role of PIN1 maxima in initiating new leaves at the margin of the shoot apical meristem (Reinhardt *et al.*, 2003; Barkoulas *et al.*, 2007).

The involvement of CUC transcription factors in the production of compound leaves was demonstrated by Blein *et al.* (2008), in a study that revealed a conserved requirement for *NAM/CUC3* genes during leaflet formation, leaflet separation and margin dissection in five distantly related plant species, including a basal eudicot (*Aquilegia caerulea*) and four core eudicots (*Solanum lycopersicum*, *S. tuberosum*, *Cardamine hirsuta*, *Pisum sativum*).

Important progress in understanding the developmental genetic underpinnings of the evolutionary transition between simple and compound leaves was obtained by comparing two closely related model species, one of which (*A. thaliana*) has leaves with nearly smooth margins, whereas the other (*Cardamine hirsuta*) has highly dissected leaves in which the leaf blade is separated into individual leaflets (Piazza *et al.*, 2010). It has been estimated that the lineage of *A. thaliana* diverged from the lineage of *C. hirsuta* approximately 14 million years ago (Koch *et al.*, 2001).

The production of the dissected leaves of *C. hirsuta* is dependent on the redeployment during leaf development of a mechanism also operating at leaf inception, i.e. the activity, along the margin of the leaf primordium, of auxin-dependent cell division foci that give rise to leaflets (Barkoulas *et al.*, 2008). Similar to its role in organ separation and delimitation in the shoot meristem of *Arabidopsis* (Reinhardt *et al.*, 2003), *PIN1* is involved in generating local auxin maxima in the *C. hirsuta* leaf primordium; its expression is also required to separate the leaflets from the rachis and to delimit the growth of the lamina (Barkoulas *et al.*, 2008).

In *C. hirsuta*, the *STM* homologue is expressed along the leaf margins; during leaf differentiation, this expression disappears at first in the apical region, then progressively from more proximal positions (Hay and Tsiantis, 2006; Blein *et al.*, 2008). *STM* expression is at its minimum also in those places along the leaf margins where there are local maxima of auxin concentration: these are the positions where leaflets are produced, similar to what usually happens at the shoot apical meristem, with production of new leaves wherever there are auxin maxima accompanied by the down-regulation of *KNOX1* gene expression (Hay and Tsiantis, 2006; Floyd and Bowman, 2010).

Piazza *et al.* (2010) demonstrated that the leaf shape in *A. thaliana* was derived from a lobed ancestral form; the loss of *STM* expression in leaf primordia contributed to the transition to an essentially simple leaf: in *A. thaliana*, *STM* prevents differentiation of the shoot apical meristem. In *C. hirsuta*, *STM* expression is found in both the shoot apical meristem and in the developing leaves. The structural cause of this difference in the expression pattern of *STM* could be localized to structural changes in its promoter.

In the Brassicaceae, leaflet development requires also the REDUCED COMPLEXITY (RCO) transcription factor, which represses growth along the margins of leaf primordia. To some extent, the presence of simple leaves in *A. thaliana* is also due to the loss of this gene in the model plant species. In different members of the family, species-specific expression patterns of the *RCO* gene in the leaf base, coupled with gene duplication and loss, are responsible for different local patterns of growth, eventually resulting in the generation of leaf shape diversity (Vlad *et al.*, 2014).

In *C. hirsuta*, the recessive mutant *rco* converts the adult leaf from the dissected phenotype of the wild type into a simply lobed leaf. In this mutant, however, number and positioning of leaves are not affected, suggesting that *RCO* is required for leaflet development but not for leaf initiation. *RCO* is part of a tandem gene triplication, of whose products only one has been retained in *A. thaliana*; this is *LATE MERISTEM IDENTITY1* (*LMI1*), known as a floral regulator (Saddic *et al.*, 2006). All three genes are present, however, in *C. hirsuta* and in *Capsella rubella*, and also in *Arabidopsis* species such as *A. lyrata* (Bailey *et al.*, 2006; Beilstein *et al.*, 2010; Couvreur *et al.*, 2010). The ascertained monophyly of *Arabidopsis* (Beilstein *et al.*, 2006) allows for robust and meaningful comparisons of leaf shape and development between the species of this genus. Major differences in leaf shape are evident in mature plants, when *A. thaliana* produces simple, unlobed leaves, whereas most of its closest relatives display lobed leaves; during primordium initiation and the early stages of morphogenesis of the first six leaves, there is no difference between the leaf margins in *A. thaliana* and in the other *Arabidopsis* species that will eventually display lobed leaves. Divergence in margin morphology is evident only around the end of the cell proliferation phase in the seventh leaf (Piazza *et al.*, 2010).

Reactivation of *KNOX1* expression during leaf development has been found in a large number of plants with compound leaves, where it marks the positions where leaflets will appear (Bharathan *et al.*, 2002; Hay and Tsiantis, 2006). Examples include representatives of the basal eudicots, Oxalidales, Brassicales and Asterales. However, production of compound leaves is not a necessary consequence of reactivated *KNOX1* expression; the latter instead is also found in plants with simple leaves, although with significant marginal elaboration, with examples among Brassicales, Gentianales and Vitales (Rosin and Kramer, 2009).

Marking more or less regularly spaced positions along the margin of the leaf primordium is thus the mechanism by which leaves may develop minute serrations or more conspicuous lobing, or produce independent leaflets, thus becoming compound leaves (Hay and Tsiantis, 2010). No surprise that, in tomato, overexpression of *KNOX1* translates into production of highly dissected leaves (Hareven *et al.*, 1996; Chen *et al.*, 1997; Janssen *et al.*, 1998). In plants with simple leaves, such as *Arabidopsis* and *Nicotiana*, overexpression of class 1 *KNOX* genes increases leaf lobing (Bharathan *et al.*, 2002; Hay *et al.*, 2003; Kessler and Sinha, 2004). However, we must refrain from generalizations (Tsukaya, 2014).

On the one hand, expression of *KNOX1* genes in leaf primordia is not necessarily conducive to complex leaves, or at least to leaves with complex margins. *KNOX1* expression has been observed in leaf primordia of species with unlobed leaves, such as *Lepidium oleraceum* (Hofer *et al.*, 2001a; Bharathan *et al.*, 2002). A *KNOX1* gene is also expressed in the indeterminate leaves of *Streptocarpus* spp. (discussed in Sections 6.6 and 8.5.4), but these leaves are simple and their margin is smooth (Harrison *et al.*, 2005). This is also the case for the indeterminate simple leaves of *Welwitschia mirabilis* (Pham and Sinha, 2003). The latter case deserves a few more words here, despite the fact that this represents an excursion outside the angiosperms, to which this book is essentially limited. The only two photosynthetically active leaves produced by a plant of *W. mirabilis* over its long life span grow continuously from a basal meristem. In the leaf primordia of this species there is no *KNOX1* gene expression, in agreement with what is observed in normal leaves with determinate growth; however, this gene starts being expressed later, in the leaf attachment region where cell proliferation occurs, when the leaf switches to an indeterminate growth pattern (Pham and Sinha, 2003).

On the other hand, in some plants the production of compound leaves is not dependent on *KNOX1* genes, as discussed in the next section.

6.4.2   Compound Leaves in Legumes

In the majority of the legumes examined (Champagne *et al.*, 2007), compound leaves are under the control of *KNOX1* gene expression, as usual in plants with dissected leaves; however, several species belonging to the inverted repeat-lacking clade (IRLC), which diverged 39 million years ago from other Fabaceae (Lavin *et al.*, 2005), form compound leaves that do not express *KNOX1* genes (Champagne *et al.*, 2007). This clade includes *Astragalus, Medicago, Trifolium, Vicia, Pisum* and *Lathyrus* (Wojciechowski *et al.*, 2004). It is notable, however, that this peculiar behaviour is not even universal among the species of the IRLC: *KNOX1* mRNA has also been detected in the developing leaves of *Medicago truncatula* (Di Giacomo *et al.*, 2008).

In the IRLC legumes, the development of compound leaves is best known in *Pisum sativum*. In this species, *KNOX1* gene expression is permanently downregulated in the incipient leaf primordium (Gourlay *et al.*, 2000; Hofer *et al.*, 2001a). The development of the (eventually compound) leaf is controlled instead by *UNI* (Hofer *et al.*, 1997). Wild-type pea leaves usually consist of two or three leaflet pairs, followed by 3–4 pairs of distal tendrils, and terminate with an apical tendril. Leaf phenotypes of *uni* mutants range from simple to trifoliate (Hofer *et al.*, 1997; DeMason and Schmidt, 2001) and are accompanied by defects in the flower.

Leaf development in pea is also regulated by *STAMINA PISTILLOIDA* (*STP*), otherwise known as a floral meristem gene. Leaf and flower, indeed, are jointly affected by mutations in this gene: in strong *stp* mutants, leaf complexity is reduced and flowers consist of sepals and carpels only (Taylor *et al.*, 2001). *STP* is homologous to *UFO* of *Arabidopsis* and *FIMBRIATA* (*FIM*) of *Antirrhinum* (Simon *et al.*, 1994; Ingram *et al.*, 1995; Taylor *et al.*, 2001). The UFO protein binds to LFY and activates it (Chae *et al.*, 2008; Souer *et al.*, 2008). It is therefore possible that *STP* controls leaf growth through activation of *UNI* (Della Pina *et al.*, 2014).

Thus, in the IRLC legumes, while *KNOX1* genes still serve functions in the shoot apical meristem itself, in compound leaf development their regulatory functions have been taken over by a *LFY* orthologue (Rosin and Kramer, 2009).

Outside of the legumes, *LFY* has a role in leaf development also in some other plants. Its tomato orthologue *FA* is expressed in vegetative meristems and in leaf primordia (Molinero-Rosales *et al.*, 1999). The leaf phenotype of *fa* mutants is not seriously impaired: the number of primary leaflets is normal, but secondary leaflets are less numerous and their differentiation is delayed, suggesting a role of this gene in controlling the length of time during which leaflets can be initiated (Townsley and Sinha, 2012).

6.4.3 Homology and Evolvability of Compound Leaves

Phylogenetic analyses reconstruct the ancestral angiosperm leaf as simple (Taylor and Hickey, 1996; Doyle and Endress, 2000). Complex leaves arose repeatedly: Bharathan *et al.* (2002) calculated an average of 29 estimated simple-to-compound transitions, followed by an average of six estimated reversals.

Easy evolvability of leaf complexity is suggested by the very common variation recorded at the intraspecific level, examples of which are given in Table 6.1.

Two hypotheses have been proposed to explain the homology of simple and compound leaves. The first hypothesis equates individual leaflets of compound leaves with simple leaves. In this model, compound leaves are seen as partially indeterminate structures that share

Table 6.1 *Evolvability of leaf complexity exemplified by variation at the intraspecific level, according to Meyen (1973).*

Simple → palmate	Ranunculaceae: *Ranunculus cassubicus, Clematis vitalba, Komaroffia integrifolia* Rosaceae: *Fragaria moschata, Rubus caesius* Cannabaceae: *Humulus lupulus* Moraceae: *Broussonetia papyrifera* Malvaceae: *Sidalcea neomexicana, Althaea ficifolia, Abelmoschus manihot* (= *Hibiscus manihot*) Convolvulaceae: *Convolvulus erubescens, C. althaeoides* Araliaceae: *Hedera helix* Apiaceae: *Pimpinella anisum*
Simple → pinnate	Ranunculaceae: *Clematis vitalba* Proteaceae: *Hakea* Rosaceae: *Rubus caesius, Prunus avium, Malus pumila* Fagaceae: *Fagus silvatica* Salicaceae: *Salix caprea* Anacardiaceae: *Pistacia vera* Sapindaceae: *Acer negundo* Brassicaceae: *Lepidium perfoliatum* Solanaceae: *Solanum dulcamara, S. tuberosum* Oleaceae: *Fraxinus excelsior* Caprifoliaceae: *Scabiosa columbaria, Knautia arvensis* Asteraceae: *Gazania linearis, Scorzoneroides autumnalis* (= *Leontodon autumnalis*), *Taraxacum officinale, Dahlia variabilis, Crepis* spp., *Sonchus* spp. Apiaceae: *Aegopodium podagraria, Oenanthe pimpinelloides*
Pinnate → palmate	Ranunculaceae: *Ranunculus repens* Rosaceae: *Rubus caesius* Adoxaceae: *Sambucus racemosa* Araliaceae: *Aralia* Apiaceae: *Sanicula*

Table 6.1 (cont.)

Simple → dichotomous	Fabaceae: *Phaseolus*
	Moraceae: *Broussonetia, Morus*
	Urticaceae: *Boehmeria biloba*
	Primulaceae: *Lysimachia vulgaris*
	Convolvulaceae: *Calystegia sepium*

properties with both shoots and leaves (Sattler and Rutishauser, 1992; Champagne and Sinha, 2004). The second hypothesis suggests that the entire compound leaf is equivalent to a simple leaf and that leaflets arise by subdivisions of a simple blade (Kaplan, 1975). Viewed in this way, leaf shape is seen as a continuum that ranges from simple leaves with entire margins to serrated, lobed and compound leaves.

These two views have both received support from recent observations: serrations of the leaf margin can be converted into structures resembling leaflets (Hay *et al.*, 2006; Barth *et al.*, 2009) but *Cardamine* leaflets appear to have a different origin because they arise from the rachis (Barkoulas *et al.*, 2008).

Different modes of compound leaf formation are indeed what we should expect, due to the multiple independent evolution of such structures (Bharathan *et al.*, 2002; Blein *et al.*, 2010).

As a rule, during compound leaf development there is (generally) a redeployment of the regulatory modules acting in the meristem; this could support the hypothesis of a shoot-like identity of compound leaves. However, compound leaf primordia do not express stem cell markers characteristic of the meristem (Hay and Tsiantis, 2006); therefore, the identity between shoot and compound leaves is only partial (Hasson *et al.*, 2010).

Complementary to the problem of homology between the compound leaves of different plants, but less often discussed, is the developmental equivalence of all simple leaves. This is in fact negated, not so much by the existence of primitively simple leaves alongside those derived by reversal from compound leaves, but especially by the diversity of ways in which this reversal has been obtained, even between species of the same genus, as suggested by Bharathan *et al.* (2002) for *Lepidium* spp. From a common ancestor with compound leaves, a number of species with simple leaves (among which *L. africanum* and *L. oleraceum*) evolved within this genus; a later reversal to compound leaves is recorded in *L. hyssopifolium*. While the compound leaves of the common ancestor depended on the usual *KNOX1* gene expression in leaf primordia, simple leaves evolved by different mechanisms: in *L. africanum* by turning off *KNOX1* genes in the primordia, in *L. oleraceum* by modifying secondary morphogenesis.

6.4.4   Dissected Leaves in Monocots

Dissected leaves are more common in eudicots than in monocots. However, taxa in which the lamina is represented by multiple leaflets are found in at least four families belonging to as many monocot orders. By far the highest diversity is found among the Araceae, of which about one-quarter have pinnately, palmately or pedately dissected leaves (Mayo *et al.*, 1997, 1998). Palmately dissected (sometimes bifid, pinnatifid or palmatisect) leaves occur in a few genera of the Dioscoreaceae (Huber, 1998; Kubitzki, 1998). The leaves of palms are more often palmately or pinnately dissected into leaflets, each with one or more longitudinal ribs (Uhl and Dransfield, 1987; Tomlinson, 1990; Dransfield and Uhl, 1998). Those of the Cyclanthaceae closely resemble those of the palmately dissected palms, but dissection of adjacent leaflets is usually incomplete (Harling *et al.*, 1998), more obviously than in palms. Complex leaves in aroids and palms deserve closer attention.

As mentioned, different kinds and degrees of dissections are found in the Araceae (which nevertheless includes also a very high number of species with entire leaves, e.g. in *Anthurium, Calla, Colocasia*; Bown, 2000):

- blade subdivided in three main divisions: e.g. *Arisaema triphyllum*
- palmate leaves, with at least five main divisions: e.g. *Anthurium pentaphyllum, A. polydactylum*

FIGURE 6.4 The unique pinnate leaves of *Zamioculcas zamiifolia*.

- same, but main leaf divisions further subdivided: e.g. *A. polyschistum*; most conspicuously however in *Amorphophallus, Anchomanes* and *Dracontium*
- pinnate leaves: e.g. *Anaphyllopsis pinnata, Zamioculcas zamiifolia* (Fig. 6.4)
- bi- or tripinnate leaves: in *Gonatopus* only

In some aroids, leaf complexity increases throughout the plant's life, e.g. first leaf with three divisions only, followed by leaves with five, seven or nine divisions.

In *Amorphophallus, Anchomanes* and *Dracontium*, the pattern of dissection is not fractal, but different in the different runs of dissections, and different in the different genera. Nevertheless, the eventual outcome at maturity is almost identical in all three genera (Bown, 2000).

In *Zamioculcas zamiifolia* the adult leaves consist of a short sheathing leaf base, a long succulent petiole and 4–8 pairs of leaflets borne on an elongate rachis. The petiole and proximal portion of the rachis are unifacial (confirmed by anatomy), while the distal region of the

FIGURE 6.5   Ontogenetic dissection of a palm leaf lamina.

rachis is bifacial (Kaplan, 1984), the opposite of peltate leaves, where the petiole is unifacial and the lamina bifacial, as described above. Leaflets and rachis of the *Zamioculcas* leaf are deciduous from the persistent petiole, and the leaflets are capable of rooting (Mayo *et al.*, 1997).

A peculiar kind of dissected leaf is found in some palms. All palm leaves have a sheathing leaf base, a distinct petiole and a corrugated blade (sometimes huge, up to 25 m in *Raphia*; Uhl and Dransfield, 1987; Tomlinson, 1990; Dransfield and Uhl, 1998). Blade morphology, either palmately or pinnately dissected, is the result of a two-step process. The blade is initially smooth and undivided, but submarginal corrugations (plications) appear soon. Later, a narrow marginal strip is removed and the remaining blade surface is progressively subdivided into individual leaflets (Fig. 6.5). This may proceed in two different ways. In some palms, leaflets separate as a consequence of tissue disintegration (schizogeny), simultaneously affecting narrow lines of cells along the lines of plication (e.g. *Chamaedorea seifrizii*; Kaplan *et al.*, 1982) and proceeding progressively from the margin (e.g. *Rhapis excelsa*). A narrow undivided zone holding adjacent leaflets together is generally saved from this process (Dengler *et al.*, 1982), but in most cases this is also broken mechanically at a later stage. In other palms, schizogeny occurs, but does not extend completely across the blade, leaving instead a narrow marginal strip of tissue that holds adjacent leaflets together (Dengler *et al.*, 1982). This narrow marginal strip is disrupted mechanically, usually late as the blade unfolds from the crown. This mode has been described, for example, for the palmate leaves of *Pritchardia filifera* and the pinnate leaves of *Cocos nucifera*.

6.5 FENESTRATED LEAVES

The cellular mechanisms involved in leaflet separation also have other functions in plant growth and development, and their recruitment into leaf development is probably secondary. The same is true of the cellular mechanisms producing fenestrated leaves in a few aroids and in two species of *Aponogeton*.

The leaves of *Monstera* (Fig. 6.6) and the related genera *Rhaphidophora, Amydrium* and *Epipremnum* are often conspicuously and elaborately perforated or, more rarely, deeply pinnatifid (Mayo *et al.*, 1997, 1998). Perforations also occur in other Araceae: *Dracontium, Cercestis* and *Dracontioides* (Bown, 2000; Gunawardena and Dengler, 2006). At inception, the leaves of *Monstera* form a complete, simple leaf blade, but holes appear soon in the lamina, due to localized histolysis.

A less popular but striking example of a plant with fenestrated leaves is provided by *Aponogeton madagascariensis* and *A. fenestralis*. Unlike many *Monstera* species, in which perforations break through the leaf margin, the entire margin of *A. madagascariensis* leaves is always left untouched. The degree of fenestration increases throughout the growth of the plant. The leaves produced by seedlings are simple, non-fenestrate; those of juvenile plants have a few

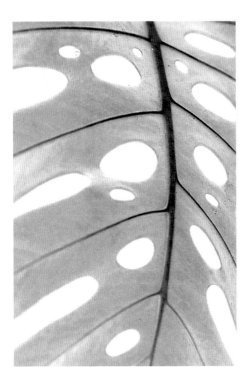

FIGURE 6.6 Detail of a fenestrated leaf of *Monstera epipremnoides*.

perforations near the midrib. In *A. madagascariensis*, perforation of the leaf lamina involves early alteration of cytoplasmic streaming, shrinkage of the cytoplasm and chromatin condensation (Gunawardena *et al.*, 2004). This sequence of events is very similar to that observed during the differentiation of tracheary elements from mesophyll cells of *Zinnia* in culture (Groover *et al.*, 1997; Fukuda, 2000). Gunawardena and Dengler (2006) therefore suggest that the cellular mechanisms involved in the production of fenestrations in the leaves of *A. madagascariensis* may represent an example of exaptation of mechanisms originally used in xylem differentiation.

6.6 LEAVES WITH INDETERMINATE GROWTH

Most leaves are determinate organs: together with the presence of an axillary bud, this is indeed one of the criteria commonly used to distinguish a 'true' leaf from a leaf-like branch. However, in some plants meristematic activity is maintained for months or even for years (Tsukaya, 2000).

There are indeed two kinds of indeterminate leaves. Some of them are provided with a perennially active basal meristem, as in *Monophyllaea* and *Streptocarpus* (Jong and Burtt, 1975; Tsukaya, 1997; Imaichi *et al.*, 2000); others have an apical meristem, as in some Podostemaceae (Imaichi *et al.*, 2005; Koi *et al.*, 2005) and in two genera of the Meliaceae, *Chisocheton* and *Guarea* (Steingraeber and Fisher, 1986; Sattler and Rutishauser, 1992; Lacroix and Sattler, 1994; Fisher, 2002; Fukuda *et al.*, 2003). In two species of *Guarea*, the leaf apex continues to initiate new leaflets periodically for more than two years (Steeves and Sussex, 1989).

7 Flowers and Fruits

7.1 THE 'ORIGINS' OF THE FLOWER

7.1.1 Origins of What?

Since Darwin declared the origin of the flower as an abominable mystery in organic evolution, a number of suggestions have been proposed as to how to bridge the gap from flowerless to flowering plants. However, most of these arguments and the accompanying scenarios, increasingly expanded to include evidence from comparative morphology, palaeontology and developmental genetics, have taken for granted that scientific questions about the emergence of organic features can be framed in terms of origins, a position about which I am very sceptical, as mentioned in Chapter 1. Philosophical questions aside, the target of this enquiry is often presented in these terms: from which pre-existing structures did the idealized textbook flower composed of four classes of organs (sepals, petals, stamens and carpels) arise? Let's approach this apparently formidable question somewhat cautiously: any theory of flower 'origins' must cope with the fact that the organs of flowers and their ancestral equivalents were (and are) not necessarily all found in association.

The probable sequence with which the flower eventually achieved its full complexity is: ovules, then stamens, carpels and tepals, in that order, with tepals eventually replaced, multiple times, by sepals and petals (Endress, 2001b). Thus, the evolutionarily oldest organs, the ovules, are in the centre of the flower while at the periphery there are the evolutionarily youngest organs, those forming the perianth. The fact that in the ontogenetic sequence (discussed in Section 9.4) the perianth is very often initiated before the reproductive organs, following the expected centripetal sequence, must not be taken as suggesting that the perianth is evolutionarily younger than the sporophylls (carpels and stamens). In other words, a recapitulatory model like Haeckel's (1866), once popular in zoology ('ontogeny recapitulates phylogeny'), does not apply here at all.

Moreover, the evolutionary sequence of ovules to tepals cannot be taken verbatim, because the definition of some flower structures is sometimes arbitrary, and the sterile organs, as we will see later, have evolved repeatedly, quite likely from different precursors.

A comparison with gymnosperms may suggest that the first step could be the emergence of bisexuality, since most reproductive branches in gymnosperms are unisexual. In addition, reproductive branches in gymnosperms are usually indeterminate, in contrast to the determinate nature of the floral axis in the angiosperms (Ambrose and Ferrándiz, 2013).

To some extent, stamens and carpels are leaf-like structures – hence the term sporophyll – but are not necessarily evolutionarily derived from conventional leaves. Although in loss-of-function mutants of *Arabidopsis* all floral organs, including stamens and carpels, appear like small and simple leaves (phyllomes) (Meyerowitz, 1994), one must take into account that early land plants may have had sporangia before there were leaves with a flat lamina.

In principle, five different possibilities for the evolution of sporophylls could be considered (Crane and Kenrick, 1997): (1) The sporangia became associated with leaves from the beginning of leaf evolution. (2) The sporangia became secondarily transformed into organs with a dorsiventral organization, that is, leaf-like structures. (3) The sporangia became synorganized with leaves that subtended them. (4) The previously autonomous sporangia became combined with leaves in consequence of ectopic gene expression. (5) Leaves may be derived from sporangia by sterilization. Unfortunately, palaeontology does not provide enough detail to help us find our way through this maze of hypotheses.

7.1.2   Angiosperms in the Fossil Record

Molecular clock estimates indicate that angiosperms arose before the Cretaceous, but the record of Mesozoic macrofossils of flowering plants is far from rich. *Archaefructus liaoningensis* was originally proposed to be from the Late Jurassic (Sun *et al.*, 1998), but more recent evidence suggests that it is in fact of Early Cretaceous age, perhaps Barremian–Aptian (130–115 million years ago). The nature of the reproductive unit of *Archaefructus* is also controversial. It has been interpreted either as a bisexual flower without perianth, with a long axis bearing paired or branched stamens proximally and carpels in more distal positions (Sun *et al.*, 2002), or as an inflorescence in which male and female flowers, each of them reduced to one or two organs only, were sequentially arranged (Friis *et al.*, 2003). The latter interpretation would make the reproductive units of *Archaefructus* comparable to those of the extant Hydatellaceae, briefly discussed in the next section.

Recently, however, Liu and Wang (2016) described *Euanthus panii* from the Middle–Late Jurassic of Liaoning, China. The reproductive unit of *Euanthus* is organized like those of extant angiosperms, i.e. as a typical flower with sepals, petals, androecium and gynoecium with enclosed ovules.

7.1.3   Euanthia, Pseudanthia: Homoplastic Origins of the Flower?

Debates on the origin of the flower have long revolved around two alternative interpretations of the reproductive units of the angiosperms. According to the unaxial or euanthial model, the flower is a single axis, typically bearing both megasporophylls (carpels) and microsporophylls (stamens), whereas the alternative pseudanthial or polyaxial models regard the flower as derived from a condensed multiaxial structure, within which each unit is per se a condensed unisexual axis (e.g. Melville, 1960; Eames, 1961; Meeuse, 1972). According to some authors (e.g. Rudall *et al.*, 2009), the two models are not necessarily mutually exclusive, but this would imply that the flower has possibly evolved repeatedly, in independent clades. This is not a necessary implication of the pseudanthial hypothesis. However, the latter suggests at least that the floral structures found in the different groups of angiosperms are not necessarily homologous, because the boundary between flower and inflorescence may have repeatedly shifted in space. This is supported (Specht and Bartlett, 2009) by the terminal structures of the racemose inflorescences in *Potamogeton* that resemble flowers but are demonstrated to be ontogenetically pseudanthial (Sokoloff *et al.*, 2006).

Problematic reproductive units combining features of flowers and features of inflorescences are found in two distantly related genera, that is, in *Trithuria*, belonging to the Hydatellaceae, one of the most basally branching clades among extant angiosperms, and in the monocot *Lacandonia*. These are the only extant angiosperms with inside-out reproductive units in which carpels surround the stamens instead of being surrounded by them. Despite their unique organization, these reproductive units are generally described as flowers, though in both taxa they have been sometimes interpreted as inflorescences (e.g. Ambrose *et al.*, 2006; Rudall and Bateman, 2006; Rudall *et al.*, 2009; Alvarez-Buylla *et al.*, 2010). Another unusual feature of the reproductive unit of *Trithuria* is that carpels develop centrifugally, in contrast to the centripetal or simultaneous development observed in typical flowers (Rudall *et al.*, 2007, 2009).

Lacandonia, with its bisexual reproductive units also showing inverted floral patterning, with stamens surrounded by carpels, evolved within the Triuridaceae, the flower of which is unisexual in all remaining genera. The Triuridaceae, in turn, may have evolved from within the Stemonaceae (Rudall, 2010). The phylogeny and especially the formal classification of the whole group are clearly in need of revision. At any rate, a few steps towards the evolution of *Lacandonia*-type reproductive units can be easily outlined. The Stemonaceae (as traditionally circumscribed) bear conventional bisexual flowers showing centripetal development, but in some species of

the Triuridaceae (these, too, intended as in their traditional circumscription) the carpels develop in fascicles, i.e. from common
primordia, in a centrifugal sequence (Rudall, 2008). According to
Rudall *et al.* (2016), the unisexual yet highly plastic flowers typical
of the Triuridaceae could have paved the way for the emergence of the
extraordinary morphology of *Lacandonia*, and the inside-out flower
of the latter could have resulted from a stabilized homeotic
transformation.

The unisexual reproductive units described by some authors as
pseudanthia are also found in other families. This is the case in the
male reproductive units of *Ceratophyllum*, consisting of up to 46
stamens (Rudall *et al.*, 2009), and in those of *Hedyosmum*, an elongate
axis bearing numerous stamens, interpreted by Endress (2015) as an
inflorescence consisting of unistaminate bractless flowers. The individual female structure of *Balanophora* appears as a minute, compact
organ bearing an embryo sac in the centre; it has been interpreted as a
reduced ovule, but it is instead an entire reduced gynoecium (Endress,
1994a).

Putative pseudanthia are also found among the monocots. In the
Hypolytreae and Chrysitricheae (two clades of the Cyperaceae), the
delimitation of flowers is unclear (Endress, 2012). Similarly, the sup-
posed flowers of some Potamogetonaceae (as mentioned above) and
Cymodoceaceae are probably the products of secondary amalgam-
ation of two or three bona fide flowers (Sokoloff *et al.*, 2006).

7.1.4   *Theories of Flower Origins*

Some so-called theories of flower origins discussed in the last two
decades are actually theories of the evolutionary emergence of a bisexual reproductive unit, irrespective of any accompanying perianth.

Among the extinct groups of land plants, the Caytoniales are
arguably the sister group of the angiosperms, and the Bennettitales, in
turn, are the sister group of angiosperms + Caytoniales (Doyle, 2008).
Among the Bennettitales, *Williamsoniella coronata* possessed bisexual flowers, with a perianth of helically arranged bracts surrounding a

whorl of microsporophylls and a terminal cluster of megasporangia (Crane, 1985). However, most other angiosperm stem-group candidates, including *Caytonia* and some of the putatively early-divergent lineages of Bennettitales, lack readily identifiable flowers; this suggests that the flowers of *Williamsoniella* evolved independently of those of angiosperms (Bateman *et al.*, 2006, 2011). Let's thus see if comparisons with gymnosperms may help us trace the origins of the flower of the angiosperms; gymnosperms are more distantly related to the latter than the Bennettitales, but in their case we are not limited to morphological evidence.

The Mostly Male theory (Frohlich and Parker, 2000; see also Melzer *et al.*, 2010) proposes that flowers are derived from the male strobili of a gymnosperm-like ancestor. The male sporophylls near the apex of the strobilus would have first become bisexual by developing ectopic ovules. At a later stage, these bisexual sporophylls would have lost the male sporangia, thus becoming functionally female. Additional evolutionary steps towards the angiosperm flower would have been the loss of the residual female strobila and the closure of the apical female sporophylls to form carpels while the basal male sporophylls evolved into stamens.

The Mostly Male theory was proposed mainly on the basis of evolutionary studies of *LEAFY* gene function. In angiosperms, only one *LFY* gene is present, whereas gymnosperms have two paralogues: one of them is the orthologue of the angiosperm *LFY*, whereas the other, of which no orthologue has been found in angiosperms, is called *NEEDLY* (*NLY*). Initial expression studies in gymnosperms showed that *LFY* was expressed predominantly in male cones, *NLY* instead in female cones. This led to the hypothesis that the major function of these genes in gymnosperms might be to control divergent differentiation into male versus female reproductive organs (Winter *et al.*, 1999). The same was hypothesized to have been true of the ancestral B-function gene of the seed plants. The absence of *NLY* in angiosperms was interpreted as suggesting that the female strobilus has been lost in the common ancestor of flowering plants, thus leaving the male strobilus as the reproductive unit out of which the flower evolved (Mouradov et al., 1998).

Albert et al. (2002) developed an alternative to the Mostly Male theory that also assumes the role of LFY and NLY genes in leading, in the gymnosperms, the development of male versus female cones. In this model, however, the segregation of male and female reproductive units in gymnosperms was considered to be due to differential regulation of LFY (male) versus NLY (female) promoters rather than to differences in the coding regions of these genes. Thus, the bisexual condition of the angiosperm reproductive units might have emerged following the loss of the NLY gene, when the previously distinct spatial expression patterns of the two LFY paralogues was replaced by a common LFY control expressed in all reproductive meristems.

However, recent data showing that LFY and NLY are in fact expressed together in both female and male cones of Picea, Podocarpus and Taxus (Carlsbecker et al., 2004; Vázquez-Lobo et al., 2007) have weakened both the original Mostly Male theory and Albert et al.'s version.

Another two theories of flower origin, the closely related Out-of-Male and Out-of-Female theories, were proposed by Theißen et al. (2002). These theories explain the bisexuality of the flower as a consequence of temporal or spatial shifts in patterns of expression of genes involved in determining the male versus female identity of reproductive organs downstream of LFY/NLY. The most important difference between these theories and the Mostly Male theory is that the former do not involve dismantling of the ancestral pathways controlling the differentiation of female reproductive units and therefore postulate that carpels are directly derived from the female organs of angiosperm ancestors rather than resulting from modified male organs, as proposed by the Mostly Male theory.

Specifically, the Out-of-Male hypothesis (Theißen et al., 2002) proposes that the expression of B-class genes in a previously male strobilus shifted towards the basal portion of this reproductive structure, leaving female structures in the apical segment. Conversely, the

Out-of-Female hypothesis postulates upregulation of B-class genes in the basal portion of a female cone; the result, a bisexual axis, would be the same.

Building upon this Out-of-Female theory, thus still regarding *LFY* as a major player, Baum and Hileman (2006) suggested that in the most recent common ancestor of extant flowering plants the reproductive shoots may have accumulated, during development, increasingly higher levels of *LFY*. Baum and Hileman (2006) hypothesize that *AG*-like genes are more responsive to *LFY* than *DEF/GLO*-like genes are, so that the top level of *AG*-like gene expression is finally much higher than that of the expression of *DEF/GLO*-like genes. Consequently, at the base of the reproductive cone, where both *DEF/GLO*-like and *AG*-like genes are expressed, male organs develop, while female organs develop instead at the top, where only *AG*-like genes are expressed.

7.2 THE PARTS OF THE FLOWER

7.2.1 *Carpels*

Unlike the perianth, the carpel is thought to have emerged only once along the lineage leading to the angiosperms (Specht and Bartlett, 2009), by synorganization between the ovule and the supporting phyllome. In the last common ancestor of the living flowering plants, the carpels were likely simple (apocarpic flower), rather than fused together into a syncarpic pistil (Vialette-Guiraud and Scutt, 2009).

Amborella is dioecious, but its female flowers produce staminodes (Endress and Igersheim, 2000a, 2000b), thus demonstrating the fundamentally bisexual nature of its flower. This is in agreement with the common assumption that the first flowers were hermaphrodite (Vialette-Guiraud and Scutt, 2009). Nevertheless, among the earliest angiosperm macrofossils, unisexual flowers are not rare (Friis *et al.*, 2000).

In spite of its phylogenetic origin, in some extant plants, including *Arabidopsis*, the ovule behaves like a constituent part of the carpel. In *Petunia*, after carpel differentiation, the residual activity of the floral meristem is exhausted in the production of a placenta from which the primordia of the ovules arise. In this plant, Colombo *et al.* (2008) regarded the ovule as a part of the carpel, based on the result of the inactivation of a petunia orthologue of the rice gene *OsMADS13*, resulting in the homeotic conversion of ovule to carpel, rather than to meristem. In contrast, ectopic carpels do not form ovules in rice mutants in which stamens are homeotically converted to carpels; in this plant, the ovule originates directly from the floral meristem, as an independent lateral organ, rather than indirectly from the carpel (Nagasawa *et al.*, 2003). This indicates that mechanisms of ovule differentiation are diverse in angiosperms (Yamaki *et al.*, 2011).

In ANA-grade angiosperms, the expression of C-function genes is mostly limited to the sepal and carpel whorls, while the E-function genes are expressed in all floral organs (Kim *et al.*, 2005b). These expression patterns are similar to those of the corresponding genes in *Arabidopsis*, suggesting extensive conservation of important elements of the control of carpel identity throughout angiosperm evolution. However, in *Amborella* and *Illicium* the expression of C-function genes extends to the perianth organs (Kim *et al.*, 2005b); this contrasts with the expression patterns of C-function genes in model eudicots (Vialette-Guiraud and Scutt, 2009).

In monocots, the C-gene clade has undergone at least one duplication prior to the separation of the rice and maize lineages; an additional duplication has occurred, in the maize lineage, in one of the two subclades generated with the first duplication. As a consequence, three C-function genes are available in maize (two in rice) to accomplish the multiple roles in which a single gene (*AG*) is involved in *Arabidopsis*, i.e. floral determinacy and stamen and carpel development. Additionally, some other factor is probably involved in the specification of carpel development, because in these monocot species this can occur independently of C-clade genes (Vialette-Guiraud and Scutt, 2009).

In the development of the characteristic fruit of the Brassicaceae, the silique, there is antagonism between the genes controlling the production of the replum which divides the fruit in two halves and the genes controlling the production of the two external valves; this antagonism mirrors the relationships between meristem and leaf, involving mostly the same genes. *STM*, *BP* and *RPL*, which have crucial roles in the meristem, are also expressed in the replum, whereas *FIL*, *YAB3* and *JAG*, with essential functions in leaf development, are expressed in the valves (Dinneny *et al.*, 2005; Martínez-Laborda and Vera, 2009).

7.2.2 Stamens and Staminodes

The genetic control of stamen production is strictly conserved: homologues of the B-function genes *AP3* and *PI* of *Arabidopsis* are uniformly required for stamen formation, and homologues of the C-function gene *AG* have been similarly shown to be required for the development of both stamens and carpels (Litt and Kramer, 2010). In some clades, including Ranunculaceae (Kramer *et al.*, 2003) and a number of Lamiales (Aagaard *et al.*, 2005), however, either *AP3* or *PI* or both genes have undergone duplication. In *Petunia*, stamen identity is controlled by two *AP* paralogues (*PhTM6* and *PhDEF*) and two *PI* paralogues (*PhGLO1* and *PhGLO2*) (for an overview, see Jaramillo and Kramer, 2007).

Sterile flower organs broadly similar to conventional stamens, except that they fail to produce spores (and thus male gametophytes, i.e. pollen grains), are the staminodes (reviewed in Ronse De Craene and Smets, 2001). Staminodes may occur in the same whorl as fertile stamens, but not necessarily.

Staminodes belonging to a staminal whorl may become secondarily petaloid; this trend is most conspicuous in the Zingiberales (Walker-Larsen and Harder, 2000). In the most derived families of this order, the number of fertile stamens (originally six, in two whorls of three) is reduced to one, less frequently two. The fertile stamen belongs to the inner stamen whorl and is petaloid. The staminodes also develop as petaloid structures. Several staminodes, 2–4 in the Zingiberaceae, five in the Costaceae, may fuse together to form the staminodial labellum, a novel structure that is the most conspicuous element of the flower in these families (Specht *et al.*, 2012). The evolution of stamens into laminar, petaloid staminodes has been accompanied by an extensive evolution of the *GLO* genes, with a family-specific sequence of duplications and losses (Piñeyro-Nelson *et al.*, 2017). This evolutionary history was probably initiated by a whole-genome duplication in the most recent common ancestor of the Zingiberales, as almost all the genes targeted by recent studies have two paralogues (*SEP3-1/2, LOFSEP-1/2, DEF-1/2, AG-1/2*) (Yockteng *et al.*, 2013; Almeida *et al.*, 2015a, 2015b); the exceptions are *AGL6*, of which a single copy has been found (Yockteng *et al.*, 2013), and *GLO*, which has instead four paralogues (Bartlett and Specht, 2010), but only the Costaceae have retained all of them.

In flowers with spiral phyllotaxis, e.g. in some Magnoliidae, staminodial structures occur frequently between tepals and stamens as well as between stamens and carpels, and are sometimes regarded as intermediates between the conventional kinds of organs. In the Ranunculales, petals have possibly arisen several times from outer staminodes (Drinnan *et al.*, 1994; Kosuge, 1994; Endress, 1995). In the Ranunculaceae, there are morphoclines leading from simple staminodes to elaborate petaloid organs bearing nectaries (Ronse De Craene and Smets, 1995).

In representatives of the Hamamelidaceae and Caryophyllaceae, petaloid staminodes are observed in positions normally occupied by petals.

In some angiosperm genera, e.g. *Linum, Hesperolinon* and *Samolus*, staminodes are present in some species, but others have none (Ronse De Craene and Smets, 2001).

7.2.3   Perianth

7.2.3.1   First Steps Towards the Differentiation of the Perianth
A preliminary discussion of what petals and sepals are is the obligate starting point from which we can move into this important chapter of plant evolution.

It is widely accepted by plant morphologists that no single character or character combination can be used to fix definitively whether a floral organ is a sepal or a petal. Usually, we call petals those floral organs that are coloured or otherwise visually striking, compared to sepals, which are usually dull and green. Moreover, petals often have a narrow base, a single vascular bundle and a delayed development (Endress, 1994a), but these criteria cannot be applied universally (Ronse De Craene, 2007; Ronse De Craene and Brockington, 2013).

In practice, in those perianths in which a distinction between petals and sepals is not possible, all perianth elements are commonly called *tepals*, but the uniform adoption of this term conceals possibly major differences in the nature or derivation of the perianth organs in different plants.

The oldest fossils of bona fide angiosperms do not help to answer the questions, whether early flowers possessed a perianth or not and, if so, whether its elements were arranged in whorls or spirals and whether they were, or not, all of the same kind. As mentioned above, *Archaefructus* lacked a perianth – but this is perhaps uninformative for a tentative reconstruction of perianth origins, because this plant, despite its age, is perhaps not basal within the angiosperms (Friis *et al.*, 2003). Among the Early Cretaceous fossils, some had an undifferentiated perianth (Crane, 1985; Friis *et al.*, 1994, 2000; Crane *et al.*, 1995), but others (*Archaeanthus*; Dilcher and Crane, 1984) had instead a more complex perianth, comparable to those of *Magnolia* species.

Among the early-divergent lineages of extant angiosperms, some have a perianth, but the delimitation between tepals and subtending bracts is sometimes uncertain (Buzgo *et al.*, 2004; Kim *et al.*, 2005c), and groups such as Chloranthaceae and Piperales lack a perianth entirely (Rudall, 2013).

In plants with two whorls of perianth organs, we can often distinguish between an external whorl of sepals and an internal whorl of petals, but the recognition of petals becomes problematic when both whorls are sepaloid, as in the Juncaceae, or when both are petaloid, as in the Passifloraceae.

A perianth with two distinct sets of organs (sepals and petals) has evolved many times independently (Zanis *et al.*, 2003; Hileman and Irish, 2009; Kramer and Hodges, 2010; Endress, 2011).

The strict distinction between sepals and petals has broken down in a number of lineages with spirally arranged perianth elements showing a progressive transition between the organs conventionally called sepals and petals. Examples are found in the Paeoniaceae, Dilleniaceae, Clusiaceae, Theaceae and Lecythidaceae (Ronse De Craene, 2008). The opposite trend is found in the Caryophyllales, where a secondary differentiation of the perianth into sepals and petals has occurred many times independently (Ronse De Craene, 2007; Brockington *et al.*, 2009).

Development does not help more than morphology in tracing a distinction between petals and sepals. In *Paeonia*, for example, all perianth organs, although morphologically differentiated into sepals and petals, are produced in a single continuous spiral; moreover, in this genus the petals have as many veins as have the sepals (Roeder, 2010).

In many monocots, the differentiation between sepals and petals is less conspicuous than in eudicots (Endress, 2011), but there are also monocot groups with strong differentiation, including some Alismatales, Commelinales and Zingiberales.

Walker and Walker (1984) presented a scheme of perianth evolution in the angiosperms, articulated into six evolutionary grades:

1 simple perianth of floral bracts, as in Austrobaileyaceae and Trimeniaceae
2 perianth entirely sepaloid or petaloid
3 differentiation of a distinct calyx and corolla of tepalar origin
4 loss of petals (primary apetaly)
5 staminodial petals acquired secondarily in descendants of these apetalous precursors by a transformation of stamens
6 further loss of petals (secondary apetaly)

In this idealized morphocline, convergence and parallelism are clearly ignored; nevertheless, this list of evolutionary transitions can still help in outlining the main events in perianth evolution.

7.2.3.2 *Andropetals and Bracteopetals*

It is widely accepted that in some angiosperm lineages petals emerged as modifications of stamen-like structures (andropetals), in other lineages as modifications of bract- or leaf-like organs (bracteopetals) (Eames, 1961; Weberling, 1989; Takhtajan, 1991). These evolutionary transformations were originally proposed based on morphological evidence, including number of vascular traces, patterns of primordium initiation and phyllotaxis, and they have been tentatively reinterpreted in terms of changes in gene expression patterns.

Baum and Hileman (2006) postulated that the corolla evolved by sterilization of the outer stamens by co-option of *WUS* in the regulation of C-class genes, causing a shift in the expression of B genes.

According to Kramer and Jaramillo (2005), the 'genetic program' responsible for petal identity in the core eudicots is essentially a 'sterilized' version of the gene expression patterns responsible for stamen identity, a version in which no C-function gene is expressed. The authors did not specify, however, whether this change in the 'stamen identity program' resulted from C-gene expression disappearing from an outer whorl of stamens or from an expansion of the B domain into pre-existing sterile organs, as suggested by Baum (1998). The former scenario would be in agreement with the idea that in many (although certainly not all) angiosperms, petals represent sterilized stamens (andropetals); the latter scenario, on the other hand, is consistent with the idea that some petaloid organs are derived from subtending leaves or bracts.

Following Ronse De Craene's (2007; also Ronse De Craene and Brockington, 2013) revisitation of the issue, however, the general opinion that the petals of the majority of angiosperms are derived from stamens seems to be an unwarranted generalization. Reading comparative data on floral ontogeny against a sound phylogenetic background, it appears that petals derive primarily by differentiation of an inner whorl of tepals. In particular, a staminodial origin of petals can be ruled out for model plants like *Arabidopsis* or *Antirrhinum*, and seems to be mainly restricted to clades (Caryophyllales, Rosales) where petals have initially been lost. In this view, petals of stamino-dial origin would be the result of a secondary repatterning of gene expression in an outer whorl of sterile stamens, as suggested, according to Ronse De Craene (2007), by the recruitment of a *TM6* homologue into the stamens of *Silene* (Zahn *et al.*, 2005).

Bracteopetals have evolved independently two or three times in the Magnoliales–Laurales clade, but apparently coexist with andrope-tals in the flower of *Fenerivia heteropetala*. The unique perianth of this plant includes very small sepals and 12 petals: the three outer oval petals are interpreted as bracteopetals, similar to the first three of the inner linear petals, whereas the six innermost linear petals are regarded as andropetals (Deroin, 2007), the only example of andrope-tals known thus far in the Annonaceae (Saunders, 2010); see Table 7.1.

Characters used to differentiate andropetals from bracteopetals are subjected to functional constraints and are thus not very reliable as indicators of homology, and a number of studies claiming the presence of andropetals in a large number of families may not stand further scrutiny: according to Ronse De Craene and Brockington (2013), the derivation of petals from staminodes remains a special case with a limited distribution.

The actual appearance of a perianth organ can indeed contrast with its origin as suggested by its relative position. In flowers with a single-whorl perianth derived from flowers with two perianth whorls by loss of one of these, the single extant perianth whorl may have petaloid appearance despite its derivation from sepals, as in apetalous Saxifragaceae (*Rodgersia, Chrysosplenium*), Rhamnaceae (*Colletia*) and Primulaceae (*Glaux*); alternatively, it may be the sepals that have been lost, in which case the remaining perianth whorl will be of petal derivation (e.g. Santalaceae, some Apiaceae). The coloured stami-nodes of *Jacquinia macrocarpa* represent an aborted stamen whorl, yet are morphologically very similar to petals (Ronse De Craene and Brockington, 2013).

Another big problem we face in our efforts to reconstruct the evolution of the flower and its parts is where to fix the topographical

Table 7.1 *Evolutionary trends in the flowers of Annonaceae; largely based on data in Saunders (2010).*

Change	Examples
Loss of one of the whorls of petals	*Anaxagorea, Annona* and *Dasymaschalon* (inner petals lost), *Annickia* (outer petals lost)
Inner and outer petal whorls united to form a single whorl with twice as many petals	*Isolona*
Inner petal development suppressed; homeotic remaining (outer) petals taking inner petal identity	*Dasymaschalon*
Compression of inner and outer whorls into a single whorl with twice as many petals; petal fusion	*Isolona* (a single whorl of six petals fused to form a perianth tube)
Evolution of an additional perianth whorl	*Fenerivia heteropetala* (innermost petals derived from stamens, i.e. andropetals, in contrast to the outermost and middle petals which are homologous with the sepals, i.e. bracteopetals)
Gain or loss of a single sepal	Species with two petal whorls typically showing 6–9 petals, but also 12 petals occasionally in *Lettowianthus* and *Toussaintia*, and up to 15 petals in *Deeringothamnus*
Congenital fusion of petals	*Miliusa* (inner petals only)
Abaxial surface of inner petals fused with adaxial surface of outer petals, leaving margins free	*Stenanona*
Unisexual flowers	Several unrelated lineages
Gradual transition between fertile stamens and sterile inner staminodes	*Anaxagorea*

Table 7.1 (*cont.*)

Change	Examples
Outer staminodes present	Common
Inner staminodes present	*Xylopia p.p.*
Syncarpy	*Isolona, Monodora*
Pseudosyncarpy (postgenital fusion of carpels during fruit formation)	*Annona, Duguetia, Fusaea*

limits of the perianth in respect to the accompanying structures (bracts and subtending leaves). These limits have often been challenged during evolution. In a number of core eudicot families, flowers are subtended by several whorls of organs, without clear separation between bracts, sepals and petals; examples are *Clusia, Reaumuria* and *Stachyurus* (Ronse De Craene and Brockington, 2013).

The perianth can evolve by inclusion of bracts as additional parts of the flower, either as an epicalyx (e.g. *Scabiosa* and *Thunbergia*) or as a replacement for a lost petal whorl (e.g. Portulacaceae) (Ronse De Craene and Brockington, 2013). *Cornus florida* is popular for its large, showy, petal-like bracts (Irish, 2009). In *Davidia involucrata*, petal identity is partially taken by the large bracts, similar to small handkerchiefs, which surround a contracted inflorescence with reduced flowers (Vekemans *et al.*, 2012b). In many flowers, e.g. *Melampyrum* and *Castilleja*, the colour of the petals expands onto the inflorescence bracts.

As a consequence of progress in developmental genetics, the angiosperm flower is sometimes depicted, rather than with reference to the traditional morphological categories (carpel, stamen, petal, sepal), in terms of 'identity programs' corresponding, for instance, to the petaloid or staminoid nature of a flower organ, irrespective of its position. Eventually, these identity programs have emerged as the 'given' in flower development and evolution. Here is an example: 'Once such a homeotic petal-identity program evolved, deactivation and reactivation of the genetic pathway could produce independent losses and gains of petaloid organs, as well as spatial shifts in their position. The existence of a homologous petal-identity program would alter the classical distinction between andropetals and bracteopetals' (Kramer and Jaramillo, 2005, p. 529). This view acknowledges the decoupling between relative position and special quality (e.g. petaloidy) of flower organs, but still implicitly accepts what Warner *et al.* (2009) describe as the unspoken rule that 'sepal' and 'petal' must refer to whole organs. The latter authors propose a novel view (they call it the 'Mosaic theory'), according to which the distinction between sepaloid and petaloid structures evolved early in angiosperm history, but these features were not fixed to particular sets of flower organs (sepals and petals, respectively) and were primarily controlled by nongenetic factors. It was only at a later stage in angiosperm evolution that sepaloidy and petaloidy became the specific traits of whole sets of organs arranged in distinct whorls, and the role of environmental control decreased in favour of predictable genetic control.

Individual perianth organs including both sepaloid (green) and petaloid (colourful) areas, occasionally delimited by sharp boundaries, are found in a number of extant plants, more characteristically in some early-divergent angiosperms such as *Nuphar*, *Nymphaea* (Fig. 7.1), *Euryale*, *Schisandra* (Warner *et al.*, 2008) and probably also *Amborella* (Buzgo *et al.*, 2004); but a similar feature occurs also in some eudicots (e.g. *Berberis*, *Hypericum*) and monocots (e.g. *Alstroemeria*). In *Impatiens*, the lower side of the adaxial petal is green, while the upper side is petaloid.

Parts of the perianth organs that are exposed in the bud are more frequently sepaloid; those which are not exposed turn out eventually to have petaloid character. In *Nymphaea*, the visible boundaries between the sepaloid and petaloid parts of each perianth organ do indeed correspond with the boundary between the areas that were exposed and covered. A mixture of green and coloured areas corresponding to uncovered and covered surfaces is also found in the sepals of the Polygonaceae and other Caryophyllales (Ronse De Craene and Brockington, 2013).

FIGURE 7.1 Among the more or less numerous sterile organs in the flower of water-lilies (*Nymphaea* spp.), some are partly sepaloid (green), partly petaloid (coloured). *A black and white version of this figure will appear in some formats. For the colour version, please refer to the plate section.*

7.2.3.3   *The Perianth: a Taxonomic Survey*

Phylogeny is the best guide to a closer approach to this tangled set of morphologies and the underlying genetic and developmental underpinnings. The following account is mainly based on Ronse De Craene and Brockington (2013).

Among the basal angiosperms, the Aristolochiaceae (except *Saruma*) and Myristicaceae have flowers with a single-whorled perianth. In some cases, the sepaloid or petaloid nature of the latter can be determined through comparison with the perianths of closely related taxa. For example, the single-whorled perianth of the Aristolochiaceae is considered a calyx (Cronquist, 1981; Takhtajan, 1991; Tucker and Douglas, 1996). But these traditional interpretations are arguably biased by the typological approach to the homology of flower organs, largely based on their relative position, that dominated the field before the last decade; see Section 7.2.3.2.

Petaloid structures of different derivation have evolved in a number of basal angiosperms, most frequently by progressive differentiation of bracteopetals (e.g. Calycanthaceae, Schisandraceae, Lauraceae), or with a growing contrast between outer (sepaloid) and inner (petaloid) tepals (e.g. Magnoliaceae, Annonaceae, Winteraceae). In the Himantandraceae and Nymphaeaceae, there is essential continuity between the innermost, narrow tepals and the coloured outer staminodes. In some basal angiosperms (Winteraceae, Monimiaceae), the outermost involucral organs of the flower are bracts that closely subtend the tepals.

In many Magnoliaceae and Annonaceae, the perianth is morphologically and functionally comparable to the typical bipartite perianth of core eudicots, despite its different derivation. In the Annonaceae (Table 7.1), the perianth is mostly trimerous, generally with two whorls of petals, often morphologically distinct. In *Asimina*, floral development appears to be controlled by a combinatorial code of gene expression broadly similar to the system operating in eudicots (Kim *et al.*, 2005c).

No perianth is present in the Chloranthaceae, except *Hedyosmum*; in the latter genus, three short scale-like organs are usually described as tepals (Endress, 2001c; Li *et al.*, 2005). In *Chloranthus*, the flower is reduced to a single stamen and a single uniovulate carpel.

In several monocots, including the basal Acoraceae and Tofieldiaceae, a number of Alismatales and clades with strongly reduced flowers, e.g. the Cyperaceae, it is not easy to determine what is a bract and what is a 'true' perianth organ (Remizowa *et al.*, 2010). Most monocots, however, have evolved a perianth with a stable two-whorl configuration. This can take different forms. In orders such as Liliales and Asparagales, both perianth whorls are petaloid; in some Alismatales and in the Poales, both perianth whorls are sepaloid; in the Alismataceae, some Melanthiaceae, Bromeliaceae and Commelinaceae, the outer, green sepals are neatly differentiated from the inner, coloured petals.

A transition from a monomorphic to a dimorphic perianth appears to have occurred independently in two monocot clades, the Commelinales and the Zingiberales (Bartlett and Specht, 2010). Within the latter, the Musaceae have a largely monomorphic perianth, with three sepals and three petals more or less similar in colour and overall aspect. The strongest differentiation among perianth organs is between one of the elements of the inner whorl, which remains free, and all other elements of the perianth, fused together into a floral tube. The family of the Zingiberales with the greatest differentiation between outer and inner perianth whorls is the Costaceae, which has sepals that are much smaller than petals and differ from them also in colour and toughness (Specht *et al.*, 2012).

Among the basal eudicots, a perianth does not seem to be present in the minute unisexual flowers of *Ceratophyllum*: male flowers in this aquatic genus contain 3–46 stamens that are initiated in a variable phyllotactic sequence; female flowers consist of a single uniovulate carpel (Iwamoto *et al.*, 2003). An alternative interpretation of the male reproductive units in this genus has been mentioned in Section 7.1.3.

Within the early-diverging eudicots, petals have evolved independently multiple times. A diversity of evolutionary paths towards a differentiated perianth is recognizable even within one order, the Ranunculales. In several genera, both of Berberidaceae (*Nandina, Podophyllum*) and of Ranunculaceae (e.g. *Anemone*), petaloid organs evolved from previously sepaloid tepals; in other taxa, e.g. Papaveraceae, the differentiation of a previously tepaloid perianth has evolved into the contrast between an outer whorl of bract-like sepals and two inner whorls of coloured petals; in many species of the Berberidaceae, Lardizabalaceae, Ranunculaceae and Menispermaceae, the unusually showy outer sepals contrast with the inner perianth organs, either petals or nectariferous leaves.

In several genera of Ranunculaceae with bipartite perianths, the organs of the second whorl, although easily classifiable as petals, strongly resemble modified stamens (Kosuge, 1994). Many taxa in this

FIGURE 7.2   The corolla of the smallflower buttercup (*Ranunculus parviflorus*) is remarkable for the instability of number and size of its elements. *A black and white version of this figure will appear in some formats. For the colour version, please refer to the plate section.*

family (e.g. *Aquilegia, Trollius*) possess two types of petaloid organs: an external whorl of large, showy sepals and an internal whorl of petals, of very different shape and sometimes provided with nectariferous spurs. However, in other genera such as *Ranunculus* (Fig. 7.2), the sepals are leaf-like and the petals do not resemble stamens. Many species, e.g. those of *Anemone* and *Caltha*, have petaloid sepals but lack a second whorl of petals altogether. In *Clematis*, some species possess petals, other species are apetalous (Rasmussen *et al.*, 2009).

Following the angiosperm tree upwards from the branching-off of the Ranunculales, in the diverse taxa forming the grade leading to the core eudicots, flowers with a perianth differentiated into sepals and petals are uncommon, and some genera are perianthless. The only taxa with a clear differentiation of sepals and petals are the Nelumbonaceae and Sabiaceae. In the remaining families of early-diverging eudicots, the perianth is simple but bipartite (Proteaceae) or undifferentiated or absent; in some taxa, even the delimitation of flowers is unclear (Drinnan *et al.*, 1994; von Balthazar and Endress, 2002; Wanntorp and Ronse De Craene, 2005; González and Bello, 2009).

A clear distinction between sepals and petals is found in the vast majority of the core eudicots (Endress, 2010a), but not in the transitional little order Berberidopsidales.

The flower of *Berberidopsis* is in fact spiral and its perianth is undifferentiated, as in many basal angiosperms. A careful study of *B. beckleri* (Ronse De Craene, 2017) and *B. corallina* (Ronse De Craene, 2004) helps to reconstruct the early steps of the evolution towards the bipartite perianth (and pentamery; see Section 7.5.3) of core eudicots. In *B. beckleri* the initiation of perianth parts is spiral, their number is variable, and their differentiation follows a gradient from the most external bracts to the following, weakly differentiated sepaloid and petaloid tepals. The androecium usually consists of 11 stamens. In *B. corallina*, the number of perianth and androecium elements is less variable and the alternation of shorter and longer plastochrons leads to an essentially whorled arrangement.

In most core eudicots, the perianth is biseriate; when it is not, this is apparently due to secondary loss of either the sepal or the petal whorl.

Most rosids have a pentamerous (less frequently tetramerous) flower with a neat differentiation between sepals and petals. However, a tendency for petals to become smaller or to be lost altogether is found in several clades, e.g. in Saxifragaceae, Rhamnaceae and Thymelaeaceae. Truly apetalous clades have evolved independently within each major order or family, but are more frequent among the Rosales, Fagales and Cucurbitales. Apetalous flowers are found even in some large taxa that usually have a showy corolla, such as the Fabaceae (*Ceratonia*) and the Rosaceae (some species in *Acaena*, *Alchemilla*, *Sanguisorba* and a few other genera) (Ronse De Craene, 2003). Petals are lost in Nepenthaceae and Polygonaceae, and possibly evolved anew in the Plumbaginaceae, where these perianth organs arise, in ontogeny, as appendages from common stamen–petal primordia.

Among the Caryophyllales, the flowers of the basal clades of the core lineage are generally pentamerous, but apetalous. At least 13 families (out of 38 in the whole order) are fully apetalous or at least include species without petals (Endress, 2010a). In other families, a bipartite perianth has evolved. This occurred, independently, along different paths: by insertion of a pair of bracts close to the petaloid sepals, as in Portulacaceae and Cactaceae, or by differentiation of outer petaloid staminodes from a common ring of stamen primordia, as in Aizoaceae and Caryophyllaceae.

In the flowers of at least some Caryophyllales, genes expressed in the petaloid structures are other than the expected *AP3* and *PI* (Brockington *et al.*, 2011); this supports the hypothesis of a novel acquisition of petals in this clade (Ronse De Craene, 2007, 2008, 2010; Brockington *et al.*, 2009). In the Aizoaceae, the perianth is either undifferentiated or differentiated into sepals and staminodial petals. In the staminodial petals of *Delosperma* there is an early expression of *AP3*, *PI* and *AG*, as expected for stamens. However, the petaloid tepals of *Sesuvium* do not show any expression of *AP3* and *PI*, suggesting in this case too a novel acquisition of petaloidy in otherwise sepaline perianth organs.

7.2.3.4 *Molecular Genetics of Sepal and Petal Specification*
In *Arabidopsis*, ectopic expression of *SEP3* accompanied by expression of A and B genes converts rosette leaves to petaloid organs (Honma and Goto, 2001; Pelaz *et al.*, 2001b); if a C gene is also expressed, leaves are transformed into staminoid organs (Honma and Goto, 2001). These results support the hypothesis that floral organs represent modifications of a leaf-like structure (Shepard and Purugganan, 2002).

In *Arabidopsis*, petal-identity specification was found to depend on the combined expression of *AP1*, *PI*, *AP3* and *SEP* (Jack *et al.*, 1994;

Krizek and Meyerowitz, 1996; Honma and Goto, 2001; Pelaz *et al.*, 2001a). Early generalizations of this mechanism to all flowering plants were, however, unwarranted. In some angiosperm clades, orthologues of some of these genes are not found. In others, such orthologues exist, but do not have the same biological function as their *Arabidopsis* (or *Antirrhinum*) counterparts (Irish, 2009). In particular, the homeotic A function of the original ABC model (see Section 4.6.3) is limited to *Arabidopsis* and its close relatives (Litt, 2007). *MADS*-box genes belonging to the same clade as the *Arabidopsis* A-function gene *AP1* are generally present in one or more copies, but do not contribute to defining sepal or petal identity, although expressed in the two outer whorls of the flower, like, e.g., *SQUA* from *Antirrhinum* (Huijser *et al.*, 1992).

In basal angiosperms the homologues of the ABC-function genes are more broadly expressed than in eudicots. In *Amborella*, where a gradual transition is seen from outer to inner perianth organs, class B gene orthologues are expressed in all of them (Kim *et al.*, 2005c). Perianth differentiation into sepals and petals has evolved in Nymphaeaceae and Cabombaceae (but see Warner *et al.*, 2009), accompanied in the latter family by a transition from spiral to whorled phyllotaxis. However, the differentiation of sepals and petals is not strong, and B-gene homologues are expressed in both sepals and petals (Kim *et al.*, 2005c; Endress, 2008a; Yoo *et al.*, 2010).

Members of the paleo*AP3* and *PI* gene lineages are expressed in all petaloid organs of the flowers of basal angiosperms, e.g. Magnoliaceae, Aristolochiaceae, Piperaceae and Chloranthaceae, and lower eudicots such as Buxaceae, Papaveraceae and Ranunculaceae; in detail, these expression patterns are quite diverse. The perianth is unipartite in some of these flowers, bipartite in others. In many species the genes are expressed, more or less strongly, at early stages of flower development, but this expression is not seen in the petaloid organs of *Liriodendron tulipifera* and *Calycanthus floridus*. Spatial restriction of these expression patterns has been recorded within the petals of several ranunculid species, in the tepals of *Magnolia* *figo* (= *Michelia figo*) and also in a monocot, *Sagittaria montevidensis* (Kramer and Irish, 1999). The actual localization of the transcripts is different in the different species, involving the tip of the developing perianth organ, or its base, or its adaxial half, or just its adaxial epidermis.

In *Magnolia*, B-class genes are extensively expressed in the perianth (represented in many species by three whorls of nearly identical tepals) as well as in staminodes and stamens (Kramer and Irish, 2000). In a few species such as *M. acuminata*, *M. liliiflora* and *M. stellata*, the organs forming the first whorl of the perianth are different from those of the following whorls. Further perianth organs in a spiral arrangement are present in *M. stellata* above the three usual perianth whorls, in the putative spatial domain of stamen primordia. In the first perianth whorl of these three species, Wróblewska *et al.* (2016) recorded the presence of *AGL6* expression, suggesting a sepaline rather than a foliar origin, despite the lack of expression of *AP3* and *PI*.

Patterns of B-class gene expression similar to *Magnolia* are seen in *Papaver* and in several representatives of the Ranunculaceae (Kramer and Irish, 2000; Kramer *et al.*, 2003).

Within the Aristolochiaceae, perianth morphology differs dramatically between *Saruma* and the remaining genera: two whorls are found in *Saruma*, but only one in the others, including the very large genus *Aristolochia*. In *Saruma*, the external whorl is leaf-like (sepaloid), the internal one petaloid, similar to the usual distinction in core eudicots. Instead, morphological aspects of both calyx and corolla are combined in the one-whorl perianth of *Aristolochia*. In *Saruma*, the homologues of the B-class genes *AP3* and *PI* are expressed in patterns suggesting a function in the development of the second and third whorl (petaloid organs and stamens), whereas the corresponding gene expression in the perianth of *Aristolochia* seems to exclude their role in specifying organ identity (Jaramillo and Kramer, 2004; Kramer and Jaramillo, 2005).

Of the many other players contributing to the regulation of perianth architecture, I will only mention the *Antirrhinum* gene

INCOMPOSITA (INCO), which represses the development of extra flower organs beneath the two lateral sepals, which are absent in the wild-type flower (Wilkinson *et al.*, 2000) and occasionally reach the size of sepals (Masiero *et al.*, 2004).

7.2.3.5 Sliding Boundaries and Fading Borders

Irrespective of the exceptions that prompted the evolution of the original ABC model into the ABCD, ABCDE or (A)BCDE versions mentioned in Section 4.6.3, the expression patterns obtained for ABC genes from basal angiosperms are generally consistent with the morphology of the flowers as predicted by the ABC model. For example, perianth organs in *Amborella*, *Nuphar* and *Illicium* expressing class B gene orthologues are always more or less 'petaloid'.

As mentioned above, B-class genes are not necessarily expressed throughout the whole perianth, but are often spatially restricted to specific areas, for example at the base, the tip or the adaxial half of the individual perianth organ (Kramer and Irish, 1999, 2000; Jaramillo and Kramer, 2004).

The spatial patterns of expression of *AP3* and *PI* homologues in *Nuphar advena* support the predictions of Warner *et al.*'s (2009) Mosaic model discussed in Section 7.2.3.2. These B-class genes are indeed expressed more strongly in the organs, or parts of organs, that are covered by other perianth organs during part of their development and eventually express petaloid character (the tips of the outer tepals) or large petaloid patches (the inner tepals) (Kim *et al.*, 2005c).

However, these findings do not support the early version of the ABC combinatorial code, predicting a precise identity of each floral organ – as sepal, petal, stamen or carpel – dependent on the specific set of ABC genes expressed in its primordium. This explains the subsequent proposals of less rigid models, beginning with the so-called Sliding Boundary or Shifting Boundary models of flower evolution (Bowman, 1997; Kanno *et al.*, 2003; Kramer *et al.*, 2003, 2006; Ochiai *et al.*, 2004; Kramer and Jaramillo, 2005; Hintz *et al.*, 2006; Soltis *et al.*, 2007; see also Tzeng and Yang, 2001; Tzeng *et al.*, 2002).

Two different versions of these models have been suggested, respectively involving outward and inward shifts of B-class gene expression.

The whorled perianths of many monocot species other than grasses, such as *Lilium* and *Tulipa*, are tepaloid. Most of these species, but not all of them, express B-class genes related to *AP3* and *PI* in the first, second and third whorls (Kanno *et al.*, 2003), supporting the idea of a sliding border for the expression domain of B-class genes: by expanding into whorl 1, they would confer to the outer tepals a petaloid identity otherwise uncommon in the external perianth whorl (Ambrose and Ferrándiz, 2013). This model has also been used to explain the differentiation of three types of perianth organs in orchids, with three outer tepals in the first whorl, two lateral inner tepals and a median inner tepal (labellum) in the second whorl. All orchid tepals are usually petaloid and express putative class B floral homeotic genes (Tsai *et al.*, 2004; Xu *et al.*, 2006). (The genetic control of the diverging identity of these three types of petaloid tepals in orchids is discussed in Section 7.4.7.)

However, it soon proved impossible to explain all perianths with petaloid first-whorl organs as the result of ectopic expression of B genes (Ronse De Craene, 2007). Throughout the angiosperms, the expression of the putative petal genes is in fact much more diverse than initially thought, and the expected correlation fails to be confirmed. For example, the perianth of *Asparagus* is formed by two whorls of petaloid tepals, but *GLO* B-class genes are only expressed in the organs of the inner whorl (Park *et al.*, 2004). In the orchid *Phalaenopsis* there is a differential expression of four different *DEF*-like paralogues in the perianth (Tsai *et al.*, 2004). In *Aristolochia* (Jaramillo and Kramer, 2004), as mentioned before, and also in Marcgraviaceae (Geuten *et al.*, 2006), the presence of petaloid tissue does not correlate with expression of *AP3* and *PI* homologues.

Evolutionary changes that have been interpreted in terms of an inward shift in expression of class B genes are less widespread. An example is *Rumex acetosa*, where these genes are expressed only in the stamens, but not in the perianth (Ainsworth *et al.*, 1995).

Shifts in the expression of B-class genes have also been suggested as responsible for the unique flower morphology of *Lacandonia schismatica* (see Section 7.1.3). In the young floral meristems of this monocot, the expression domain of the B-class *PI* homologue is extensive, but is subsequently restricted to the area where stamens and carpels will form, coextensive with the expression of the *Lacandonia* homologue of the C-class gene *AG*, whereas the expression of the other B-class gene, the *AP3* homologue, is restricted to the central zone of the meristems, where stamens will differentiate. As a consequence, in the central whorl (eventually, stamens), all three homologues (*AP3*, *PI* and *AG*) are simultaneously expressed; in the surrounding whorl (eventually, carpels) only *PI* and *AG*; and none of these genes is expressed more externally, where the perianth will form (Alvarez-Buylla *et al.*, 2010).

An interesting example of a shifting boundary of gene expression correlated with intraspecific variation in the specification of floral whorls is provided by the two floral morphs of *Nigella damascena*. These differ in the identity and number of perianth organs and in the position of the perianth–androecium boundary on the flower meristem. In one of the morphs, the expression domain of *AG* paralogues extends outside the fertile organs, correlating with the change in identity of the inner perianth organs (Jabbour *et al.*, 2015).

Buzgo *et al.* (2004) extended the Sliding Boundary model, initially proposed for B-class genes and for the sometimes uncertain divide between sepals and petals, or petals and stamens, to other genes and to the gynoecium. Stressing the frequent patterns of gradual transitions among floral organs, Buzgo *et al.* (2004) suggested a 'fading borders' view of gene expression, in which each organ identity gene is broadly expressed across the floral meristem but only weakly at the outer and inner limits of its expression (see also Soltis *et al.*, 2006; Theissen and Melzer, 2007).

Theissen and Melzer (2007) suggested that in the course of evolution the borders of the domains of gene expression have became sharper because of positive autoregulatory control leading to the amplification of small differences in expression levels, so that eventually only domains of no expression or full expression remain.

The relationships between the patterns of gene expression in magnoliids, especially those of the *AP3*, *PI* and *AG* homologues, and the corresponding floral phenotypes, are in agreement with the Fading Borders model: in this clade, the expression domains of these genes are broader and less constrained than in core eudicots. In the more differentiated flowers of *Asimina*, *AP3* and *PI* are expressed in petals, *AP3* (marginally) also in sepals (Kim *et al.*, 2004), but in *Magnolia grandiflora* the B genes are strongly expressed in all perianth whorls (Kim *et al.*, 2004).

In basal angiosperms, B genes are broadly expressed across the floral meristem with a gradient in the level of gene expression that has been suggested to correspond to a similarly gradual transition from one floral organ kind to another (Kramer *et al.*, 2003; Kim *et al.*, 2005c; Shan *et al.*, 2006; Chanderbali *et al.*, 2009). However, a correspondence between the expression of B-class *MADS*-box genes and the presence and absence of petaloidy is far from granted (Ronse De Craene and Brockington, 2013). Here are a few examples.

In *Arabidopsis* and *Antirrhinum*, continuous expression of B-class genes is required to maintain petal development, but in some Ranunculaceae the initial expression of these genes in the petals is turned off early in development (Kramer and Irish, 1999). Homologues of *AP3* but not *PI* are expressed in late stages of petaloid organs of *Impatiens hawkeri* (Geuten *et al.*, 2006). More importantly, there are many examples of petaloid organs developing independently from B-class *MADS*-box gene expression. *AP3* and *PI* homologues are not expressed in the external perianth whorl in *Asparagus*; nevertheless, these tepals are petaloid (Park *et al.*, 2003, 2004). A *PI* homologue is expressed in both perianth whorls in *Aquilegia*; nevertheless, this expression is required for petal identity in the second whorl only, not in the organs of the first whorl which are also petaloid (Kramer *et al.*, 2007). In *Rhodochiton atrosanguineum*, the sepals are petaloid and closely resemble the petals of the same flower, but do not show any B-gene expression (Landis *et al.*, 2012). On the other hand, there are examples of B-gene expression in non-petaloid perianth parts such as the lodicules of grasses.

A different and more articulated reconstruction of the possible evolution of genetic control in the specification of the identity of floral organs has been proposed by Irish (2009). The model differs conceptually from the Sliding Borders or Fading Borders models because it denies a direct role for *AP3/PI* genes in specifying petaloid identity. According to Irish (2009), the expression of *AP3* and *PI* orthologues specifies instead a regional domain of the flower, rather than a kind of floral organ (see also Drea *et al.*, 2007; Whipple *et al.*, 2007). This domain corresponds to the inner perianth, not necessarily differentiated as a whorl of petals, plus the androecium. Organs other than stamens that arise in this domain can be inner tepals, petals or – in grasses – the lodicules, as mentioned above.

A second point in Irish's (2009) interpretation is that the alternative between staminoid versus petaloid nature of the flower organs formed in this area specified by the expression of *AP3* and *PI* may not depend so much on these genes per se, but on nature of pre-existing cascades of genetic control that have eventually became regulated by them. Irish supposes that in some lineages *AP3/PI* homologues may have evolved the capacity to regulate genes responsible for stamen differentiation, in others, instead, the capacity to regulate genes previously involved in the differentiation of leaf-like lateral organs. This would imply that different perianth types have evolved as a consequence of a difference in the genetic cascade regulated by *AP3/PI*.

7.2.3.6 Compound Petals

Are there compound sepals or petals, comparable to compound leaves? The answer is possibly no, but to some extent it depends on whether one adopts a stricter or broader definition of 'compound'. To avoid unwanted confusion, I recommend not using this adjective to denote 'compound origin', that is, for flower organs deriving from the confluence of two primordia, as in the case of the 'compound' sepals described by Tucker (2000) for the caesalpinioid legume *Brownea coccinea* subsp. *capitella* (= *B. latifolia*). Sepals with marginal elaborations are found in a number of genera scattered across the angiosperm tree; examples are found in *Fumaria*, *Hypericum*, *Crataegus* and *Pedicularis*. The following remarks are restricted to petals.

Flowers with fringed petals are present in several unrelated families (Endress and Matthews, 2006a), such as Caryophyllaceae (e.g. *Dianthus* spp.), Malpighiaceae (*Hiptage benghalensis*) and Cucurbitaceae (*Trichosanthes* spp.). No gene expression study is currently available for the development of this marginal patterning, which is occasionally very conspicuous, especially in *Trichosanthes*. Different kinds of morphological elaboration are common in the Sapindaceae, where the petal is frequently provided with an internal basal ligule, or has an overall (asymmetric) peltate shape (reviewed in Leinfellner, 1958). Conspicuous thread-like fimbriae are present on the adaxial side of the corolla lobes of *Menyanthes trifoliata* (Fig. 7.3) and the related *Nymphoides indica* (but not in other species of the latter genus).

The shape closest to that of a hypothetical compound petal is found in some or all species in four genera of the Saxifragaceae (*Lithophragma*, *Tiarella* and especially *Mitella* and *Tellima*) in some species of which a very narrow, straight 'rachis' is flanked on either side by equally narrow, long lobes (Soltis, 2007).

7.2.3.7   *Perianth Reduction*

The following remarks largely follow Endress (2011). Strong reduction and eventual loss of sepals without a parallel reduction of the other flower organs is quite rare; in fact, what gets lost in this case is the external flower whorl, but sepaloidy (at least in the form of protective function in respect to the internal parts of the flower) is still usually expressed, either in the petals (Apiaceae, Araliaceae, Rubiaceae) or in

FIGURE 7.3   The bogbean (*Menyanthes trifoliata*) is one of the very few plant species in which the whole internal (adaxial) surface of the petals is covered with long appendages. *A black and white version of this figure will appear in some formats. For the colour version, please refer to the plate section.*

the subtending bracts, as in the Acanthaceae Thunbergioideae. Petals are more commonly lost in lineages where sepals are, instead, conserved.

Flowers without a perianth are more common in basal clades in which a divergent differentiation into sepals and petals has not developed; examples are found among the Hydatellaceae, Chloranthaceae, Piperales, Eupomatiaceae and Lauraceae. Within monocots, perianthless flowers are found in some Alismatales, especially among the Araceae, and in Pandanaceae, Typhaceae, some Restionaceae and Cyperaceae. Perianthless eudicots include representatives of basal lineages such as *Euptelea*, *Achlys*, *Trochodendron* and a number of Saxifragales. In the groups currently known as the higher core eudicots (whatever 'higher' may eventually mean; see Rigato and Minelli, 2013), perianthless flowers are limited to some Betulaceae and Myricaceae, *Callitriche* and the female flowers of the parasitic

Balanophora. In the clades in which the entire perianth is occasionally lost, flowers are nearly always characterized by a low degree of complexity and synorganization.

Developmental instability is widespread in lineages in which there is a trend towards a reduction of floral parts. Unusual variation in the size of petals, both in the same flower and between flowers of the same plant, is found is some members of the Ranunculaceae (Fig. 7.2). In the flowers of *Galopina* (Rubiaceae), no trace of sepals is usually found, except for a few flowers in which the four sepals are reduced, but are still large enough to frequently show, at the stage of primordia, a difference in size between the adaxial and abaxial sepal, while the lateral ones are equal in size (Ronse De Craene and Smets, 2000).

With the loss of the perianth, position and number of stamens often become highly labile (Endress, 1990b). Examples include *Lindera*, *Litsea*, Pandanaceae, Typhaceae, *Euptelea*, *Achlys*, *Styloceras* and *Trochodendron* (Endress, 2011). In some instances, this trend extends to carpels.

At least seven independent transitions from petalous to apetalous flowers have been ascertained in the Ranunculaceae. In some instances, petal loss was strongly associated with decreased or lost expression of *AP3-3*. In apetalous genera such as *Thalictrum*, *Beesia* and *Enemion*, the gene has either been lost altogether or disrupted by deletions affecting either coding or regulatory regions. This condition has been demonstrated in an apetalous mutant of *Nigella*, in which the insertion of a transposable element into the second intron of this gene has led to its silencing, followed by the transformation of petals into sepals. In *Clematis*, the *AP3-3* homologue is likely a pseudogene, and thus the genus is essentially apetalous, although a number of *Clematis* species have evolved secondary petaloidy (Zhang *et al.*, 2013a).

7.2.3.8 Grass Flowers

In the flowers of the grasses (Fig. 7.4), the typical perianth organs are replaced by three organ types termed lodicules, palea and lemma. At anthesis, lodicules expand to separate the lemma and palea and

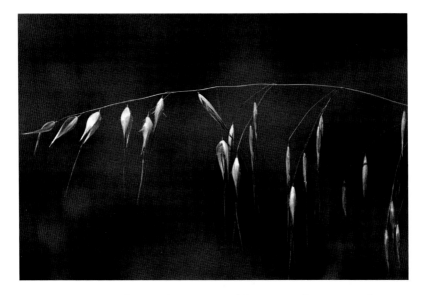

FIGURE 7.4 The reproductive units of the Poaceae deviate strongly from conventional flowers, to the extent that the homology of their characteristic elements (lodicules, palea, lemma) are problematic. Here, an inflorescence of common wild oat (*Avena fatua*). *A black and white version of this figure will appear in some formats. For the colour version, please refer to the plate section.*

expose the anthers to the wind. It was suggested in the past that the palea may represent a prophyll and the lemma a true bract (Clifford, 1987); however, gene expression patterns favour the alternative inter-pretation of palea and lemma as homologues of eudicot sepals.

Lodicules in maize and rice have been shown to express petal-identity genes (Ambrose *et al.*, 2000; Kyozuka *et al.*, 2000; Nagasawa *et al.*, 2003; Xiao *et al.*, 2003; Whipple *et al.*, 2004, 2007). This seems to support the idea that lodicules are the grass homologues of petals; in agreement with this interpretation are also the position of the lodicules with respect to the reproductive organs of the flower and the fact that in the basal grasses the lodicules occur in threes, as is typical of perianth whorls in monocots (lodicules are reduced to two in the remaining grasses). A different evolutionary origin has occa-sionally been suggested, however, based on the fact that the third lodicule, when present, is attached higher on the floral axis than the other two. Adding to these uncertainties is the fact that in the most basally branching clade of the Poaceae, which includes *Anomochloa* and *Streptochaeta*, no petals or lodicules are present, just a ring of hairs outside the stamens in *Anomochloa*. Whether lodicules originated at the base of the grasses and were lost in *Anomochloa* and *Streptochaeta*, or evolved after the splitting of the Anomochlooideae, is unknown (Kellogg, 2001).

In grasses, *DEF*- and *GLO*-like genes are involved in specifying the identities of stamens and lodicules. Two *GLO*-like genes have been reported from rice (Chung *et al.*, 1995) and three from maize (Münster *et al.*, 2001). Since petals are replaced by sepaloid organs in eudicot class B gene mutants, the loss-of-function phenotypes of *DEF*- and *GLO*-like genes in maize and rice suggest that lodicules are homologous to petals, and palea and lemma are homologous to sepals (Kang *et al.*, 1998; Ambrose *et al.*, 2000).

7.3   FLORAL PHYLLOTAXIS

7.3.1   *Patterns of Floral Phyllotaxis*

Floral phyllotaxis is labile in basal angiosperms, sometimes even within the individual plant (Endress, 2006). Irregularly whorled phyllotaxis is found in perianthless flowers, e.g. *Euptelea* and *Achlys*, and in flowers with numerous stamens and carpels, as in a number of Ranunculaceae and Papaveraceae (Endress, 2006). In *Anemone*, the arrangement of floral organs can be spiral, whorled or irregular. The sepals, commonly five, are often initiated in spiral sequence, but in *A. chinensis* the six tepals are initiated in two whorls. Both irregularly spiral and irregularly whorled floral phyllotaxis are even present in the same species (*A. taipaiensis*) (Ren *et al.*, 2010).

In many Nymphaeaceae the phyllotaxis of the perianth organs varies between different floral series of the same flower (Warner *et al.*, 2008). In eudicots, phyllotaxis is spiral in the calyx, but becomes whorled from the petals onwards (Endress, 2006).

The floral organs of one whorl often develop in a spiral sequence, e.g. in several Apiaceae studied by Leins and Erbar (2004), but this has virtually no effect on their final arrangement, due to the very short plastochrons (Endress, 2006).

Superposed whorls have their organs arranged in the same radii and violate Hofmeister's rule (see Section 5.7). Superposition of whorls is observable in several flowers, e.g. between tepals and stamens in *Basella* and *Berberis* (Lacroix and Sattler, 1988; Endress, 1992, 1994c).

Instead of one organ in an expected position, two organs are sometimes found side by side (Endress, 1994a, 1994b). The site with the first double positions in a flower is usually somewhere in the perianth or the androecium (Endress, 2011). Double positions occur especially in basal angiosperms, basal monocots and basal eudicots, more often in the androecium, less frequently in the perianth or in the gynoecium. Examples are also known from the core eudicots, e.g. in the Brassicaceae.

7.3.2 Taxonomic Distribution of Phyllotactic Patterns

As a general trend, floral phyllotaxis is mainly spiral in the early-branching angiosperm lineages, and mainly or totally whorled in the more derived ones (Endress, 2011), but the distinction between spiral and whorled phyllotaxis is not always clear. For example, floral phyllotaxis in *Amborella* is usually described as spiral, but carpels are actually initiated in a nearly whorl-like manner (Buzgo *et al.*, 2004).

All flower organs are arranged in a spiral in the Austrobaileyaceae, Schisandraceae, Eupomatiaceae, and in some Monimiaceae, Calycanthaceae, Ranunculaceae and Menispermaceae. However, flowers are whorled in the Nymphaeales, except for the perianth in *Nuphar* (Endress, 2001c, 2004). Spiral and whorled patterns may coexist in the same family and even in the same species (*Endiandra montana*; Endress, 1994a).

According to Endress (1987b), in basal angiosperms the switch between whorled and spiral floral phyllotaxis is easy because of the lack of synorganization between floral organs.

All monocots have whorled flowers, as do most of the core eudicots. In some groups, the perianth and the fertile organs are arranged according to a different phyllotaxis; for example, some Magnoliaceae have whorled tepals but spiral stamens and carpels (Endress and Doyle, 2015).

Floral phyllotaxis is sometimes irregular, and this has been explained as a consequence of the small size and large number of organs crowded in the flower (Endress, 2006). In the Annonaceae, genera with lower organ numbers such as *Monanthotaxis* have a whorled phyllotaxis (Ronse De Craene and Smets, 1990), whereas in *Annona*, with a comparatively high number of fertile organs, stamens and carpels are arranged irregularly (Endress, 1990a).

Irregular phyllotaxis occurs repeatedly in polystemonous taxa within most major clades, including basal angiosperms (e.g. in Nymphaeales and Magnoliales), monocots (Arecales), basal eudicots (Ranunculales) and core eudicots (Endress, 2011). Stamens and carpels are also arranged irregularly in flowers lacking a perianth (Endress, 1978, 1989, 1990a; Tucker, 1991).

7.4 FLOWER SYMMETRY

7.4.1 Basic Patterns of Flower Symmetry

In terms of overall symmetry, most flowers are either actinomorphic (radially symmetrical) or zygomorphic (bilaterally symmetrical). Asymmetric flowers are rare and occur mostly in basal angiosperms (e.g. Winteraceae; Endress, 1999) and within clades with predominantly zygomorphic flowers (e.g. Orchidaceae, Zingiberales, Fabaceae and Lamiales; Marazzi and Endress, 2008). Asymmetry is sometimes limited to features such as the size of the anthers (*Labichea*), or the spatial orientation of the style (enantiostyly: *Cassia* and other legumes). These restrictions, although corresponding to minor forms of asymmetry, are nevertheless interesting from an evo-devo perspective, as expressions of developmental modularity.

In some flowers, e.g. in the Apocynaceae, the petals are asymmetric, and thus the flower has no plane of symmetry, but five equivalent sectors overlap after rotation of 72° around the flower axis (Fig. 7.5).

A few flowers are disymmetric, that is, they possess two non-equivalent planes of symmetry; examples are found in the Papaveraceae Fumarioideae such as *Hypecoum* and *Dicentra* (Jabbour

FIGURE 7.5 The large periwinkle (*Vinca major*) is a typical example of the peculiar symmetry of the flowers of many Apocynaceae: the individual petal is asymmetric, but the whole flower is symmetrical in respect to a rotation of 72° around its axis.

et al., 2009). Multiple forms of bilateral flower symmetry are derived from disymmetry; examples are *Iberis* and *Corydalis* (Hileman, 2014a).

In describing flower symmetry, morphologists generally focus on the perianth, often disregarding differences between the symmetry of the sterile parts of the flower and those of the fertile organs, which are not always identical. Some interesting examples of diverging patterns of symmetry between perianth and fertile organs (further examples of modularity) will be mentioned in the following.

Flowers of basal eudicots (e.g. Menispermaceae and Ranunculaceae) with spiral phyllotaxis may attain the same symmetry patterns as whorled flowers: flowers with a large number of organs are almost actinomorphic, as in *Adonis* and *Nigella*; flowers with dorsiventral differentiation (e.g. *Aconitum* and *Delphinium*) are almost zygomorphic; flowers with few organs (e.g. *Hypserpa decumbens*) are asymmetric. Flowers of basal angiosperms with irregular phyllotaxis and a small number of organs (e.g. *Zygogynum*) are also asymmetric (Endress, 1999).

7.4.2   Floral Aestivation and Symmetry

A fairly consistent relationship exists between perianth aestivation and flower symmetry.

Aestivation patterns are defined by the mutual positions of organs of a whorl when in the bud (Endress, 2011). Several patterns are recognized: in some of these (collectively called imbricate aestivation patterns) one or more organs cover one or both neighbours either on one side or on both sides; in valvate aestivation, organs are in lateral contact with their two neighbours, without overlap; in open aestivation, organs are more extensively spaced and their margins are nowhere in contact. Among the imbricate patterns, different degrees of symmetry are possible. In decussate aestivation, external (covering) organs alternate with internal (covered) organs, in a radially symmetrical pattern. In contorted aestivation, one side of each organ covers one of the neighbouring organs but the other side is covered by the other neighbour. Symmetry is lost in the two remaining patterns: quincuncial, with two internal organs, two external organs, and one with an internal and an external side; and cochleate, with one internal organ, one external organ, and three with an internal and an external side.

As mentioned in Section 7.2.3.2, organ parts that are covered by other parts when in the bud may attain different properties (thickness, colouration, hairiness) than parts that are not covered (Endress, 2008b; Warner *et al.*, 2009).

If five sepals (tepals) are present they are mostly quincuncial. This is also true for the rare pentamerous monocots, such as *Pentastemona*.

Contort petal aestivation determines left–right asymmetry. The direction of coiling is fixed in some species, anticlockwise as in *Datura stramonium* or clockwise as in *Nerium oleander*; in others it fluctuates randomly. Flowers with contort petal aestivation are frequent in rosids and in the Gentianales. In the Lamiales and some other plants with strongly zygomorphic flowers, both perianth whorls are often cochlear (Endress, 2011). The aestivation pattern of the petals is stable within some larger taxa: for example, it is characteristically contort in Gentianales, Malvaceae, Oxalidaceae, Linaceae and Plumbaginaceae. In many families, however, aestivation patterns are different in different genera or groups of genera. Dextrorse and sinistrorse morphs may be equally common within a species, or even between flowers of an individual plant (e.g. some Oxalidaceae, Linaceae, Plumbaginaceae; Bahadur *et al.*, 1984), but a single morph may be exclusive for a large taxon (e.g. among the Apocynaceae, the Plumierioideae are sinistrorse, the Apocynoideae and Asclepiadeae dextrorse; Endress, 1994a). Contort petal aestivation is enantiomorphic in many rosids, whereas in almost all asterids one of the two contort morphs is fixed at genus or even family level (Endress, 2012).

Unstable petal aestivation has been reported for *Cadia purpurea*, a legume with zygomorphic flowers. The 21 different patterns reported in this species (van der Maesen, 1970) contrast strongly with the highly canalized petal aestivation found in most Fabaceae. In *Cadia*, all petal margins have a similar marginal growth pattern; therefore, when adjacent petal margins meet, it is purely a matter of chance which margin grows outside the other (Tucker, 2002).

7.4.3   Morphological Patterns in Zygomorphic Flowers

Zygomorphic flowers differ widely in the nature and conspicuousness of the floral parts whose number, position or form are responsible for the bilateral symmetry of the flower. The morphological diversity in the expression of floral zygomorphy is summarized in Table 7.2.

Elaborate zygomorphy occurs especially in flowers with highly synorganized organs (Endress, 2006) and is therefore uncommon among the basal angiosperms; nevertheless, it is conspicuous in *Aristolochia* and in some Papaveraceae and Ranunculaceae. Flowers of the most basal monocots (*Acorus*) are also zygomorphic (Buzgo and Endress, 2000). The elaborate zygomorphic flowers of the Orchidaceae are discussed in Section 7.4.7.

Among the eudicots, classical cases of conspicuous zygomorphy are the bilabiate flowers, which have evolved in two extreme forms as the keel flowers of most Fabaceae and Polygalaceae and the lip flowers of the Lamiales. In the latter, the largest clade with almost exclusively zygomorphic flowers, the upper median stamen is reduced or absent; in many taxa belonging to different families, one of the two pairs is also reduced, either the upper or the lower (e.g. Endress, 1999). In other asterids (e.g. Balsaminaceae, Apocynaceae, Rubiaceae, Asteraceae), floral zygomorphy is commonly restricted to the perianth, with at most a slight additional expression in the androecium (Endress, 2012).

Oblique zygomorphy is found in many genera, mostly in Sapindales, Brassicales and Solanales (Endress, 2012).

Strongly reduced flowers with one stamen or one carpel only are zygomorphic (Endress, 2012), their plane of symmetry being determined by the attachment of the anther in the first case, by the sealed margin of the carpel in the second. The zygomorphy of the gynoecium

Table 7.2 *Diversity in the expression of flower zygomorphy, mainly after Endress (2012), with non-exhaustive examples.*

Flower parts affected	
Calyx	Rubiaceae: *Mussaenda, Warszewiczia*
Corolla	Apocynaceae: *Isonema, Rauvolfia*
Calyx and corolla	Balsaminaceae
Androecium	Lythraceae: some *Lagerstroemia* species
	Solanaceae: *Solanum lidii*
Gynoecium	Annonaceae: *Isolona, Monodora*
All organs	Lamiales, Fabaceae, Orchidaceae
Kind of differential modification	
Organs differentially shaped on either side of symmetry plane	Lamiales
Organs curved in one direction	Cleomaceae
Organs reduced on either upper or lower half	Gesneriaceae and other Lamiales (odd stamen)
Organs lost on either upper or lower half	Lamiaceae (stamens)
Number of organs increased on either upper or lower half	Lecythidaceae (stamens)
Degree of organ union different on either upper or lower half	Lamiaceae: *Teucrium* (petals)
Organs neofunctionalized on either upper or lower half	Lecythidaceae (stamens)
	Bignoniaceae: *Jacaranda* (odd stamen)
	Plantaginaceae: *Penstemon* (odd stamen)
Zygomorphy by reduction	
Flowers with a single stamen or a single carpel	Lacistemataceae (one stamen)
	Lauraceae (one carpel)
	Chloranthaceae: *Sarcandra* (one stamen and one carpel)

Table 7.2 (cont.)

Orientation of the plane of symmetry with respect to the axis of the next higher order	
Median	Lamiales, Orchidaceae etc.
Transversal	Papaveraceae Fumarioideae
	Dilleniaceae: *Hibbertia* p.p.
Oblique	Some Malpighiales, Brassicales,
	Sapindales, Solanales

is sometimes the result of selective reduction or suppression of carpels in an originally syncarpous gynoecium, only one carpel remaining fertile (pseudomonomery) (Rudall and Bateman, 2004). This condition has evolved several times, in both monocots and eudicots. In monocots, the pseudomonomerous condition is mostly confined to the Araceae and the commelinid clade (e.g. *Sparganium*, *Pontederia*, Restionaceae and Arecaceae). In many cases (e.g. in *Pontederia*), gynoecial zygomorphy consequent to pseudomonomery is correlated with the bilateral symmetry of the perianth. However, carpel abortion does not always occur in this symmetry plane; for example, different carpels are expressed in different species of the Restionaceae (Ronse De Craene *et al.*, 2002).

As a rule, in plants with determinate inflorescences, both the lateral flowers and the terminal flower are actinomorphic. An exception is *Capnoides sempervirens* (= *Corydalis sempervirens*), the only species of the *Corydalis* lineage with determinate inflorescence; here, both the terminal flower and the lateral flowers are zygomorphic (Troll, 1964).

7.4.4 *Phylogenetic Distribution of Zygomorphic Flowers*

The oldest bisexual flowers known from the fossil record, dated around 120–125 million years ago, were actinomorphic; the first zygomorphic flowers appeared only 30–40 million years later (Crane et al., 1995).

At the family level, about 70 independent transitions from actinomorphy to zygomorphy, 23 in monocots and 46 in eudicots, have been suggested by Citerne et al. (2010), but these figures underestimate the true number of evolutionary changes in flower symmetry. For example, within the Lamiales, Citerne et al. (2010) recovered only one transition from radial to bilateral symmetry, and one reversal; but Hileman (2014b), scoring at the species (rather than family) level for symmetry of corolla and androecium, based on the same phylogeny (Schäferhoff et al., 2010), recovered possibly two transitions from radial to bilateral flower symmetry early in the Lamiales, followed by multiple reversals. Similarly, multiple transitions from actinomorphic to zygomorphic flowers have been identified in the Ranunculales (Damerval and Nadot, 2007), Solanaceae (Olmstead et al., 2008; Knapp, 2010) and Brassicaceae (Busch et al., 2012). In the Malpighiales, a single transition to bilateral flower symmetry was followed by multiple independent reversals (Davis and Anderson, 2010; Zhang et al., 2010).

Zygomorphy is all but non-existent in basal angiosperms and is rare in early-diverging eudicots (Endress, 1999). In these groups, many deviations from actinomorphy are the result of the extreme reduction of the flower, although this is not the case for the showy perianth of Aristolochia (Ronse de Craene et al., 2003).

Monocot flowers are also usually actinomorphic, but zygomorphy has evolved at least four times in this clade, by suppression of either the abaxial or the adaxial stamen (in the latter case, often associated with the evolution of a labellum) (Rudall and Bateman, 2004). Distinct floral zygomorphy occurs in more species of Asparagales and commelinids than in other monocots (Rudall and Bateman, 2004). In Pentastemona, zygomorphy probably evolved from an actinomorphic trimerous condition by loss of the adaxial outer petal and stamen (Fukuhara et al., 2003). Zygomorphy is shown most commonly by the inner tepals and stamens, although gynoecial zygomorphy (including pseudomonomery by selective abortion of all but one fertile carpel) is common in some monocot groups.

In the Ranunculales, zygomorphy evolved twice, once from disymmetry (in the Fumarioideae, where the symmetry plane is transversal rather than median as usual; Endress, 2012), once from actinomorphy (in the tribe Delphinieae: *Aconitum* and *Delphinium* s.l.; Jabbour *et al.*, 2009).

Among rosids, zygomorphic flowers are particularly frequent in the Fabales and in malvids (Endress and Matthews, 2006b), among asterids especially in Lamiales, Asterales and Dipsacales. Zygomorphic flowers are especially elaborate in many subclades of the Fabaceae and in most Polygalaceae. Floral zygomorphy is also common in the Melastomataceae; extreme examples are found in the Vochysiaceae, with transitions to asymmetry.

A graded range of conditions between actinomorphy and zygomorphy is found in the Geraniaceae. The flowers of *Geranium* are actinomorphic, but in a few species the stamens are sigmoidally curved, and in *G. arboreum* the corolla is also zygomorphic; the flowers of *Erodium* are weakly zygomorphic and those of *Pelargonium* strongly zygomorphic. Zygomorphy is unusual in the perianth of the Brassicaceae, but is occasionally conspicuous, either in the calyx (*Streptanthus*) or in the corolla (*Iberis*, *Teesdalia* and *Erysimum*) (Endress, 2012).

7.4.5 Asymmetric Flowers

Asymmetric flowers have evolved many times, more frequently in clades with highly elaborated and otherwise zygomorphic flowers (Endress, 1999). For example, three genera with asymmetric flowers like *Pedicularis*, on the one side, and *Phaseolus* and *Lathyrus*, on the other, are respectively nested within the Lamiales and the Fabaceae Faboideae, two angiosperm clades with mainly zygomorphic flowers.

The simple, unordered flowers of a few basal angiosperms are asymmetric; for example, in some *Zygogynum* species, the innermost perianth parts and the stamens are arranged irregularly (Endress, 1999).

Asymmetry based on modification of elaborate zygomorphy is found among the monocots (Asparagales, Commelinales and Zingiberales), rosids (Myrtales and Fabales) and asterids (Lamiales, Asterales and Dipsacales) (Endress, 2012). Several legumes have asymmetric flowers, e.g. *Lathyrus* and the Phaseolinae, in which the keel is curved or coiled, whereas in species of *Chamaecrista* and *Senna* the gynoecium and the stamens are curved to one side (Endress, 1999).

Asymmetric petals are found in many flowers with contort aestivation such as *Malva, Hibiscus, Linum austriacum, Oxalis floribunda, Plumbago capensis*; this pattern is more conspicuous in *Vinca* and other Apocynaceae. However, other plants with contort aestivation, e.g. *Linum usitatissimum, Gypsophila paniculata, Geranium sanguineum* and *Ceratostigma plumbaginoides* (= *Plumbago larpentae*), have perfectly symmetrical petals (Endress, 2012).

Asymmetric flowers may also evolve by simplification, as in the extreme example of *Centranthus*, with a single stamen; these flowers are enantiomorphic (Endress, 1999).

Molecular systematic studies have revealed that *Hippuris* and *Callitriche*, two genera of water plants with extremely reduced flowers, evolved within the Plantaginaceae (Olmstead and Reeves, 1995). In *Hippuris*, each flower has a single stamen and a single carpel, in addition to a very small perianth.

In the Cannaceae and Marantaceae, only one theca develops on the single fertile stamen, because the other side is petaloid (Rudall and Bateman, 2004).

Hidden physiological asymmetry is more widely distributed than morphological symmetry. For example, in legume fruits asymmetry can be revealed by desiccation (Goebel, 1930).

7.4.6   Genes and Flower Symmetry

Most of the multiple independent transitions from actinomorphic to zygomorphic flowers are based on changes in the expression of *CYC*-like genes (reviewed in Busch and Zachgo, 2009; Preston and Hileman, 2009; Martín-Trillo and Cubas, 2010).

Within the Ranunculaceae, as mentioned above, zygomorphy has evolved in the stem lineage of the Delphinieae. In this family, two main *RANUNCULACEAE CYCLOIDEA*-like (*RANACYL*) gene lineages have been identified, but an additional duplication, possibly predating the emergence of the Delphinieae, resulted in up to four gene copies. The *RANACYL* paralogues are expressed early in floral buds, and flower symmetry correlates with the duration of their expression. In species with actinomorphic flowers, at most one *RANACYL* paralogue is expressed during late floral development, but all paralogues are expressed at this stage in species with zygomorphic flowers (Jabbour *et al.*, 2014).

In the Papaveraceae Fumarioideae, the shift from disymmetry to zygomorphy is triggered by three developmental events: the elongation of stamen filaments, the development of the nectary and the emergence of the spur. A duplication predating the divergence between Fumarioideae and Papaveroideae produced two *CYCLOIDEA*-like (*CYC*-like) paralogous lineages (Kölsch and Gleissberg, 2006; Damerval *et al.*, 2007). Both paralogues have an asymmetric pattern of expression in the outer petals of *Lamprocapnos spectabilis*, stronger in the basal region of the outer petal that will form the nectar spur than in the basal region of the other petal, which will not form a spur (Damerval *et al.*, 2013). In the bilaterally symmetrical flowers of *Capnoides*, derived from disymmetric flowers, the expression of two *CYC*-lineage paralogues is asymmetric (Kölsch and Gleissberg, 2006; Damerval *et al.*, 2007).

In core eudicots, correlations between changes in *CYC*-like gene expression patterns and evolutionary shifts in floral symmetry have been documented in many genera including *Mohavea* (Hileman *et al.*, 2003), *Antirrhinum* (Luo *et al.*, 1996, 1999), *Linaria* (Cubas *et al.*, 1999), *Veronica* and *Gratiola* (Preston *et al.*, 2009) (all Plantaginaceae); *Cadia* (Citerne *et al.*, 2006), *Pisum* (Wang *et al.*, 2008), *Lotus* (Feng *et al.*, 2006) and *Lupinus* (Citerne *et al.*, 2006) (all Fabaceae); *Chirita* (Gao *et al.*, 2008) and *Oreocharis* (= *Bournea*) (Zhou *et al.*, 2008) (both Gesneriaceae); *Byrsonima* and *Janusia* (both Malpighiaceae) (Zhang

FIGURE 7.6 In the Brassicaceae, the flower is zygomorphic, but this is essentially due to the arrangement of the (usually six) stamens. The corolla, however, is generally actinomorphic, with four identical petals. The main exceptions include species of the genus *Iberis*. Here, the unequal development of the petals is best visible in the external flowers of the inflorescence, as in this *Iberis semperflorens*, quite similar to the model species *Iberis amara*.

et al., 2010); and *Iberis* (Brassicaceae) (Busch and Zachgo, 2007), representing at least five lineages that evolved zygomorphy independently.

In the Brassicaceae, as also mentioned above, a zygomorphic corolla is found in only a small number of species, but this phenotype is conspicuous in *Iberis*, owing to the very large size of the two ventral petals (Fig. 7.6). This differential elongation is associated with relatively late expression of a *CYC2* homologue in the smaller adaxial petals (Busch and Zachgo, 2007; Busch *et al.*, 2012). Comparison of early and late *CYC2* expression in different species of Brassicaceae, including both zygomorphic and actinomorphic ones, suggests that an early adaxial *CYC2* expression in floral meristems is likely ancestral in this family. However, this pattern of expression is not a prerequisite for the establishment of zygomorphy in the corolla of the

Brassicaceae, because it got lost in all zygomorphic species studied thus far. Instead, flower zygomorphy evolved in this family via a heterochronic shift of *CYC2* expression from the ancestral condition (early expression in the adaxial part of the floral meristem) to an accumulation of *CYC2* transcript in the adaxial petals at a later stage of their development (Busch *et al.*, 2012).

Another group of rosids in which zygomorphy has independently evolved is the Fabaceae. The zygomorphic corolla typical of legumes includes a zygomorphic adaxial petal (standard), two asymmetric lateral petals (wings) and two fused abaxial petals (keel).

The expression of *CYC2* and *CYC3* orthologues is required in *Lotus corniculatus* var. *japonicus* (= *L. japonicus*) (Feng *et al.*, 2006) and in *Pisum sativum* (Wang *et al.*, 2008) for adaxial petal identity. The *CYC2* homologues are expressed only in the abaxial petal, while the expression domains of the *CYC3* homologues are broader and include also the lateral petals. However, the actinomorphic corolla of *Cadia purpurea* is not dependent on a loss of expression of this *CYC*-like gene, but on its expression in the whole developing flower (Citerne *et al.*, 2006)

Detailed studies of the genetic underpinnings of zygomorphy are available for *Antirrhinum majus*, where three dorsal identity genes (*CYC*, *DICH* and *RADIALIS* (*RAD*)) and one ventral identity gene (*DIVARICATA* (*DIV*)) are involved. At an early developmental stage, *CYC* reduces the growth of the dorsal half of the floral meristem; later on, it increases the growth rate of the petals (Luo *et al.*, 1996). *DICH* expression is restricted to the dorsal half of the dorsal petals (Luo *et al.*, 1999). *CYC* positively regulates the transcription factor *RAD* in petals; in turn, *RAD* negatively regulates *DIV* (Galego and Almeida, 2002; Corley *et al.*, 2005; Preston and Hileman, 2009).

Despite the well-documented correlation between the patterns of expression of *CYC2*-like and *RAD*-like genes and the zygomorphic symmetry prevailing in lamiid flowers, it is still unclear whether the asymmetric expression of these genes during the development of the floral meristem is ancestral or derived in the Plantaginaceae (Zhong and Kellogg, 2015).

In the floral meristems of early-diverging Lamiales with actinomorphic corollas, *CYC2*-like and *RAD*-like genes are expressed broadly; this contrasts with the restricted expression of the same genes in adaxial/lateral regions of the floral primordia in the mainly zygomorphic core Lamiales. According to Zhong and Kellogg (2015), the expression pattern of *CYC2*-like genes has evolved stepwise. In the floral meristem of representatives of the Oleaceae (one of the most basal clades within the lamiids) studied by these authors, *CYC2*-like genes are expressed only during a short segment of flower development. A first change is reconstructed at the level of the next node in lamiid phylogeny, that is, in the common ancestor of Tetrachondraceae and core Lamiales, in which the expression of *CYC2*-like genes in petals is prolonged. Finally, in the common ancestor of the core Lamiales, the expression *CYC2*-like genes in adaxial/lateral petals becomes asymmetric. Zhong and Kellogg (2015) reconstruct a similar evolution for the expression patterns of *RAD*-like genes: this expression first appeared in petals in early-diverging Lamiales or earlier, and eventually became asymmetric in adaxial/lateral petals in core Lamiales.

In the zygomorphic flower of *Lonicera*, duplicate *CYC*-like genes are expressed in the dorsal petals, sometimes also in the lateral ones; the orthologues of these *CYC*-like genes show no pattern of differential expression across the floral axis in a related taxon (*Viburnum*) in which the flower is radially symmetrical (Howarth *et al.*, 2011).

Developmental origins of zygomorphy are diverse among the monocots; in one lineage at least (the orchids) these are independent of control by *CYC*-like genes (Rudall, 2013).

In rice, however, the *CYC* homologue *RETARDED PALEA1* (*REP1*) is implicated in floral zygomorphy by regulating the structure of the palea in early flower development (Yuan *et al.*, 2009).

TEOSINTE BRANCHED (*TB*) homologues are responsible for flower symmetry in the monocot orders Zingiberales and

Commelinales. In *Heliconia stricta* and *Commelina communis*, both with zygomorphic flowers, *TBL2/TB1a* are expressed ventrally at mid stages of flower development (Mondragón-Palomino and Trontin, 2011), whereas in *Costus spicatus* and *Tradescantia pallida*, both with actinomorphic corollas, the expression of these genes is either homogeneous or undetectable (Bartlett and Specht, 2011; Preston and Hileman, 2012).

7.4.7 Flower Symmetry in the Orchidaceae

A different genetic mechanism involved in patterning zygomorphic flowers, based on the differential expression of B-class (*DEF*-like) genes, has been postulated for orchids (Mondragón-Palomino and Theißen, 2009).

The zygomorphic flowers of orchids include three types of perianth organs: an external whorl of three tepals, often described as sepals, plus a second whorl of two lateral inner tepals (petals) and a median inner tepal called the lip or labellum. In orchids, flower zygomorphy is not the consequence of specific patterns of expression of *CYC* genes. Instead, both the symmetry and the extraordinary diversity of orchid flowers are largely due to the evolution of B-class genes. Various numbers of *AP3*-like and *PI*-like genes have been isolated from a number of orchid species (reviewed in Tsai *et al.*, 2014). The *AP3* duplication predated the split of the Orchidaceae from their sister group and may even be older than the origin of Asparagales. A secondary duplication probably occurred at the beginning of orchid radiation or closely thereafter (Tsai *et al.*, 2014).

An actinomorphic flower with a perianth comprising two undifferentiated whorls of tepals, as in *Lilium*, is the condition from which the orchid flower quite probably evolved. Observations on other representatives of the Asparagales, *Agapanthus praecox* (Nakamura *et al.*, 2005) and *Muscari armeniacum* (Nakada *et al.*, 2006), reconstruct such a perianth as controlled by the uniform expression of *DEF*- and *GLO*-like genes in both whorls of tepals.

FIGURE 7.7 Moth orchids (*Phalaenopsis*) are among the most popular representatives of their family in the lab.

Four clades of *DEF*-like genes have been identified in orchids, based on studies on *Phalaenopsis equestris* (Fig. 7.7), *Dendrobium crumenatum* and *Habenaria radiata*. These are known as *PeMADS2*-like genes (clade 1), *OMADS3*-like genes (clade 2), *PeMADS3*-like genes (clade 3) and *PeMADS4*-like genes (clade 4), respectively. In analogy to the combinatorial model of floral organ identity (ABC model and later versions), the combined expression of the four *DEF*-like genes is regarded as the mechanism responsible for the divergent morphologies of orchid perianth organs. According to a first version of this 'orchid code', organs with outer tepal identity develop where only clade 1 and 2 genes are expressed; lateral inner tepals are instead under the control of the overlapping expression of genes of clades 1, 2 and 3; finally, expression of genes of all four clades specifies lip development.

Clade 1 and clade 2 genes, expressed in all tepals, are perhaps responsible for a generalized 'petaloid' identity, in agreement with one of the ancestral functions of *DEF*-like genes in monocots, in addition to their role in specifying stamen identity (Mondragón-Palomino and Theißen, 2008). In a revised version of this 'orchid code', Mondragón-Palomino and Theißen (2011) suggest that the lateral inner tepals are specified by overlapping high levels of expression of clade 1 and clade 2 genes, together with low levels of expression of clade 3 and clade 4 genes, whereas the development of the labellum requires low levels of expression of clade 1 and clade 2 genes and high levels of expression of clade 3 and clade 4 genes.

7.4.8   Floret Zygomorphy in the Capitula of the Asteraceae

Different degrees of zygomorphy are expressed by the individual florets in the capitula of the Asteraceae. In ligulate capitula, as in *Taraxacum*, all florets are zygomorphic. In capitula with differentiation between central (disc) florets and peripheral (ray) florets, disc florets are more or less strictly actinomorphic, while ray florets are zygomorphic (see Fig. 5.4). The evolution of such different degrees of modularity within the capitulum, which creates huge variation in flower head morphology, is under simple genetic control: one or two major genes are apparently involved, in addition to a number of modifiers (reviewed in Gillies *et al.*, 2002; Broholm *et al.*, 2014).

Two types of zygomorphic florets are distinguished: bilabiate florets, with three outer and two inner petals, and ray florets, with strap-shaped fused ligules with a maximum of three apical teeth (Stuessy and Urtubey, 2006). Phylogeny suggests that the shape of the corolla evolved from bilabiate to ray: only bilabiate corollas are found in the most basal subfamilies Barnadesioideae and Mutisioideae, whereas typical ray flowers seem to have evolved only after the split of those two clades from the rest of the family (Broholm *et al.*, 2014). Within a heteromorphic capitulum, the change in symmetry can be very sharp, as observed in *Helianthus*, where the strongly bilateral (and functionally neutral) ray flowers contrast with the tubular, radially symmetrical (and hermaphrodite) disc flowers. In *Gerbera*, instead, there is a gradual change in symmetry along the radius of the capitulum.

The zygomorphic florets develop in different ways, unlike the actinomorphic florets, all pentamerous, which are radially symmetrical throughout development. In *Tagetes* and *Chrysanthemoides*, five petal primordia are initiated, but only three persist; four petal primordia are initiated in *Galinsoga*, whereas in *Rudbeckia* and *Gazania*, only two are initiated per flower, which eventually fuse to form the ray (Tucker, 1999).

The evolution of the complex inflorescence architecture in Asteraceae is associated with expansion and apparent functional conservation of the *CYC2* genes: both in *Gerbera* and in *Helianthus*, this gene clade contains the largest number of gene duplicates. These are all expressed in the primordia of the future ray flowers.

In *Helianthus*, the key gene in specifying ray versus disc identity is *HaCYC2c* (Chapman *et al.*, 2012). In *Gerbera*, where the orthologue of *HaCYC2c* is probably *GhCYC5*, rather than *GhCYC2*, there seems to be functional redundancy between the members of this gene family: in addition to *GhCYC2*, *GhCYC3* and *GhCYC4* are also involved in regulating ray flower identity, but this is probably not true of *GhCYC5*. This would indicate that in this case orthologous genes do not have the same function in these two plants.

At the level of single flowers, *GhCYC2* is expressed in the large ventral ligule but not in the dorsal domain of the flower; this suggests the evolution of a novel function of this gene in petal fusion and growth of the ligule (Broholm *et al.*, 2008). In *Gerbera hybrida*, as mentioned above, there is a gradient of increasing zygomorphy of the florets from the centre to the periphery of the inflorescence; central disc flowers are completely radially symmetrical with five short but separate petals. The external ray flowers are zygomorphic, with three large and fused abaxial petals and two rudimentary adaxial petals; their stamens are incompletely developed. Flowers positioned between the two main types (trans flowers) have intermediate morphology.

Similar is the control of floret morphology by the two *CYC2* clade paralogues in *Senecio vulgaris* (*RAY1* and *RAY2*): *RAY1* discriminates between disc and ray florets, while *RAY2* is apparently involved in organ fusion (Kim *et al.*, 2008).

7.4.9 Evolvability of Flower Symmetry

Easy evolvability of flower symmetry, coupled with variation in the number of perianth organs, is exhibited by the basal eudicot genus *Stephania*, in the species of which both actinomorphic and zygomorphic flowers are present.

The actinomorphic flowers of *S. japonica* and *S. longa* have perianth whorls of three or four organs: two whorls of sepals in the male flowers but only one whorl in the female flower; one whorl of petals in both sexes. *S. kuinanensis* and *S. mashanica* are sexually heteromorphic: the actinomorphic male flower has usually two whorls of sepals and a whorl of petals (all whorls of three), whereas female flowers, always zygomorphic, have mostly one sepal and two petals only. In some species, trimerous and tetramerous perianths coexist frequently in the same inflorescence. Variation in the number of perianth organs and loss of perianth parts in the female flowers result in flower symmetry switching from actinomorphy to zygomorphy (Meng *et al.*, 2012).

A clade with unusually unstable flower symmetry is the basal eudicot family Proteaceae, within which zygomorphy has independently evolved 10–18 times from actinomorphic ancestors, with at least four reversals to actinomorphy. Phylogenetic analysis suggests a duplication of *CYC*-like genes prior to the diversification of Proteaceae, followed by loss or divergence of one of the two paralogues in lineages which include more than half of the 35 species in 31 genera studied by Citerne *et al.* (2017). The genus *Grevillea* includes species with zygomorphic flowers and species with actinomorphic flowers: the *CYC* genes are expressed in developing flowers of all of them, but at a later stage the spatial pattern of expression becomes asymmetric in the perianth of the zygomorphic ones.

The first asterids probably had actinomorphic corollas; zygomorphy evolved in this clade at least eight times independently, with at least nine reversals, especially among the lamiids (Donoghue *et al.*, 1998; Ree and Donoghue, 1999; Donoghue and Ree, 2000).

7.5   MEROSITY

7.5.1   *Number and Spatial Arrangement of Flower Organs*

A most conspicuous aspect of flower diversity is the number of organs, sometimes variable within the species and even among the individual flowers of the same plant, sometimes fixed in whole large clades. In flowers with regularly whorled phyllotaxis, the number of organs (merosity; also called, less properly, merism) is often the same in all whorls, but exceptions are common, e.g. with carpels less numerous than sepals, petals and stamens. In flowers with spiral phyllotaxis, and/or with a continuous transition between different kinds of organs (most frequently, sepals vs. petals), determining merosity retains a degree of arbitrariness. Flower merosity has been studied in detail by Ronse De Craene (2016); this work is the single main source of information for the following account.

To some extent, floral merosity is related to floral phyllotaxis and to the ratio between the width of the organ primordium and the diameter of the floral tip at the time the organs are formed: the number of floral organs in a whorl or series is inversely proportional to that ratio (Soltis *et al.*, 2005).

7.5.2   *The Range of Numbers*

In most monocots and core eudicots, the number of flower organs is quite small, but within the whole of the angiosperms numbers range between 1 and ca. 10 000 (Endress, 2006). Extreme meristic diversity is observed in a few families, e.g. palms (see below) and Ranunculaceae (Tamura, 1995). The organs that reach the highest number in a flower are the stamens and, to a lesser degree, the carpels.

Very high numbers are found, for example, in *Gustavia*, with 500–1200 stamens in flowers with a diameter of up to 20 cm (Endress, 1994a); *Tambourissa* spp., with up to 2000 stamens and carpels

(Endress, 1994a); *Annona montana*, with ca. 2000 stamens, and *A. muricata*, with ca. 500 carpels (Leins and Erbar, 2008); *Laccopetalum giganteum*, with ca. 10 000 carpels (Tamura, 1995).

At the opposite end of the range there are, among others, *Euphorbia* with one stamen, *Ascarina* with one carpel and *Hippuris* with one stamen and one carpel (the latter sunken into the flower axis) (Leins and Erbar, 2008).

7.5.3 Number of Perianth Organs in the Main Groups of Angiosperms

In the absence of further specification, the following description refers to the perianth; the merosity of the other floral parts is often the same, but not necessarily, as illustrated in some detail in Section 7.5.6. In the majority of the pentamerous core eudicots, the number of carpels and/or stamens is lower or higher than the merosity of the perianth. Carpels are often three or two.

Overall, trimerous and pentamerous perianths dominate angiosperm flowers. Monocots (with a number of satellite groups in basal angiosperms and the Ranunculales) are the major clade with mainly trimerous flowers, while core eudicots are the domain of pentamerous flowers (Remizowa *et al.*, 2010; Ronse De Craene, 2010; Doyle and Endress, 2011). Tetramerous and dimerous flowers are subordinate conditions, derived from the two major types; however, tetramerous flowers are frequent in several clades of the core eudicots, and dimerous flowers are not rare in basal eudicots and basal angiosperms.

Indeterminate merosity, associated with spiral phyllotaxis, characterizes some early-branching lineages like *Amborella* and several Austrobaileyales. Other basal lineages, however, have trimerous or tetramerous perianths (Endress, 2001c). Here are a few examples of merosity in the flowers of basal angiosperms:

- *Brasenia schreberi*: sepals 3–4, petals 3–4, stamens 18–36, carpels 4–18
- *Cabomba* spp.: sepals 2–3, petals 2–3, stamens 3 or 6, carpels 1–7

- *Euryale ferox*: outer tepals (sepals) 4, inner tepals (petals) 20–35, stamens 78–92, carpels 8–16
- *Austrobaileya scandens*: tepals 12–24, stamens 6–11, inner staminodes 6–16, carpels 4–14
- *Takhtajania perrieri*: petals 11–15, stamens 12–16, carpels 2

In the most basal angiosperm groups, merosity is often variable or even indeterminate. For example, within the Winteraceae, some species have a perianth arranged in tetramerous and even pentamerous whorls, but in others the outermost floral organs are in dimerous whorls (Endress *et al.*, 2000).

Perianths with 2, 3, 4, 5 or 8 organs are found in the flowers of different species of Nymphaeaceae and Monimiaceae. Nevertheless, even in quite basal lineages there is a trend towards stable or dimerous (Monimiaceae, Winteraceae) or trimerous flowers (Piperales, Magnoliaceae, Annonaceae, Lauraceae). In basal angiosperms, pentamerous flowers are very rare and tetramerous flowers occur as occasional exceptions in generally trimerous families, e.g. in Annonaceae, *Nymphaea* and Monimiaceae (Ronse De Craene, 2016). The perianth of some species of *Magnolia* is indeterminate, but the perianth organs of other species of *Magnolia* are arranged in three trimerous whorls, as in other taxa of the same family, e.g. *Liriodendron*. The indeterminate merosity found in several magnoliid taxa, including *Calycanthus*, *Daphnandra*, *Hortonia* and some species of *Magnolia*, is regarded as a reversal from an already established condition of a trimerous perianth (Soltis *et al.*, 2005).

Among the monocots, which are basically trimerous, flowers with a dimerous perianth have evolved e.g. in *Maianthemum* and the Restionaceae; dimerous and tetramous flowers are present in the Triuridaceae; and pentamerous flowers have evolved only once (*Pentastemona*) within a dimerous clade (Rudall *et al.*, 2005a). In the Orchidaceae, a reduction in the number of stamens is associated with the evolution of zygomorphy (Rudall and Bateman, 2004). Only a few genera show high variation in the number of floral organs: *Paris*: sepals 3, 4–5, 7 or 10; petals 0, (3)4–5, 7 or 10; stamens (3)–8, 15–20

(21); *Aspidistra*: tepals 6–8 (rarely 4–10); stamens (4)6–12; carpels 3–5 (Fischer, 2015).

Many early-diverging eudicots (examples in Buxaceae, Proteales, Ranunculales and Trochodendraceae) have dimerous or trimerous perianths, but in core eudicots, as stated above, pentamerous flowers predominate. The developmental canalization that yielded the pentamerous condition of core eudicots occurred after the node leading to Gunnerales, which are dimerous (Soltis *et al.*, 2003).

In a number of basal angiosperms and in some early-diverging eudicots, especially among the Ranunculaceae, floral merosity is highly variable within a species, sometimes even among the flowers produced by the same individual plant. In contrast, in more derived groups, especially in the asterids, floral merosity is essentially fixed, in some cases even at the level of family (e.g. Apocynaceae) or order (Lamiales) (Soltis *et al.*, 2005).

Among the basal eudicots, remarkable meristic diversity is found in the Ranunculaceae. Their flowers are trimerous, pentamerous or dimerous. This diversity is likely correlated with the variable floral phyllotaxis, which is a mix of spiral and whorled patterns. Pentamerous flowers, in particular, are widespread and linked with a spiral initiation, except in *Aquilegia* (Endress, 1987b; Endress and Doyle, 2007). In *Clematis*, typically with two dimerous perianth whorls, a pentamerous organization may evolve following the division into two organs of one of the sepals of the inner pair (Ren *et al.*, 2010).

The reduced flowers of other basal eudicots (e.g. Proteaceae, Buxaceae, *Tetracentron*) are mainly dimerous, those of the Platanaceae trimerous or tetramerous.

The vast majority of core eudicots share a common pentamerous ancestor and are therefore known as the Pentapetalae (Chanderbali *et al.*, 2017), although – as we will soon see – many of them have subsequently evolved a different merosity. Pentamerous flowers, however, evolved independently in some non-core eudicots, i.e. in

Meliosma and in the Ranunculales. In these plants, floral organs are arranged in whorls, but floral development is spiral (Ronse de Craene *et al.*, 2003; Damerval and Nadot, 2007; Ronse De Craene and Wanntorp, 2008).

Within the Pentapetalae, flowers with alternative merosity have evolved many times. Most importantly, several families are almost exclusively tetramerous (e.g. Ebenaceae, Onagraceae, Oleaceae, Brassicaceae) or at least contain a large number of tetramerous genera (e.g. Rubiaceae, Melastomataceae, Rutaceae, Crassulaceae). Trimerous and dimerous flowers are much less frequent (Ronse De Craene and Smets, 1994; Endress, 1996). Some dimerous flowers, e.g. those of *Circaea lutetiana*, are fully developed (Ronse De Craene, 2016), but truly dimerous flowers are more often small and reduced, as in *Salix*, *Dorstenia*, *Betula* and *Myriophyllum*.

Compared to reduction, increase in merosity is much more widespread (Table 7.3) and occasionally characterizes whole tribes or genera. Hexamerous flowers are common in some families that also include species with pentamerous flowers, but not less than two-thirds of the Lythraceae have hexamerous flowers, as have a number of Polygonaceae, Lecythidaceae, Fagaceae and Loranthaceae. Heptamerous flowers are less frequent (examples: *Lysimachia*, *Fothergilla*), but sometimes this condition exists as a variant of hexamerous ones. Octomerous flowers are more widespread and characterize whole species-rich clades of the Sapotaceae (Kümpers *et al.*, 2016). Flowers with still higher merosity are uncommon; examples are found among the Crassulaceae (*Sempervivum* and *Aeonium*) and in *Lysimachia* (9-merous, unstable) (Ronse De Craene, 2016).

7.5.4 Stamen Number

In general, the number of stamens in a flower is more variable than the number of sepals, petals and carpels. In many groups with whorled floral phyllotaxis, stamens occur in two whorls, each of which is formed by as many organs as there are sepals or petals in the same flower (diplostemony).

Table 7.3 *Core eudicot families from which flowers with trimerous, hexamerous, heptamerous or octamerous flowers have been recorded; based on data in Endress (1996) and Ronse De Craene (2016).*

Merosity	3	6	7	8
Achariaceae	x			
Actinidiaceae		x		
Aextoxicaceae		x		
Alzateaceae		x		
Amaranthaceae	x[a]			
Anacardiaceae	x			
Anisophylleaceae	x			
Aphloiaceae		x		
Araliaceae		x		x
Balanopaceae	x			
Balanophoraceae	x[a]			
Barbeyaceae	x[a]			
Begoniaceae	x[b]	x		
Bixaceae		x		
Boraginaceae		x	x	
Brunelliaceae		x		x
Burseraceae	x			
Caryocaraceae		x		
Celastraceae	x[c]	x		
Cistaceae	x[c]			
Clethraceae		x		
Cornaceae	x	x	x	
Crassulaceae			x	
Crossosomataceae	x			
Crypteroniaceae		x		
Cunoniaceae	x[a]	x		
Datiscaceae	x[b]	x		x
Dirachmaceae			x	
Droseraceae				x
Ebenaceae	x[b]			
Ericaceae	x[a,c]			
Escalloniaceae		x		

Table 7.3 (*cont.*)

Merosity	3	6	7	8
Euphorbiaceae	x	x		
Fabaceae	x[a]			
Fagaceae	x		x	
Gentianaceae		x	x	x
Haloragaceae	x			
Hamamelidaceae	x[a]		x	
Hydrangeaceae		x		x
Lecythidaceae	x[b]			x
Limnanthaceae	x			
Loasaceae				x
Lophopyxidaceae		x		
Loranthaceae			x	
Lythraceae	x		x	x
Marcgraviaceae	x			
Melastomataceae	x	x	x	
Meliaceae	x[(c)]			
Misodendraceae	x			
Moraceae	x[a]			
Muntingiaceae			x	
Myricaceae	x[a]			
Myrothamnaceae	x			
Olacaceae	x	x	x	
Oleaceae		x	x	x
Onagraceae	x			
Opiliaceae		x	x	
Passifloraceae	x	x		x
Penaeaceae		x		
Penthoraceae				x
Phellinaceae		x		
Phyllanthaceae			x	x
Picramniaceae		x		
Picrodendraceae	x	x		
Platanaceae	x			
Podostemaceae	x			
Polemoniaceae		x		
Polygonaceae	x			

Table 7.3 (cont.)

Merosity	3	6	7	8
Primulaceae	x	x		x
Putranjivaceae	x	x	x	
Resedaceae		x		x
Rhamnaceae	x	x	x	
Rhizophoraceae				x
Rosaceae	x^{a,b}		x	
Rubiaceae	x^a	x	x	x
Rutaceae	x	x		
Salicaceae	x^{a,b}	x		
Santalaceae	x			x
Scrophulariaceae	x^a			
Setchellanthaceae			x	
Simaroubaceae	x$^{(a)}$			
Simmondsiaceae		x		
Sladeniaceae		x		
Tapisciaceae		x		
Ulmaceae		x	x	x
Urticaceae	x^a			
Vitaceae			x	

^{a} Flower trimerous (limited to some species, in the families for which the superscript is given in parentheses) except for the gynoecium, which commonly has fewer than three carpels, rarely more than three.
^{b} Flower trimerous except for polymerous androecium.
^{c} Flower trimerous (limited to some species, in the families for which the superscript is given in parentheses) except for pentamerous calyx.

Nearly all Brassicaceae have flowers with four petals and six stamens, but in *Lepidium* this trait is unstable: flowers with four or two stamens have evolved several times, in parallel with the loss of the corolla (Endress, 1992; Bowman *et al.*, 1999; Karoly and Conner, 2000; Mummenhoff *et al.*, 2001).

The highest diversity in the number and arrangement of stamens is found in the Caryophyllales, with stamen numbers ranging from 1 up to 4000. Caryophyllales are the only core eudicots in which a ring meristem gives rise to stamens and petals, with centrifugal development and according to a diversity of ontogenetic paths (Ronse De Craene, 2013).

7.5.5 Carpel Number

The number of carpels in a flower is diverse in the ANA-grade angiosperms, in magnoliids and in basal eudicots. The number becomes essentially fixed (but with exceptions) in the more advanced groups. In monocots, the number of carpels is most often three, and in core eudicots between two and five (Endress, 1990a).

Higher carpel numbers are often correlated with apocarpy. However, a huge number of carpels (up to 200) occur in *Schefflera pueckleri* (= *Tupidanthus calyptratus*), a plant with a syncarpous gynoecium (Sokoloff *et al.*, 2007).

Unicarpellate gynoecia are relatively common in the Lauraceae and other basal angiosperms (Hydatellaceae, Chloranthaceae, Ceratophyllaceae, Myristicaceae, Degeneriaceae, Hernandiaceae). Basal eudicots with unicarpellate gynoecium are Berberidaceae and Proteaceae; among the core eudicots, this condition characterizes the Fabaceae but is otherwise uncommon. Pseudomonomerous gynoecia, with a single functional carpel accompanied by the sterile remnants of other carpels, are found in some monocots (e.g. in Typhaceae, Poaceae, Restionaceae) and eudicots (e.g. in Anacardiaceae and Acanthaceae) (Endress, 2011).

In a few families, the number of carpels is higher than the number of sepals, petals and stamens, which are isomerous (Ronse De Craene and Smets, 1998; Endress, 2014).

7.5.6 Evolvability of Merosity

The genetic basis for meristic change is still poorly known. In *Arabidopsis*, mutants that control just floral organ merosity are rare. A loss-of-function mutant of the gene *PERIANTHIA* (*PAN*) is pentamerous, a reversal from the derived tetramerous flowers characteristic of Brassicaceae. *PAN* is a transcription factor that acts independently of pathways controlling meristem size and meristem identity (Chuang *et al.*, 1999). Otherwise, the majority of the floral organ number mutants identified in this plant are related to the control of meristem size. An example is the gene coding for the transcription factor *ULTRAPETALA* (*ULT*) (Carles *et al.*, 2005). In its loss-of-function mutant *ult1*, the number of floral organs (but also the number of flowers in the inflorescence) is higher than in the wild type, possibly owing to an increase in the size of floral and inflorescence meristems.

In the different plant clades, the evolution of merosity of flower organs uncovers different degrees and patterns of modularity of flower construction. The most fundamental distinction is between isomerous change, i.e. concerted change affecting all whorls, and anisomerous change, where the number of elements in a whorl evolves independently from the number of elements in other whorls.

Strictly isomerous change is frequently found in the Crassulaceae. Flowers of most species in this family are 4–5-merous, but in *Sempervivum* between 6-merous and 18-merous and in *Aeonium* from 6-merous to 32-merous (Ronse De Craene, 2016).

Isomerous tetramerous flowers seem often to derive from pentamerous flowers by loss of one organ in each whorl. In the Rutaceae, positional relationships of organ whorls are conserved irrespective of merosity, including genera with pentamerous, tetramerous and trimerous flowers. In such cases, the transition e.g. from pentamery to tetramery is sometimes described (e.g. Goebel, 1933) as the loss of a flower sector.

Intraspecific variation in merosity (Fig. 7.8) is infrequent, except for some clades. Indeed, in some Winteraceae, variation in floral organ number can be very remarkable. An example is *Drimys piperita* (= *Tasmannia piperita*), with 0–15 petaloid perianth organs, 7–109 stamens, 1–15 carpels and 2–46 ovules per carpel (Endress, 1994a).

In the palm subtribe Ptychospermatinae, variation in the number of stamens is amazingly high, at all levels: interspecific, intraspecific and even intra-individual. Stamen number is fixed in some species, e.g. 20 in *Ptychosperma pulleni*, 30 in *P. furcatum*,

FIGURE 7.8 A flower of cleavers (*Galium aparine*) with five sectors instead of the usual four. *A black and white version of this figure will appear in some formats. For the colour version, please refer to the plate section.*

40 in *Balaka macrocarpa*, 50 in *B. diffusa*; highest in *Veitchia pachyclada* (174–320), also very high in *V. subdisticha* (135–219), *Ptychococcus paradoxus* (100–213) and *Brassiophoenix schumannii* (130–200); very variable also in *Drymophloeus oliviformis* (30–66) (Alapetite *et al.*, 2014).

Some plant genera are extremely unstable, with great variation between species, e.g. *Lysimachia* with 5- to 9-mery; some *Hydrangea* with 6- to 12-mery; *Alangium* with 4- to 10-mery; *Diospyros* with 3(–5)- to 8-mery; *Lafoensia* with 8- to 16-mery; *Sanguisorba* with 3- to 7-mery (Ronse De Craene, 2016).

In some families (e.g. Styracaceae, Polemoniaceae, Gentianaceae) meristic variations are more frequent than in others. In some families where the large majority of species are pentamerous, individual species with different merosity may occur. Examples are the tetramerous *Potentilla erecta* and *Rhodotypos scandens* and the

8-merous *Dryas octopetala* in the Rosaceae; the 7-merous *Trientalis europaea* in the Primulaceae; the 12-merous *Blackstonia perfoliata* in the Gentianaceae (Ronse De Craene, 2016).

Flower merosity is highly conserved in some clades, for example the Malvaceae, most of which are pentamerous and isomerous throughout all flower whorls. The Caryophyllales are mainly pentamerous to tetramerous. Reductions to trimery are rare in core eudicots (e.g. *Polylepis* in the Rosaceae; *Empetrum* in the Ericaceae). Increases above hexamery are also rare and often unstable. Merosity is more variable in other large families, e.g. in the Lecythidaceae, which are mainly either hexamerous or tetramerous, and in the Sapotaceae, which can be pentamerous, hexamerous or octomerous (Ronse De Craene, 2016).

In some families, and even within lesser clades, the ontogenetic basis for changes in floral merosity may differ in different genera. For example, in the Detarieae, a tribe of the Fabaceae, a reduced number of floral organs may result from fusion, suppression or loss of organs during development. A saltational increase in the number of organs may result from developmental innovations such as the presence of a ring meristem or the production of multiple organs from common primordia, e.g. in stamen fascicles (Tucker, 1999).

In the Sapotaceae, octomery has evolved at least twice, once from tetramery, once from pentamery, and hexamery has evolved at least three times. Different paths leading to increase in merosity have been identified in this family, corresponding to different degrees of developmental modularity: (1) isomerous change affecting all organ whorls more or less equally; (2) isomerous increase in the number of petals, stamens and (mostly) carpels, without effect on sepals, as in *Madhucca* and *Payena*; (3) independent increase in the number of carpels, while the other organs are not affected. In a number of genera, the number of stamens has increased secondarily, independent of changes in the overall merosity of the flower (Kümpers *et al.*, 2016).

Variation in the number of organs in a whorl is not necessarily continuous and with normal distribution (Bachmann and Gailing,

2003). Some numbers are clearly preferred, and this often suggests a correlation with phyllotactic patterns. In *Salix pentandra*, the number of stamens varies between three and eight, but is most often five. These numbers (3, 5, 8) are the lowest (not counting 2) and most common in the Fibonacci sequence. A similar constraint in plasticity is documented in leaf phyllotaxis: in *Elatine alsinastrum*, the leaf whorls above water level are mostly trimerous, those under water pentamerous.

Polemoniaceae and Fouquieriaceae are sister groups. In the former, the flower has five stamens, in the latter 10–23; perhaps, the 'duplication' (that is, moving from 5 to 10, but not from 10 to 23) shows that the five-stamen whorl is not so precise, that is, it is rather a spiral – errors only show up when magnified, as there are no flowers with a fractional number of elements. Similarly, there are no animals with a fractional number of vertebrae or body segments, but duplication may uncover an 'excess of morphogenetic potential' that could not be expressed before as the addition of an integer to the series.

Constraints other than phyllotactic are probably revealed by Huether's (1968, 1969) experiments on *Leptosiphon androsaceus* (= *Linanthus androsaceus*). Here, selection for a number of petal lobes higher than the usual five resulted in the emergence of a new stable condition with 10 petal lobes, this being always the most frequent number larger than six recorded in the selected lines. In this case, the synorganization of petals probably canalized variation in petal lobe number, eventually leading to duplication of the original number (Fig. 7.9) rather than to a further exploration of Fibonacci's ratios.

7.6 SYNORGANIZATION WITHIN THE FLOWER AND AMONG FLOWERS

In his many fundamental contributions to flower morphology and evolution, Peter Endress has repeatedly discussed patterns and trends of synorganization within the flower and among the flowers in the

FIGURE 7.9 In some members of the Caryophyllaceae, as in this chickweed (*Stellaria media*), each petal is very deeply dissected into two lobes (laciniae), suggesting evolvability from 5-mery to 10-mery through a possible saltational transition. *A black and white version of this figure will appear in some formats. For the colour version, please refer to the plate section.*

inflorescence. A large part of what is summarized in this section is based on his work, mainly Endress (2006, 2011).

7.6.1   Synorganization Among Organs of the Same Whorl (Connation)

7.6.1.1   Perianth

Synorganization of perianth organs without actual fusion of the latter is produced by interdigitation of epidermal cells between the tepals (Proteaceae), sepals (*Oenothera* and Vitaceae) or petals (Loranthaceae,   Santalaceae,   Apocynaceae   Asclepiadoideae, Goodeniaceae,   Menyanthaceae, Campanulaceae and Asteraceae). In some Campanulaceae the sepals are connected laterally by long, curly hairs, functioning like Velcro.

In many basal angiosperms, tepals are united, at least basally. The flowers of *Amborella*, *Cabomba* and *Hedyosmum* are syntepalous. In many magnoliids the outermost 2–3 tepals are united, and more rarely the fusion occurs among the inner tepals. In the Myristicaceae there are only three tepals, which are united, as are, basally, the outer tepals (also three) of the Degeneriaceae. In many genera of the Annonaceae, only the three outer tepals are united, in others also the inner three or six tepals. Fusion occurs also between the inner 3–6 tepals of *Cinnamosma*, which form a bell-shaped structure, and the two or three outer tepals in many Winteraceae. In the Piperales, syntepaly is most pronounced in the flowers of *Aristolochia* and in the parasitic *Hydnora*.

In the Ranunculanae, there are almost no instances of synsepaly and sympetaly, but synsepaly occurs in *Cyclea* and in some Papaveraceae, while sympetaly is known from *Antizoma*, *Cyclea*, *Cissampelos* and *Consolida* (Endress, 1995).

Among core eudicots, sympetalous corollas evolved several times, e.g. in the Plumbaginaceae, Caricaceae and Cucurbitaceae. The great majority of the sympetalous eudicot species belong to the asterids, but the fusion among the petals of their corollas may arise through different ontogenetic pathways, suggesting a possible multiple derivation of sympetaly (at least four times) within this large clade (Olmstead *et al.*, 1992).

Erbar and Leins attached much systematic importance to the pattern and timing of initiation of sympetalous corollas, based on which they distinguished between early and late sympetaly. In early sympetaly the petals arise on a ring primordium or are otherwise connected already at initiation, while in late sympetaly the petals arise separately and become connected only at a later stage (Erbar and Leins, 1996, p. 428). But the distinction is of limited phylogenetic importance (Ronse De Craene and Smets, 2000).

Little is known about the control of sympetaly, but genetic pathways involved in the establishment of organ boundary (e.g. the *CUC1–3* genes) are strong candidates (Zhong *et al.*, 2016).

7.6.1.2 Androecium

Synandry, the congenital union of stamens, occurs in representatives of more than 10 families (Matthews and Endress, 2008), but is much less common than the fusion of sepals, petals or carpels. In most examples of synandry, only the filaments are united, but in rare instances the anthers are also united. Synandry is more prominent in unisexual flowers, where the stamens can be united in a central structure that involves both filaments and anthers, as in the Myristicaceae and in a few species of the Menispermaceae, Euphorbiaceae and Cucurbitaceae. Cases in which only filaments are united are more common in bisexual flowers, e.g. of Amaranthaceae, Malvaceae and Meliaceae (Soltis *et al.*, 2005).

Postgenital fusion of the filaments occurs in *Iris unguicularis* and in *Sisyrinchium* (Weberling, 1989). Postgenital fusion of the anthers occurs in a few genera of the Gesneriaceae, Linderniaceae, Scrophulariaceae, Lamiaceae and Bignoniaceae; in some species, anthers are fused pairwise, but more often all four anthers are fused together (Endress, 1994a).

7.6.1.3 Gynoecium

Unsurprisingly, angiospermy is the most prominent synapomorphy of angiosperms. In flowering plants, the ovules are borne on carpels. At some time during flower development, the carpels become closed, with the ovules hidden inside. In some clades, the carpels close early and the ovules are initiated only after the closure of the carpels. Early carpel closure occurs in all members of the ANA grade and in almost all magnoliids studied to date. In monocots, early carpel closure is present in early-diverging lineages, that is, in the Acoraceae and in part of the Alismatales (Tofieldiaceae, Araceae and Butomaceae). In early-branching eudicots, early carpel closure occurs in the Eupteleaceae and in the majority of the Papaveraceae and core Ranunculales (Endress, 2006, 2015). Late carpel closure occurs in many representatives of the Alismatales and in at least some Papaveraceae, Berberidaceae and Ranunculaceae.

Delayed development of carpel wall accompanied by precocious ovule development has been documented in a few Alismatales and some Poales (including part of the Poaceae and Cyperaceae) and in some Ranunculales. Among core eudicots, this behaviour is only known from a few fabids and malvids and several Malpighiales and Caryophyllales.

In some flowers, the closure of the ventral slit is not synchronous along its length. In representatives of the Calycanthaceae, Annonaceae, Winteraceae, Lardizabalaceae, Fabaceae and Araliaceae, the longitudinal slit closes first in the lower part, whereas in other plants, such as *Degeneria vitiensis* and *Penthorum sedoides*, the longitudinal slit closes first in the upper part (Endress 2015). In *Sagittaria trifolia*, the middle parts of the carpel margins undergo postgenital fusion during late carpel development, but at maturity the lowest and uppermost parts of the carpel margins are still open (Huang *et al.*, 2014). In exceptional cases, the carpels are not sealed at all at anthesis. In *Tiarella*, carpel margins are rolled inwards, ensuring that the ovules are hidden from the outside world even in the absence of carpel sealing; in *Reseda*, the entrance to the gynoecium is not sealed, but its margins are covered by protective hairs.

In the most basal angiosperms, carpel sealing is predominantly by secretion, in others by postgenital fusion. Carpels that are sealed only by secretion take often a tubular (ascidiate) form, which is probably the ancestral carpel form of angiosperms (Doyle and Endress, 2000; Endress and Igersheim, 2000a).

In syncarpous gynoecia in which the styles remain free, e.g. in *Laurus nobilis*, *Dillenia alata*, *Illicium anisatum*, *Hernandia nymphaeifolia* and *Aquilegia vulgaris*, carpel closure begins at mid-length and extends upwards; the closure is somewhat delayed at the lower and upper end of the slit. Fusion extends to the stigma only in a limited number of lineages (Rutaceae, Simaroubaceae and Apocynaceae) (Tucker, 1984a).

In syncarpous gynoecia, the separate pollen-tube transmitting tracts of the individual carpels are united together to form a compitum which extends at least for part of their length. The advent of syncarpy and the development of a compitum represent key innovations that evolved independently at least twice, once in monocots (Buzgo and Endress, 2000; Igersheim et al., 2001) and once in eudicots. Apocarpy is present in the majority of basal angiosperms, both in the ANA grade and in the magnoliids; many basal eudicots are also apocarpous, but others are unicarpellate (Endress, 2011).

The gynoecium of early angiosperms probably had numerous free carpels (Takhtajan, 1969; Stebbins, 1974). Most recent basal angiosperms and many basal eudicots, including *Amborella*, Austrobaileyales, *Ceratophyllum*, Laurales, Magnoliales and most Ranunculales, have apocarpous gynoecia. Among the Nymphaeaceae, *Cabomba* and *Brasenia* are also apocarpous, but syncarpy has evolved in the remaining genera of the family (Soltis et al., 2005). Other basal angiosperms are also syncarpous, including Piperaceae, Saururaceae and Aristolochiaceae (Igersheim and Endress, 1998). Syncarpy is found in more than 80% of extant angiosperm species and is the rule in core eudicots and monocots, but secondary apocarpy occurs in a couple of monocot clades (Alismatales; Triuridaceae) and in a number of core eudicots (e.g. Rosaceae, Sapindales, Malvaceae, Apocynaceae) (Endress, 1982).

Overall, during angiosperm evolution there have been at least 17 independent transitions from apocarpy to syncarpy (Armbruster et al., 2002), probably based on different molecular mechanisms. In *Arabidopsis*, a number of mutations, e.g. *aintegumenta*, *crabs claw*, *ettin*, *leunig*, *spatula* and *tousled*, are known to produce more or less complete carpel separation (Alvarez and Smyth, 1999; Liu et al., 2000), but it is not clear whether these genes were actually involved in the evolution of syncarpy in this lineage (Vialette-Guiraud and Scutt, 2009).

One genetic system potentially responsible for the generation of congenital syncarpy, however, deserves special attention.

This involves negative regulation of *CUC1* and *CUC2* expression by microRNAs of the *miR164* family. In *Arabidopsis*, this family is represented by three genes. The negative regulation of *CUC2* by miR164 in the process of carpel fusion is demonstrated by the pheno-type with complete carpel separation corresponding to the triple mutant *miR164abc* (Sieber *et al.*, 2007). The *CUC* and *miR164* system has been recruited many times independently over the course of angiosperm evolution; its role in controlling leaf dissection (Niko-vics *et al.*, 2006) has been mentioned in Section 6.4.1. It is possible that the recruitment of this developmental module to generate syn-carpy has also occurred in distinct angiosperm lineages independently (Vialette-Guiraud and Scutt, 2009).

7.6.2 *Synorganization Between Whorls (Adnation)*

In some eudicots, e.g. Tropaeolaceae and Myrtaceae (especially *Eucalyptus*), sepals and petals are fused and form a synorganized perianth. In some Asparagaceae and a few other monocots, the outer and inner perianth whorls are fused (Soltis *et al.*, 2005).

The simplest form of synorganization between perianth and androecium occurs in trimerous flowers with two whorls of perianth organs and two whorls of stamens. These whorls alternate with each other, thus dividing the flower into sectors, each with a stamen and a closely associated perianth organ. In monocots (Endress, 1995), but also in a number of basal eudicots, especially core Ranunculales with trimerous flowers (Endress, 2010a), these two organs are united basally. This link is retained in cases where the flower switches from trimery to pentamery, despite the fact that with the latter change, the positional alternation previously existing between organs of two sub-sequent whorls has been lost. This is the case in *Pentastemona* (Rudall *et al.*, 2005a), in the terminal flowers of *Berberis* (Endress, 1987b) and in the pentamerous flowers of the Sabiaceae.

In sympetalous flowers, stamens may fuse with the petal tube. This form of synorganization is widespread among the asterids, espe-cially in representatives of the Ericales, Gentianales, Lamiales, Sola-nales, Boraginaceae and Asterales (Endress, 2010a). In *Petunia*, a *PI* paralogue is required for the fusion of stamens to the corolla tube (Vandenbussche *et al.*, 2004).

A further organ complex involving perianth and androecium is the corona, typical of a number of Amaryllidaceae and Velloziaceae among the monocots and Apocynaceae among the core eudicots (Endress, 2011).

In two families, Orchidaceae and Apocynaceae, the intimate synorganization of androecium (specifically, the anthers) and gynoecium (specifically, its distal part) has independently culminated in the evolution of pollinaria, complex structures that participate in unique pollination mechanisms. In a pollinarium, two or four pollinia (the compact mass of pollen grains produced in a pollen sac) are held together by a translator that attaches them to the body of a pollinator.

In the Orchidaceae, synorganization occurs by congenital fusion of gynoecium and androecium, resulting in a gynostemium. In the Apocynaceae, instead, postgenital fusion between anthers and stigma has resulted in a gynostegium (Endress, 1994a; Endress M. E., 2001; Fishbein, 2001; Endress, 2011); see Section 7.6.4.

Away from the orchids, a gynostemium is also present in *Aristolochia* (Igersheim and Endress, 1998). In this genus, the perianth consists of three petaloid sepals congenitally fused into a zygomorphic calyx; the reproductive organs are represented by a gynostemium formed by congenital fusion between the stamens and the stigmatic region of the carpels. In *A. fimbriata*, the expression of the *AG* homologue is spatiotemporally restricted to stamens, ovary and ovules, suggesting conserved function in the control of stamen and carpel identity, consistent with the function of these genes in monocots and core eudicots. Expression of the *STK* homologue has been found in the anthers, stigmas, ovary and ovules, and later on in both fruit and seeds, suggesting roles in stamens, ovary and fruit development not found in other angiosperms, in addition to conserved roles of this gene in controlling ovule and seed identity. In *A. fimbriata*, *STK* expression has been additionally found in areas of organ abscission and in dehiscence zones (Suárez-Baron *et al.*, 2017).

7.6.3 Synorganization Between Flowers in an Inflorescence

Synorganization may occur also between the flowers of an inflorescence, especially when these are tightly packed together to form a pseudanthium, which resembles a single flower. The capitula of the Asteraceae, the second-largest of angiosperm families, are the best demonstration of how successful this organization can be. A further level of synorganization has evolved in a few genera of this family, e.g. in *Oedera*. Here, a number of small capitula are packed together in the same way as florets are grouped together in an ordinary capitulum; ray flowers are present only at the margin of the peripheral capitula (Classen-Bockhoff, 1992, 1996).

Despite the less close packing, synorganization is also present among the flowers in the umbels of the Apiaceae, ordinarily grouped in turn in umbels of umbels. Here again, the peripheral flowers of the peripheral umbels may have conspicuously enlarged petals on their peripheral side, an example of synorganization particularly conspicuous in *Orlaya* (Endress, 2012).

Fusion of the two lateral flowers occurs in the inflorescence of some species of *Lonicera* and other Caprifoliaceae, a modified dichasium retaining the two lateral flowers, but with loss of the terminal flower. The fusion starts with the ovary and may extend to the rest of the flower, as observed also in plants of other families, e.g. in *Cornus mas* and in *Mitchella*, where varying degrees of fusion of two tetramerous flowers are occasionally observed, resulting in trimerous, pentamerous or hexamerous forms (Blaser, 1954).

7.6.4 Evolvability and Evolution of Synorganization

The patterns of aestivation have an impact on the possibility of evolution of synorganization. In asymmetric configurations, as in imbricate aestivation (see Section 7.4.2), it is difficult for two contiguous organs to fuse because their edges grow in different directions; opportunities for closer interaction of organs of the same whorl eventually leading to fusion are provided instead by symmetrical, e.g. valvate, aestivation patterns (Endress, 1994a). As noted by Endress (2006), there are strict morphological preconditions for the evolution of synorganization. Among these are a moderate organ number per whorl, commonly not more than five, and a precise localization of organs or organ parts, as obtained only when floral organs are arranged in whorled phyllotaxis (Endress, 1987a, 1990a, 2006). In these flowers, synorganization may evolve in two directions: as tangential synorganization among the organs of a whorl, as usual, but also as sectorial (radial) synorganization between each element of a whorl and the element of a contiguous whorl which grows in the same radial sector of the flower (e.g. between perianth organs and stamens in many monocots, and between inner perianth organs and stamens in several basal eudicots). Tangential and radial synorganization may occur in the same flower: in many monocots there is both syntepaly and fusion between each tepal and the stamen in the same radius; similarly, in many asterids and some rosids (e.g. Galipeinae of Rutaceae) synorganization involves all petals and stamens (El Ottra *et al.*, 2013). The most integrated structures produced by combined radial and tangential synorganization are those that involve androecia and gynoecia in the Orchidaceae and Apocynaceae (Endress, 2015).

In the large clade of the Apocynaceae formed by the Asclepiadoideae plus the Secamonoideae (overall, more than 3000 species), the number of floral organ numbers is fixed, without exception, with five sepals, five petals, five stamens and two carpels (Endress, 2015). Except for the sepals, the organs of all floral whorls are tangentially synorganized by congenital fusion; in addition, corona and stamens are also radially synorganized, again by congenital fusion. On top of this, postgenital fusion occurs between the anthers and the style head and between the distal parts of the two carpels. As mentioned in Section 7.6.2, the resulting gynostegium is formed by postgenital union of the anthers, whereas in the gynostemium there is congenital fusion of the filaments (sometimes also the anthers) with the style (Rudall and Bateman, 2002). Gynostegia are also present in a number of monocots, e.g. *Burmannia*, *Avetra* and *Trichopus*, in which apical processes of the anthers fuse postgenitally with the stigma before anthesis (Caddick *et al.*, 2000).

In flower evolution, the advent of a new synorganization is often per se a key innovation (Endress, 1997, 2001b) – witness the success of syncarpy in core eudicots and monocots, the dominance of sympetaly in asterids, the widespread fusion of petals and stamens in euasterids and the fusion of stamens and carpels in orchids.

Synorganization also paves the way for essentially novel structures (Endress, 2006, 2015), an example of which is the corona of the Apocynaceae.

Synorganization is accompanied by a stabilization of floral phyllotaxis and a reduction of variation in the number of floral organs. This is remarkable in comparison with the variation, or plasticity, shown by plants with no evidence of synorganization in their flowers. In the same plant, for example in *Drimys* (Doust, 2001) or *Trochodendron* (Endress, 1990a), flowers may have spiral or whorled phyllotaxis, depending on their terminal versus lateral position in the inflorescence; in other plants, such as *Distylium*, the different position of the flower translates into a different number of flower organs (Endress, 1970, 1978). No parallel variation is found in flowers with any degree of synorganization among their parts. Similar differences between flowers with different degrees of synorganization are observed in starved individuals compared to normal ones: starved individuals tend to produce fewer flower organs, e.g. stamens, in plants whose flowers have a low degree of synorganization (e.g. *Papaver*), but smaller flowers with the full number of organs in plants whose flowers have a high degree of synorganization (e.g. asterids).

7.7 THE FRUIT

7.7.1 *Fruit Patterning Genes*

A large amount of information is available on the identity, expression patterns and control of expression of genes involved in the growth, differentiation, patterning and ripening of fruits, and also in a number of mechanisms of seed dispersal, especially when this is a consequence of fruit dehiscence or is preceded by predictable detachment of the fruit from the plant. This evidence is particularly detailed for plants with fruits or seeds of economic importance, while a comparative genetics of fruit development is still to be produced. Overviews of the genetic network involved in fruit development in *Arabidopsis* were offered by Roeder and Yanofsky (2006) and Seymour *et al.* (2013).

The genes responsible for fruit patterning include genes additionally involved in a number of different morphogenetic processes. In the case of *Arabidopsis*, for example, the list includes, among others, genes with a role in leaf development, such as *AS1, AS2* and *BP* (Alonso-Cantabrana *et al.*, 2007).

A number of *MIKC^c*-type genes, which are collectively better known because of their roles in the specification of floral organs, are also involved in the control of fruit development and seed dispersal mechanisms in many groups of flowering plants. For example, in *Arabidopsis*, two members of the *AG* subfamily (*SHP1, SHP2*) are responsible for the specification of the replum, the septum that divides the silique longitudinally in two halves and to which seeds are attached. Their tomato orthologue *TAGL1* also has a function in fruit development, but its actual role is very different, being involved in controlling fruit expansion and ripening (Itkin *et al.*, 2009; Vrebalov *et al.*, 2009; Giménez *et al.*, 2010).

In different plants, the expression of genes belonging to the same subfamily may result in the production of very different fruit types (Smaczniak *et al.*, 2012), for example the true fruit (a berry) of *Vaccinium*, the monocarpic accessory fruit (a pome) of *Malus*, the aggregate accessory fruit of *Fragaria* (Cevik *et al.*, 2010; Jaakola *et al.*, 2010; Seymour *et al.*, 2011).

Very promising examples of what will eventually contribute to a large fruit chapter in plant evo-devo are already available, as the two following examples will show.

The first is the Chinese lantern, the unique fruit of *Physalis* species (Fig. 7.10). The genus *Physalis* is closely related to *Solanum*

FIGURE 7.10 Mature fruit of the Chinese lantern (*Physalis alkekengi*).

(Olmstead and Palmer, 1992). Species with an inflated calyx that increases in size while the fruit is growing and eventually ensheaths it at maturity are restricted to six of the 96 genera of the Solanaceae, namely *Anisodus*, *Nicandra*, *Physalis*, *Physochlaina*, *Przewalskia* and *Withania* (D'Arcy, 1991). However, sepals approximately as large as the petals of the same flower and occasionally able to resume growth after fertilization are also known in a small number of species within the huge genus *Solanum*, such as *Solanum aetiopicum* and *Solanum macrocarpon* (in some populations at least). The production of the very showy 'lantern' of *Physalis* species seems to depend on the joint presence of an unidentified signal produced by the fertilized ovules and the ectopic expression in the floral structures of the *MADS*-box gene *MATURATION PROMOTING FACTOR2* (*MPF2*), which in *S. tuberosum* is expressed vegetatively (He *et al.*, 2004; He and Saedler, 2005). The complex evolution of sepal morphologies in Solanaceae is arguably dependent on duplication of *MPF2*-like genes followed by functional diversification at the regulatory and protein levels (Khan *et al.*, 2009).

A peculiar kind of fruit has evolved in the tribe Brassiceae, within which it is shared by most species. This is the heteroarthrocarpic fruit, characterized by a joint that divides it into proximal and distal segments of different nature and composition and which, unlike the usually dehiscent fruits of the Brassicaceae, is partially or completely indehiscent. The proximal segments of the heteroarthrocarpic fruits of *Erucaria erucarioides* and *Cakile lanceolata*, studied by Hall *et al.* (2006), correspond to the valves of a typical silique. Only elements of the ovary and the style contribute to the indehiscent distal segments, because valve tissue is not differentiated from the ovary wall of this segment. In *Cakile* there is also a transverse dehiscence zone that allows the detachment of the two parts. Changes in lignification patterns and novel relationships between valve, style, ovary and mesocarp are involved in the evolution of these unusual and complex fruits.

7.7.2 Evolvability of Fruits

In many angiosperm lineages, the correlation between flower morphology and fruit morphology is quite loose. Despite the fact that it develops out of the flower, in a renewed burst of morphogenesis normally released by pollination and subsequent fertilization of ovules, the fruit can evolve, sometimes to a remarkable extent, following its own path. This is an example of temporal (sequential) modularity, comparable to the large autonomy with which larva and adult evolve in many animal groups. To be sure, in the production of fruits are involved tissues with different ploidy levels and different genotypes, and this will contribute to the divide between flower and fruit morphogenesis: to be compared, again, with some animals, e.g. *Drosophila*, where bulky larval structures, including the musculature, are polyploid, whereas the cells of the histoblasts and the imaginal discs out of which adult organs such as eyes, antennae, wings, legs and genitalia will be formed are diploid. The parallelism may stop here (during the stage of pupa, polyploid tissues are subjected to lysis, from which diploid cells escape, but this is only remotely comparable to the abscission of a number of flower organs that generally accompanies the production of the fruit); however, to the extent to which it applies, it would deserve closer study.

In evolutionary terms, there are several different paths potentially leading to a diversity of fruit types within a family, or a group of closely related families. In the case of the Ranunculaceae, this is largely a consequence of the variation in composition and organization of the gynoecium, but this correlation does not apply in other cases. In the Myrtaceae, fruits are primitively dehiscent, but indehiscent fruits have evolved four times, including three independent derivations of fleshy fruits (Soltis *et al.*, 2005); similarly, in the Melastomataceae, berries evolved from capsules at least four times (Clausing and Renner, 2001). Other lineages with inconstant fruit types include Malvaceae, Fagales, Myrtales and Celastrales (Soltis *et al.*, 2005).

Within the asterids, a huge clade (ca. 35 000 species) of mainly herbaceous plants including Campanulaceae, Asteraceae, Caprifoliaceae, Adoxaceae, Araliaceae, Apiaceae and a few minor families is sometimes recognized under the name Campanulidae (Cantino *et al.*, 2007). The transitions among the major types of dry and fleshy fruit types found in this group are largely unidirectional: once the ancestral capsular condition was abandoned, whether through the loss of dehiscence or through a reduction to a single seed, there was no reversal. In particular, the evolvability of the achene (dry, single-seeded and indehiscent; the most common fruit type within the campanulids) into a different fruit type is close to zero (Beaulieu and Donoghue, 2013).

The fruits of bamboos contain one seed only, like the ordinary caryopses of grasses, but are green, fleshy and often very large, compared to the fruits of the other Poaceae, up to 300 g in the case of *Melocanna baccifera* (Govindan *et al.*, 2016). These unusual caryopses may have evolved independently up to seven times (Ruiz-Sanchez and Sosa, 2015).

Multiple independent evolution of indehiscent from dehiscent fruits has been demonstrated in the Brassicaceae (Appel and Al-Shehbaz, 2003; Koch *et al.*, 2003; Mummenhoff *et al.*, 2005; Al-Shehbaz *et al.*, 2006), even within a genus, or group of closely related genera (*Lepidium* s.l.) (Mummenhoff *et al.*, 2009).

7.7.3   *Heterocarpy*

In some species, different types of seeds (heterospermy) or fruits (heterocarpy) are produced by the same individual plant (Hannan, 1980).

Heterospermy has been described in a number of families, including monocots (Commelinaceae, Poaceae), basal eudicots (Papaveraceae) and core eudicots (Fabaceae, Euphorbiaceae, Thymelaeaceae, Cistaceae, Brassicaceae, Polygonaceae, Caryophyllaceae, Nyctaginaceae, Rubiaceae, Plantaginaceae, Caprifoliaceae, Apiaceae), but especially in Amaranthaceae and Asteraceae (Imbert, 2002).

Examples of heterocarpy are *Calendula*, in which up to four achene morphs are produced by the same plant (Heyn *et al.*, 1974), and *Ceratocapnos heterocarpa*, which produces short monospermous and indehiscent fruits and long dehiscent fruits, each of which contains two seeds, of different shapes. The three types of seed differ also in germination patterns. Short and long fruits, respectively adapted to short- and long-distance dispersal, are produced in seasonally varying proportions (Ruiz de Clavijo, 1994).

8 Architecture and Syntax of the Plant Body

The geometrical regularities of plant architecture are obvious, fascinating to the human eye and often, but perhaps not always, of adaptive value. A good degree of robustness in producing flowers with either radial or bilateral symmetry is arguably important, in so far as flower geometry contributes to the fidelity of pollinator visits; a regular phyllotactic positioning of leaves, in turn, may play a role in obtaining a spatial distribution and orientation of their blades such that suitable exposure to the sun is obtained. However, in the absence of a specific selection pressure, there is no reason to expect that plant organs will show a higher degree of regularity, in their shape and distribution, than a very basic set of growth rules may be able to provide.

In fact, in the absence of disturbance, highly symmetrical structures are easier to produce than structures with lower degrees of symmetry, or overtly asymmetric ones. If we look for regular forms among the green algae, we find them most easily among the paucicellular kinds, e.g. *Pediastrum*. Similarly, regular actinomorphic flowers require, in principle, less genetic control than zygomorphic flowers, and still more complex is the production of robustly asymmetric ones. The same is true of snails, whose left or right chirality is usually fixed at the species level (and above), and vertebrates, with their visceral asymmetry – in either case, early expression of lateralizing factors is required to generate directional deviation from mirror symmetry.

On the other hand, plants can adjust their occupancy of the space in different ways, the most curious being perhaps resupination (see Section 5.2.2), the most common a differential bending or torsion of leaf petioles such that the original distribution determined by

FIGURE 8.1 In the imparipinnate leaves of *Rosa* spp., the degree of pairing between leaflets ranges from very precise to non-existent. *A black and white version of this figure will appear in some formats. For the colour version, please refer to the plate section.*

phyllotaxis is eventually immaterial in determining the exposure of photosynthetic organs to the light.

Thus, rather than representing the results of subtle adaptations, minor deviations from geometrical precision in symmetry or other kinds of spatial relationships (Fig. 8.1) perhaps tell us more about development than about evolution, and it is mainly in this sense that I look at them throughout this chapter, in Sections 8.2 and 8.4 especially.

Emblematic, in this respect, is *helical anisoclady*, as found in several Rubiaceae (Rutishauser, 1981; Rutishauser *et al.*, 1998) and Caryophyllaceae (Troll, 1964) (see Fig. 5.2). The phenomenon is better observed in the latter: the two decussate leaves of one whorl are not equivalent in regard to their axillary buds, one of which is more vigorous than the other one. Ideally, joining all favoured buds from whorl to whorl, we obtain a helix with a divergence of 90° or less.

Ideal versus functional symmetry must often be distinguished also at the level of the individual leaf. Some leaves (e.g. *Tilia, Corylus*) are obviously asymmetric along the axis that joins the petiole to the leaf tip (directional asymmetry, again); strongly asymmetric also are

FIGURE 8.2 In the compound leaf of barrenwort (*Epimedium* sp.), the terminal leaflets have approximately symmetrical laminae, whereas the other leaflets are strongly asymmetric, the left side prevailing in the leaflets on the left, the right side in those on the right.

the leaflets of some pinnate leaves, except for the terminal one, but in this case left and right leaflets are reciprocally mirror images (Fig. 8.2). However, most leaves are described as 'truly' bilaterally symmetrical, minor deviation from the ideal shape (fluctuating asymmetry, in this case) being attributed to noise experienced during development (Miller, 1995; Kozlov *et al.*, 1996; Hódar, 2002).

Nevertheless, sophisticated geometrical analyses have revealed that even plants such as *Arabidopsis thaliana* or *Solanum lycopersicum* produce leaves with either a left- or a right-handed bias (Chitwood *et al.*, 2012). In this case, the asymmetry of leaves correlates with the handedness of the individual plant and is a consequence of the auxin distribution also responsible for the spiral phyllotaxis (Koenig *et al.*, 2009; Scarpella *et al.*, 2010). In *Hedera helix*, a plant with distichous phyllotaxis, leaves are slightly asymmetric, and left-biased leaves alternate from node to node with right-biased ones (Martinez *et al.*, 2016).

8.2 POSITIONAL EFFECTS

8.2.1 *Plant Architecture and Ontogenetic Contingency*

In a series of important papers analysed in this section, Pamela Diggle has put the flowers back in their natural context. Instead of the idealized reproductive units suggested by the laboratory studies of model species, identical copies of a species-specific blueprint varying only through the effects of mutations and environmental disturbance, flowers appear, in her studies, in all the variation of their multiple copies occurring within one inflorescence. In nature, flowers are not more or less identical units borne on the individual plant in an arbitrary number. The structural variation between individual flowers must be addressed because, as remarked by Diggle (2014), it is not the 'average flower morphology' that evolves, but the whole range of subtly different flower phenotypes that each plant can ordinarily produce.

Differences among the flowers of the same inflorescence are not limited to the common singularities of the terminal flower, described below, and the obvious contrast between disc and ray florets in the capitula of many Asteraceae (see Sections 5.10.3 and 7.4.8). A number of elegant studies, especially by Diggle (1994, 1995, 2003), have demonstrated variation in floral traits consistently correlated with the relative position of the individual flowers within the inflorescence.

In many plants, features of the corolla, androecium and/or gynoecium show significant intra-inflorescence variation attributable to flower position alone. This is demonstrated by a literature review involving studies on 65 species in 27 families (Diggle, 2003). The most common pattern is proximal-to-distal decline in size or mass, but in some species positional variation follows the opposite trend. Within a flower, the different organ whorls can vary independently. Results of this literature review are summarized in Table 8.1. Eight studies detected a pattern of greater fruit production in distal flowers. In three species with proximal-to-distal increase in fruit maturation, i.e. *Narcissus dubius* (Worley *et al.*, 2000), *Ipomopsis aggregata* (Brody

Table 8.1 *Summary of the results of Diggle's (2003) literature review of studies on 65 species in 27 families in which features of the corolla, androecium and/or gynoecium show significant intra-inflorescence variation attributable to flower position alone.*

	Number of species showing intra-inflorescence variation (proximal to distal)			
Feature	Increase	Decrease	No change	Variable (non-directional)
Corolla	2	9	2	—
Androecium	1	3	2	1
Pollen	—	6	2	—
Gynoecium	—	5	1	—
Ovule number	2	32	1	—
Fruit length or mass	1	11	1	—
Fruit maturation	8	12	14	—

and Morita, 2000) and *Euphorbia characias* (Espadaler and Gómez, 2001), distal flowers reach anthesis earlier than basal flowers.

These effects are due in part to developmental plasticity, a consequence of the fact that individual flowers develop in different microenvironmental conditions – including the developmental stages of the flowers closest to them within the same inflorescence – but this does not explain all the differences. To some extent, 'architectural effects', to use Diggle's (2003, 2014) aptly descriptive term, are also involved.

In *Solanum hirtum*, phenotypic plasticity in the production of staminate flowers (i.e. hermaphrodite flowers with reduced, non-functional gynoecia) has been demonstrated by experimental manipulation of fruit-set. The normal or aborted development of the gynoecia in distal flowers of the inflorescence is a plastic response to the fruiting status of the plant: individuals with no fruit-set produce only, or predominantly, hermaphrodite flowers, whereas the inflorescences of the plants with high fruit-set produce staminate flowers in distal positions (Diggle, 1991, 1993, 1994). However, analyses that included appropriate architectural controls (Diggle, 1991, 1995), while confirming plasticity in declining petal length, nevertheless demonstrated that declining anther length is an architectural effect: anthers of distal flowers are smaller than those of proximal flowers, irrespective of treatment. Both plasticity and position are responsible for the decline in the size of the ovary in staminate flowers (Diggle, 1991, 1995).

In *Nicotiana alata* and *N. forgetiana*, the size of the flowers declines both from basal to apical position within an inflorescence and among the inflorescences successively produced by a plant (Bissell and Diggle, 2008); measures of corolla limb and mouth decline in size with flower position, while the lengths of corolla tube, stamens and gynoecium are invariant with flower position. Significant association between the variation in the size of flowers and their position within the inflorescence was also found in *Delphinium glaucum*; sepal size also declined with the distance of the flower from the base of the inflorescence (Ishii and Harder, 2012). In this study, petal development was shown to be more robustly controlled: both individual petal size and correlations among different petal traits were invariant with position.

In the corolla of *Dactylorhiza* flowers, the length of the lip declines with flower position, but the length of the spur does not (Vallius, 2000). In *Aquilegia canadensis*, the size of the sepals declines with position, while the lengths of nectar spur, gynoecium and stamen are invariant (Kliber and Eckert, 2004).

Structural or functional factors that might explain these architectural effects are still unclear, but do not seem to involve hormones (Granado-Yela *et al.*, 2017). The amount of vasculature available to the individual flowers and the fruits and seeds deriving from these may play a role (Carlquist, 1969). Distal reproductive structures may receive lower nutrient supply because of the smaller diameter of the supporting stem, which may contain smaller amounts of vascular tissue, compared to those in proximal positions.

Granado-Yela *et al.* (2017) tested the relative influence of competition for resources and architectural constraints in *Olea europaea* var. *sylvestris*. In this andromonoecious tree, hermaphrodite flowers occur mainly on the apical and the most proximal positions in determinate inflorescences. The authors experimentally modified the resources available to individual inflorescences, either by removing half of the inflorescences per twig, thus increasing the resources available to each one, or by removing leaves, thus reducing resource availability. An increase in resources increased the probability of hermaphrodite flowers developing in any position, while removal of the apical flower increased the probability of producing hermaphrodite flowers in proximal positions but not in subapical positions. This suggests an interaction between resource competition and architectural constraints.

8.2.2 Positional Effects: the Terminal Flower

In most determinate inflorescences, the activity of the apical meristem comes to an end with the production of a terminal flower. However, the presence of the latter is sometimes facultative, as in *Aesculus hippocastanum*. Under serious limiting conditions, many plants that usually produce determinate inflorescences with numerous flowers are reduced instead to producing terminal flowers only. An opposite behaviour is seen in others, e.g. *Agrimonia eupatoria*, *Campanula rapunculoides* and *Actaea cimicifuga*, where a terminal flower appears only in the case of less than vigorous growth (Troll, 1964).

In *Daucus carota*, the terminal flower is often recognizable by its deep purple, rather than white, colour (Fig. 8.3). In several plants, the terminal flower differs in merosity from the other flowers in the same inflorescence: in *Berberis vulgaris*, the terminal flower has five sepals and five petals whereas the lateral flowers have six (Weberling, 1989) (but are actually trimerous; Troll, 1964); in *Menyanthes trifoliata*, the terminal flower is hexamerous, the lateral flowers pentamerous (Weberling, 1989); in *Adoxa moschatellina* (Weberling, 1989)

FIGURE 8.3 The terminal flower of the carrot (*Daucus carota*) umbel is frequently conspicuous because of its deep purple colour.

and *Phellodendron* (Zhou *et al.*, 2002), the terminal flower is mostly tetramerous, the lateral flowers mostly pentamerous; in *Ruta* (Wei *et al.*, 2011), the terminal flower is mostly pentamerous, the lateral flowers mostly tetramerous.

In other plants, lateral and terminal flowers differ in phyllotaxis. In several genera of the Berberidaceae (*Berberis, Jeffersonia, Plagiorhegma*), the terminal flowers of the main inflorescence axis may have a spiral perianth, in contrast to the whorled floral phyllotaxis otherwise found in that family (Endress, 1989).

In determinate inflorescences, the symmetry of the terminal flower is often in agreement with the plant's phyllotaxis: for example the 'decussate' perianth in the terminal flower of *Clematis*; the same applies, based on dominant non-decussate arrangement of flower parts, to species of *Adonis, Nigella* or *Ranunculus* (Troll, 1964). In some species of *Dianthus* and *Hypericum*, in the production of the terminal flower there is a change from decussate to non-decussate phyllotaxis (Troll, 1964).

Species with a radial terminal flower and zygomorphic lateral flowers have also been reported, including *Impatiens balsamina* and *Nepeta racemosa* (= *N. mussinii*) (Tucker, 1999). This corresponds to the phenotype of the *centroradialis* mutant of *Antirrhinum* (Bradley *et al.*, 1996; Cremer *et al.*, 2001; Gillies *et al.*, 2002). Mutants with terminal peloric flowers are common in Lamiaceae and other Lamiales, such as *Linaria* and *Digitalis*, and occur also in Malpighiales (*Viola tricolor*), but none has been recorded in monocots (Rudall and Bateman, 2003).

8.2.3 Positional Effects: Anisophylly

Anisophylly (reviewed in Dengler, 1999) is a special case of dorsiventral shoot asymmetry in which leaves on the dorsal and ventral sides of the stem differ in size, shape, or both. In some extremely anisophyllous shoots, such as the lateral branches of *Anisophyllea disticha*, dorsal leaves are reduced to scales.

In *Aucuba japonica*, anisophylly only occurs along the reproductive shoots. In *Strobilanthes glomerata*, buds are present in both dorsal and ventral leaf axils but the preferred buds are associated with the dorsal leaves. In species of *Columnea* and other anisophyllous taxa of the Gesneriaceae, axillary buds associated with ventral leaves are reproductive, while those in the axils of smaller dorsal leaves are vegetative.

Goebel (1928) recognized two broad categories of anisophylly, based on their pattern of occurrence within the plant's body: habitual anisophylly, expressed throughout the adult shoot system, and lateral anisophylly, restricted to horizontally growing (plagiotropic) lateral branches.

In plants with habitual anisophylly having a creeping or climbing growth habit, as in *Procris repens* (= *Pellionia daveauana*), the transition to plagiotropy and anisophylly occurs simultaneously with the switch from juvenile to adult growth (Troll, 1937).

8.2.4   Challenging Position: Epiphylly

Epiphylly is the occurrence of leaves, branches, inflorescences or other structures upon a leaf or leaf homologue, in any position, although those sprouting out of the abaxial side of a leaf should be actually called hypophyllous; see below in this section. The topic was reviewed by Dickinson (1978), the main source for the following account.

Leaf-like epiphyllous structures are observed on the leaves of some *Begonia* species. Those of some varieties of *B. hispida* are in the form of stalked laminae, others are funnel-shaped, still others peltate. Remarkable leaf-like appendages are also found in *Eriospermum*: in *E. paradoxum*, the epiphyllous structure is larger than the leaf supporting it (Troll, 1939).

Dickinson (1978) divided epiphyllous vegetative shoot systems into two groups, based on whether they remain attached to the parent leaf as *epiphyllous branches*, or function as *epiphyllous propagules*. In the latter case, a period of dormancy is usually required before they start growing.

Epiphyllous branches intergrade with 'axillary' buds that in fact sprout from the leaf petiole, as in *Stephanopodium peruvianum* and *Tapura latifolia* (= *T. pedicellaris*). Many epiphyllous propagules, e.g. those of *Hammarbya paludosa* (= *Malaxis paludosa*), *Pinguicula vulgaris*, *Cardamine pratensis* and many species of *Kalanchoe*, *Drosera* and *Ornithogalum*, are discrete, dispersible units. Other epiphyllous propagules, e.g. those of *Scadoxus cinnabarinus* (= *Haemanthus cinnabarinus*), *Nymphaea micrantha* and *Tolmiea menziesii*, establish new plants only while still attached to the parent plant.

Epiphyllous inflorescences are a peculiar feature of a number of species belonging to distantly related families, most of which belong to the superrosids, although the most popular examples are found in a family of the superasterids (Gesneriaceae). Most epiphyllous inflorescences sprout out of the adaxial (upper) side of the leaf, often in characteristic positions: at the junction of petiole and lamina in some species of *Begonia*; along the petiole and the midrib of fertile leaves in *Phyllobotryon* and *Mocquerysia*; in association with the leaflet pairs of the up to 2 m long pinnate leaves of *Chisocheton*; at the end of the midrib, at the base of a sinus resulting from a cleft at the leaf's apex, in *Polycardia phyllanthoides*; on the midrib at the base of a lateral sinus in *P. lateralis*. Interestingly, these two species also produce sterile leaves without apical or lateral sinuses.

Inflorescences are very rare on the abaxial side of the leaves; this condition is found in *Erythrochiton hypophyllanthus* and is obtained secondarily in *Phyllobotryon spathulatum*, where the inflorescences become hypophyllous by growing through the lamina.

In the Gesneriaceae, epiphyllous inflorescences have evolved multiple times. Among the Cyrtandroideae, a number of species are either normally or occasionally unifoliate, with inflorescences formed at the base or along the midrib of the phyllomorph, the single enlarged cotyledon representing the whole photosynthetic organ of the plant (Jong 1973; Jong and Burtt, 1975).

8.3   SEX AND BODY SYNTAX

Whenever staminate or pistillate flowers occur in an inflorescence together with unisexual flowers of the opposite sex, or with hermaphrodite flowers, the spatial distribution of sexes in the inflorescence is generally fixed at the species level, but quite diverse in different taxa. The mechanism by which these spatial patterns are determined, their evolvability and their actual evolution are poorly known, with a few exceptions such as the capitula of the Asteraceae. The mechanisms underlying these patterns and the possible constraints on their evolvability deserve closer scrutiny.

In the inflorescences (cyathia) of *Euphorbia*, the terminal flower is female and all other flowers are male (e.g. Prenner and Rudall, 2007). In *Buxus*, the terminal flower is female and the lateral flowers are male; in contrast, in *Pachysandra*, the lowermost lateral flower is female and all other flowers are male (von Balthazar and Endress, 2002).

In *Callitriche*, female flowers are commonly found lower down the inflorescence than male flowers, except for *C. autumnalis*, where several male and female regions may alternate (Arber, 1920). In the tiny *Littorella uniflora*, two or more female flowers are found in the lower part of the inflorescence, followed by one male flower (Troll, 1964).

In *Gunnera*, the flowers are often unisexual. Typically, the basal flowers of the inflorescence are pistillate, the upper ones staminate and the middle flowers bisexual, although all flowers of an inflorescence may be either unisexual or bisexual (Soltis *et al.*, 2005). Similarly, in *Myriophyllum spicatum*, the upper flowers of the spike are generally staminate and the lower pistillate, while perfect flowers occur in the intermediate region (Arber, 1920).

In *Kirkia*, the cymes of the thyrsoid inflorescences exhibit rhythmic changes from one branch order to the next between functionally male and female flowers (Endress, 2010b).

Based on the distribution of sexes among the individual florets, the capitula of the Asteraceae are either homogamous or heterogamous. In homogamous capitula of monoecious composites, all florets are hermaphrodite; in those of dioecious species, all florets are either male or female. Most heterogamous species are gynomonoecious, with central hermaphrodite florets surrounded by one or more rows of female florets; other conditions are possible, of which the monoecious heterogamous is the most frequent, with functionally male flowers surrounded by functionally female ones.

8.4 FRACTAL PATTERNS AND PARAMORPHISM

Since the early 1980s, when Benoit Mandelbrot's fractal geometry became a fashionable subject in popular science books and magazines, examples of its possible application were frequently taken from the plant and animal world. The structure of a cauliflower, indeed, is amazingly similar to some patterns generated by simple algorithms that iterate (infinitely, in principle) a short series of generative rules producing a self-similar pattern repeated at every scale. Of course, any seemingly fractal pattern we may discover in living organisms is

FIGURE 8.4 The outline of the leaf lamina in plume poppies
(*Macleaya* spp.) is strongly reminiscent of the fractal object known as the
Koch snowflake.

necessarily limited to a few iterations only, spanning between the
finest details at the level of the spatial arrangement of a local cluster
of cells and the overall extension of a leaf (Fig. 8.4), an inflorescence,
or a suture line on the inner surface on an ammonite's shell.

Fractal patterns in animals and plants are obviously attractive
from a developmental point of view, because the analogy to fractal
geometrical objects generated by simple algorithms suggests that
living organisms are perhaps able to produce those patterns quite easily
in a mathematically elegant way. However, we cannot take for granted
that biological forms are always produced in the algorithmically

simplest way. Living nature is often less elegant than our models inspired by mathematics. A cautionary story comes from the development of the respiratory system in the larva of *Drosophila*. In this insect, there are 10 pairs of lateral spiracles, through which open as many primary tracheae. In the body of the larva, these branch in a stereotyped way, giving rise to secondary tracheae which branch in turn into tertiary tracheae, of increasingly smaller diameter. In principle, we might expect that an identical process, iterated at two different levels, is responsible for producing the whole arborescence, but in fact the branching of primary into secondary tracheae is under different genetic control than the branching of secondary into tertiary tracheae (Metzger and Krasnow, 1999).

In plants, the architectural diversity of grass inflorescence has been described as generated in a combinatorial and iterative manner which reveals the modularity of the underlying developmental processes (Kellogg, 2000; Malcomber *et al.*, 2006). At any step along the process, three options are available to the local meristem: to terminate in a spikelet, to continue producing lateral meristems, or to simply stop developing. In principle, the fate of meristems at the different orders of branching is independent of the fate of the previous order, manifesting a kind of 'temporal modularity' which largely increases the diversity of morphogenetic paths available to these inflorescences.

Some degree of independence between subsequent levels of branching opens the way to a potentially huge diversity of resulting structures.

As a rule, structural similarity at different levels of organization is likely to result from developmental processes in which a largely conserved and iteratively expressed pathway is differentially modulated by a small number of additional factors, responsible for the remaining differences. But largely conserved and iteratively expressed pathways may also be responsible for less tight but often intriguing similarities in the patterning of plant structures as distantly related, in terms of anatomy and function, as a leaf and the stem to which it is attached – expressions of correlations suggested long ago by

FIGURE 8.5 Comparison between the branch architecture of a *Tilia* tree and the branching vein pattern in its leaf – suggestive of the paramorphism described by M'Cosh (1851).

Uittien (1928) in a largely forgotten article. A principle of iteration involving both shoot and leaf was formulated over 70 years ago by Agnes Arber (e.g. Arber, 1941, 1950) and more recently discussed and elaborated by Rolf Rutishauser (e.g. Rutishauser and Isler, 2001). Virtually unknown, however, were it not for a short article in *The Linnean* newsletter (James, 2009), is the much earlier contribution of James M'Cosh (1851; see also M'Cosh and Dickie, 1856), from which the following excerpts are taken. To illustrate the concept, I provide here a visual example (Fig. 8.5).

> To us it appears possible to reduce a plant by a more enlarged conception of its nature to a unity. According to our idea, it consists essentially of a stem sending out other stems similar to itself at certain angles, and in such a regular manner, that the whole is made

to take a predetermined form. The ascending axis for instance sends out at particular normal angles in each tree, branches similar in structure to itself. These lateral branches again send out branchlets of a like nature with themselves, and at much the same angles. The whole tree with its branches thus comes to be of the same general form as every individual branch, and every branch with its branchlets comes to be a type of the whole plant in its skeleton and outline.

Taking this idea of a plant along with us, let us now inquire whether there may not be a morphological analogy between the stems and the ribs or veins of the leaf. As these veins are vascular bundles, proceeding from the fibro-vascular bundles of the stem, they may be found to obey the same laws.

Some trees ... send out side branches along the axis from the root, or near the very root, and the leaves of those trees have little or no petiole or leaf stalk, but begin to expand from nearly the very place where the leaf springs from the stem. There are other trees ... which have a considerably long unbranched trunk, and the leaves of these trees will be found to have a pretty long leaf stalk. ...

Some plants ... send off leaves which have a tendency to become whorled, and their branches have also a tendency to become verticillate.

In most plants the angle at which the side stems go off will be found to widen as we ascend to the middle, and thence to decrease as we ascend to the apex, and the venation of the leaves will be found to obey a similar law.

Generally we shall find a correspondence between the angle of the ramification of the tree, and the angle of venation of the leaf. ... All that we argue for is a general correspondence between the tendency of the direction of the branches, and the tendency of the direction of the veins of the leafage. ...

We are not prepared to say what is the special law of order in plants of the monocotyledonous class. Some of these, such as our ordinary lilies and grasses, send off no branches, and the leaves of these plants have their veins parallel or nearly parallel to the stem, and have no ramified venation.

I think it proper to add, that while strongly convinced that there is truth in this doctrine, I am at the same time prepared to believe that it may have to submit to modification, which may correct, but will not destroy, the general view.

Independent from this botanical tradition, years ago I proposed the notion of axis paramorphism, according to which animal appendages can be regarded as evolutionary duplicates of the main body axis, originated by a re-expression on a secondary (lateral) axis of the developmental routines responsible for the production of the main body axis (Minelli, 2000). In terms of gross morphology, this is suggested for example by the segmented structure of the appendages in arthropods (and to some extent also in vertebrates), animals in which the main body axis is similarly segmented, compared to the unsegmented appendages of other animals, e.g. molluscs, whose main body axis is correspondingly unsegmented. In plants, intriguingly similar correspondences are quite frequent. The examples provided in the following lines are complementary to those listed by M'Cosh (1851).

In plants with pinnate leaves there is a frequent agreement between the precision in the pairing of leaflets within the compound leaf and the kind of phyllotactic distribution of the compound leaves along the stem. In many species with decussate phyllotaxis, indeed, the pairing of the leaflets is very strict. Examples are found in a number of families, such as Zygophyllaceae (e.g. *Guaiacum, Tribulus*), Oleaceae (e.g. *Fraxinus*), Bignoniaceae (e.g. *Campsis*) and Adoxaceae (e.g. *Sambucus*; Fig. 8.6). This is not the same as the reiteration of the same pattern of branching within a bi- or tripinnate leaf, where the branching pattern of the first-order pinnae is repeated in the second (and eventually further) order, as noted e.g. in *Sambucus nigra* and *Phellodendron amurense*; in some instances, the module repeated at different scale in a fractal pattern includes the stipules, which are repeated (as stipels) at the base of the leaflets, as in *Phaseolus* (Rutishauser and Isler, 2001).

FIGURE 8.6 Paramorphism in elder (*Sambucus nigra*): decussate arrangement of the compound leaves and precise pairing of their leaflets.

In the examples on which I am focusing, the symmetry relationship is conserved *beyond* the limits of the leaf and expresses a commonality of pattern between the architecture of the shoot and the organization of the leaf.

Whatever its causal explanation(s) may eventually be, this circumstantial pattern agreement between phyllotaxis and arrangement of leaflets leads to the expectation that in plants with pinnate leaves and sparse phyllotaxis, there will be no precise pairing of leaflets. This is indeed what we observe, for example, in *Juglans* and in a large number of Fabaceae, e.g. *Robinia*.

To the best of my knowledge, these paramorphic patterns have never been studied from an ontogenetic point of view. As in the case of many other recurrent patterns discussed in this book, there is no need to expect that the underlying mechanisms will be always the same; however, it is sensible to expect that paramorphism is a kind of 'modulated fractal pattern', where the iteration of a small set of rules over different body axes (the longitudinal axis of shoot and leaf) is to some extent constrained by the different context but nevertheless results in recognizable repetitions of a basic regularity.

This said, it is fair (and important) to mention that this relationship between phyllotaxis and leaflet pairing is far from general. But it is also interesting to remark that the relationship is consistently manifested in some taxa, and consistently (or at least frequently) absent in others. In the Rosaceae, the phyllotaxis is nearly always sparse, but pinnate leaves with a good pairing of the leaflets are common, e.g. in the genus *Sorbus* (e.g. *S. aucuparia*), which includes species with pinnate leaves (subgg. *Cormus* and *Sorbus*) and species with undivided leaves (subgg. *Aria* and *Torminaria*). Among the many *Rosa* species and cultivars there is an amazing diversity in the degree of leaflet pairing, including some of the worst I have ever observed in a pinnate leaf (see the right panel in Fig. 8.1).

Besides the arrangement of leaflets in compound leaves, there are other aspects of plant architecture whose symmetry sometimes correlates with phyllotaxis. My first example is again from the Rosaceae. As just mentioned, in this family the phyllotaxis is usually sparse. Species with decussate leaves are limited to three genera, *Rhodotypos* (Fig. 8.7), *Coleogyne* and *Lyonothamnus*; the affinities of the last genus are uncertain, whereas the first two belong to the Kerrieae, the other two genera of which (*Kerria, Neviusia*) have alternate leaves (Byng, 2014). Interestingly, in *Rhodotypos scandens*, the only species in this genus, the flower is vaguely zygomorphic (Fig. 8.8) rather than actinomorphic, as otherwise in the family.

A correlation between floral zygomorphy and inflorescence architecture has been sometimes reported (Coen *et al.*, 1991; Gillies *et al.*, 2002); in particular, many species with strongly zygomorphic flowers have racemose inflorescences (Rudall and Bateman, 2003).

Another trait that may recur in a paramorphic fashion over different body axes is organ fusion (or lack of disjunction), resulting in some form of synorganization. Two genera in which this is observed are

FIGURE 1.1 A bisexual reproductive unit with a perianth (commonly differentiated into calyx and corolla) ensheathing androecium and gynoecium is characteristic of the angiosperms, but deviations from this idealized model are frequent. In the small burnet (*Sanguisorba minor*), bisexuality is more a property of the inflorescence than of the individual, all apetalous, flowers. Within the inflorescence, the basal flowers are staminate, the intermediate often hermaphrodite, the top ones carpellate. *A black and white version of this figure will appear in some formats.*

FIGURE 3.1 Brassicaceae of the genus *Cardamine* are increasingly studied as model plants, partly – although not exclusively – because of the diversity of their leaf shapes, which include pinnate and palmate. Their strict affinities to *Arabidopsis thaliana* (a plant with entire leaves) allows for interesting evolutionary comparisons. Here, large bitter-cress (*Cardamine amara*). *A black and white version of this figure will appear in some formats.*

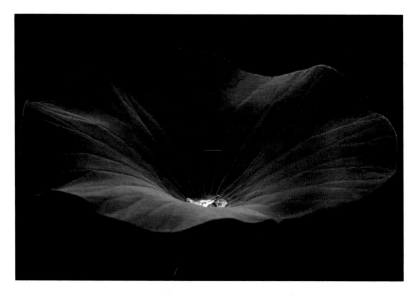

FIGURE 6.1 The sacred lotus (*Nelumbo nucifera*) is one of the most popular examples of plants with peltate leaves. *A black and white version of this figure will appear in some formats.*

FIGURE 7.1 Among the more or less numerous sterile organs in the flower of water-lilies (*Nymphaea* spp.), some are partly sepaloid (green), partly petaloid (coloured). *A black and white version of this figure will appear in some formats.*

FIGURE 7.2 The corolla of the smallflower buttercup (*Ranunculus parviflorus*) is remarkable for the instability of number and size of its elements. *A black and white version of this figure will appear in some formats.*

FIGURE 7.3 The bogbean (*Menyanthes trifoliata*) is one of the very few plant species in which the whole internal (adaxial) surface of the petals is covered with long appendages. *A black and white version of this figure will appear in some formats.*

FIGURE 7.4 The reproductive units of the Poaceae deviate strongly from conventional flowers, to the extent that the homology of their characteristic elements (lodicules, palea, lemma) are problematic. Here, an inflorescence of common wild oat (*Avena fatua*). *A black and white version of this figure will appear in some formats.*

FIGURE 7.8 A flower of cleavers (*Galium aparine*) with five sectors instead of the usual four. *A black and white version of this figure will appear in some formats.*

FIGURE 7.9 In some members of the Caryophyllaceae, as in this chickweed (*Stellaria media*), each petal is very deeply dissected into two lobes (laciniae), suggesting evolvability from 5-mery to 10-mery through a possible saltational transition. *A black and white version of this figure will appear in some formats.*

FIGURE 8.1 In the imparipinnate leaves of *Rosa* spp., the degree of pairing between leaflets ranges from very precise to non-existent. *A black and white version of this figure will appear in some formats.*

FIGURE 8.10 A leaf whorl of cleavers (*Galium aparine*). Of the six lateral organs in this whorl, only two possess axillary branches, as in typical leaves, whereas the remaining four, without axillary buds or branches, are, by definition, more similar to stipules. Otherwise, the morphology of all six lateral organs is uniform. *A black and white version of this figure will appear in some formats.*

FIGURE 9.3 Flower colour change in lungwort (*Pulmonaria officinalis*). *A black and white version of this figure will appear in some formats.*

FIGURE 9.5 In the common poppy (*Papaver rhoeas*), the abscission of the hairy sepals precedes the full expansion of petals. *A black and white version of this figure will appear in some formats.*

FIGURE 10.2 Flower of a large-cupped daffodil (*Narcissus*) with conspicuous corona. *A black and white version of this figure will appear in some formats.*

FIGURE 10.4 Flowers of a hybrid cultivar of columbine (*Aquilegia*) showing the five long spurs, an unmistakeable innovation of this genus of the Ranunculaceae. *A black and white version of this figure will appear in some formats.*

FIGURE 11.3 Dandelion (*Taraxacum* sp.), capitulum. *A black and white version of this figure will appear in some formats.*

FIGURE 8.7 Decussate leaves in *Rhodotypos scandens*.

FIGURE 8.8 Approximate zygomorphy in the flower of *Rhodotypos scandens*.

Lonicera and *Eucalyptus*. In each genus, this form of paramorphism is somehow 'distributed', in the sense that the characteristic fusion is not necessarily expressed in different axes in the same individual, or species, but is nevertheless recognizable as a common syndrome arguably dependent on shared forms of genetic/developmental control, in the different species of the genus. In *Lonicera*, the fruits of *L. alpigena* and *L. nigra* are fused at the base while the leaves are free, while in *L. caprifolium* the fruits are free, but the leaves of the same pair are united through a margin that becomes larger and larger moving up from the oldest to the most recently formed leaves, until the last ones form a disc pierced by the stem. In *Eucalyptus lehmannii*, numerous fruits are fused by the walls of the hypanthium to form a syncarpic structure; in other species of the same genus, other organs are fused, e.g. stamen filaments in *E. synandra* and opposite juvenile leaves in *E. uncinata* (Rudall, 2010).

No targeted study has so far addressed the issue of how these paramorphic patterns are controlled. However, similar effects on multiple axes have been reported in a number of studies of plant developmental genetics.

Floyd and Bowman (2010) suggested that class III *HD-ZIP* and *KANADI* genes previously playing a role in the radial patterning of stems (Emery *et al.*, 2003) may have been co-opted to control, additionally, leaf polarity. The argument was based on the fact that land plants evolved vasculature before the emergence of the seed plants and the origin of their megaphylls. In *Arabidopsis*, some mutants show the same defect in the stem and in the leaf petiole; in the *acaulis 2* mutant, the inflorescence axis, the flower stalks and the leaf petioles are all shortened, whereas leaf blades are of normal size (Tsukaya *et al.*, 1995). Again in *Arabidopsis*, *ORGAN BOUNDARY1* (*OBO1*) is expressed at the boundaries between the apical (shoot and root) meristems and lateral organs at all stages, from late embryo to seedling to mature plant; its overexpression may cause petal–stamen fusion and irregularities in the number and size of petals (Cho and Zambryski, 2011).

8.5   CHALLENGING THE IDENTITY OF BODY PARTS

8.5.1   Continuum Morphology and Process Morphology

The notion of paramorphism applies to comparisons between body axes of different order or between an axis (stem, branch) and its lateral organs. However, the distance is not necessarily great between the iteration of a pattern of growth and differentiation along the different axes of the plant and its systemic expression, such that even the conventional identity of the plant's parts may be challenged. We can speculate that a shallow distinction between e.g. stem and leaf may have characterized a segment of the early evolution of land plants. In this scenario, a simple organization mirroring the iteration or the extensive pervasiveness of morphogenetic processes and their systemic effects would have progressively evolved towards higher degrees of complexity under the effect of spatially restricted expression patterns of regulatory factors. Evidence from palaeontology and phylogenetics on the one hand, from developmental genetics on the other, converges towards this reconstruction (e.g. Langdale and Harrison, 2008; Harrison, 2017). Most of this history, however, is considerably older than the emergence of the flowering plants, which are the subject of this book. But organs of uncertain identity are not rare among the extant plants: in the most extreme instances, virtually all categories of plant organography are inapplicable. This asks for developmental explanations, while opening at the same time new vistas into the evolvability and evolution of plant organization.

Unease with the assignment of a number of plant organs to one of the traditional categories of plant morphology has been repeatedly expressed: among modern authors, for example, by Tsukaya (1995), Sinha (1999), Cronk (2001), Hofer *et al.* (2001b), Bharathan *et al.* (2002), Fukuda *et al.* (2003) and Rutishauser and Moline (2005).

In a number of papers, occasionally in collaboration with other researchers (e.g. Rutishauser and Sattler, 1986; Sattler and Jeune, 1992; Lacroix and Sattler, 1994; Rutishauser, 1995, 1999, 2016a; Rutishauser and Isler, 2001; Rutishauser *et al.*, 2008), Rolf Sattler and Rolf Rutishauser have developed an approach to plant comparative morphology alternative to what they define as Classical Morphology. A distinct anticipation of their perspective is found in several works by Agnes Arber (especially Arber, 1941, 1950); for this reason, Rutishauser and Isler (2001) refer to it as the Fuzzy Arberian Morphology of vascular plants. Following their own characterization of these two concepts of plant morphology, in Classical Plant Morphology the traditional categories of vascular plant organization, such as root, shoot, leaf and stipule, are regarded as concepts with clear-cut borders and, in principle, without intermediates; in Fuzzy Arberian Morphology, these structural categories are treated instead as concepts with fuzzy borders, allowing for the occurrence of virtually any kind of intermediate. In other words, we can therefore characterize the two views of plant morphology as following a Discrete Model versus a Continuum Model. Rutishauser and Isler (2001) regarded the two approaches as complementary rather than strictly alternative. Only the Continuum Model, however, accepts that there can be developmental mosaics of plant organs and mixed homologies between root, shoot, leaf and their parts (Rutishauser, 1995, 1999; Sattler, 1996; Baum and Donoghue, 2002; Hawkins, 2002; Rutishauser et al., 2008) (Fig. 8.9).

From a phylogenetic perspective, the Continuum Model is consistent with the hypothesis that both leaves (e.g. Cronk, 2001; Friedman et al., 2004) and roots (e.g. Raven and Edwards, 2001; Schneider et al., 2002) are derived from shoot-like organs, at least in ferns and seed plants.

Eventually, rather than insisting on a static approach to plant morphology, in which clear-cut archetypes are replaced by fuzzy descriptive categories, Sattler and Rutishauser have opted for *process morphology* (Sattler and Rutishauser, 1990, 1997; Sattler, 1992, 1994; Lacroix et al., 2003; see also Jeune and Sattler, 1992, 1996; Weston, 2000; Classen-Bockhoff, 2001; Kirchoff, 2001). The traditional analysis in terms of structural categories is thus replaced (or at least complemented) by a description in terms of developmental routines and their combinations.

FIGURE 8.9 A compound leaf of mahonia (*Berberis* × *hortensis* '*Charity*') and a fragment of an old, decaying leaf which shows that the rachis disarticulates in segments between two subsequent points of insertion of leaflets.

Four examples of plants with organs that cannot be classified according to the criteria of traditional morphology are given in the following sections; a most extreme case, the duckweeds, where the whole non-reproductive part of the body is represented by a nondescript mass of cells, except for the roots still present in most species of this taxon, will be discussed in Section 10.1.2 as an example of miniaturization.

8.5.2 Leaves with Indeterminate Growth

In some plants, organs that morphologically do not differ from ordinary leaves have persistent apical meristems. Those of *Chisocheton* and *Guarea* may show seasonal bursts of apical growth repeated for several years. The apical meristem of the very large compound leaves of these trees gives off cells and leaflets in the same way that any shoot apical meristem does while producing leaves. Moreover, each time the shoot apical meristem adds new leaves, the apical meristem of older leaves also initiates new leaflets.

At the cellular level, this leaf apical meristem is similar to the shoot apical meristem, except for its dorsoventral, rather than approximately circular structure; moreover, no lateral buds are given off. A further similarity between these unusual leaves and ordinary shoots is the production, by some species of *Chisocheton*, of epiphyllous shoots (*Ch. tenuis*) or epiphyllous inflorescences (*Ch. pohlianus*) (Stevens, 1975; Steingraeber and Fisher, 1986; Fisher and Rutishauser, 1990; Fisher, 2002); see Section 8.2.4.

A meristematic zone has been described at the base of the phylloclade (cladode) of *Asparagus asparagoides*, a leaf-like homologue of the lateral shoot, ectopically expressing some leaf genes involved either in the identification of the adaxial domain of leaves (members of the *HD-ZipIII* gene family) or the abaxial domain (*miR166*), or in the establishment of various leaf axes (*AS1*) (Nakayama *et al.*, 2012).

A leaf apical meristem has been described also in the Podostemaceae, but this characterization is justified only if we are willing to identify as leaves the unusual green structures where these meristems reside (see Section 8.5.6).

8.5.3   Stipules versus Leaf

In standard plant morphology, stipules are accessory and usually minor laminae accompanying a leaf, usually present in pairs, more or less identical and regularly disposed at the sides of the leaf or its petiole. No axillary bud is subtended by a stipule. In respect to the leaf with which they are associated, stipules behave as quite independent modules: a stipule's growth often stops before the growth of the leaf stops, and stipules often abscise independently of the leaf (Rutishauser, 1999). But in a number of eudicots, stipules are quite similar to the associated leaf, and in a few taxa the distinction between leaf and stipule becomes difficult and perhaps arbitrary. Extreme cases are provided by a few *Acacia* species with lateral organs arranged in whorls, and a number of species of the Rubiaceae (Rutishauser, 1984, 1999; Rutishauser and

Table 8.2 *Number of leaves per whorl in some species of* Galium *and related genera, in agreement with Wörz's (1996) descriptions. Recorded variation is given.* **Six** *is the number of organs corresponding to two opposite 'true' leaves accompanied by two stipules each. See also Fig. 8.10.*

	Number of leaves per whorl									
	Lower whorls					Upper whorls				
Galium palustre	4					4				
Asperula cynanchica	4					4 5 **6**				
Sherardia arvensis	4					5 **6**				
Asperula arvensis	4					**6** 7 8				
Galium uliginosum	5 **6** 7					5 **6** 7				
Galium lucidum	5 **6** 7 8					5 **6** 7 8				
Galium saxatile	**6**					**6**				
Galium mollugo	**6** 7 8					**6** 7 8				
Galium odoratum	**6** 7 8 9 10					**6** 7 8 9 10				
Galium glaucum	8 9 10					8 9 10				
Galium verum	8 9 10 11 12					8 9 10 11 12				

Sattler, 1986), in particular in *Rubia, Galium* and related genera. In the latter plants, there are 4–12 lateral appendages per whorl (Rutishauser, 1984). The first whorl of many Rubieae seedlings consists of four leaves. In several species in this tribe the number of leaves increases along the stem, as shown in Table 8.2. Usually, axillary buds are associated with only two leaves per whorl (Fig. 8.10) (occasionally three, in a few species, e.g. *Phuopsis stylosa* and *Rubia fruticosa*; Rutishauser, 1984). All additional whorl members lack an axillary bud and have therefore often been described as stipules. However, this interpretation would easily work for whorls with six leaf-like organs (two decussate 'true' leaves, each with two associated stipules), less easily with other numbers, especially odd ones. Cronquist (1968) suggested that in the Rubieae the stipules have undergone a homeotic transition to leaves, but this does not solve the numerical mismatch.

FIGURE 8.10 A leaf whorl of cleavers (*Galium aparine*). Of the six lateral organs in this whorl, only two possess axillary branches, as in typical leaves, whereas the remaining four, without axillary buds or branches, are, by definition, more similar to stipules. Otherwise, the morphology of all six lateral organs is uniform. *A black and white version of this figure will appear in some formats. For the colour version, please refer to the plate section.*

8.5.4 The Acaulescent Gesneriaceae

In those members of the Gesneriaceae that are popularly known as the one-leaf plants, the entire vegetative structure consists of a single enormous cotyledon quite similar to a conventional leaf and a stem + petiole-like organ called a petiolode (Jong, 1973; Jong and Burtt, 1975). There is no conventional shoot apical meristem and no new leaf primordium is produced next to the cotyledons, only one of which starts to grow after germination (Tsukaya, 1997). Its growth is indeterminate. No mutants with morphologies like those of one-leaf plants are known in *Arabidopsis thaliana* or other conventional model species.

One-leaf plants evolved independently in *Monophyllaea* and *Streptocarpus* (Smith *et al.*, 1997; Mayer *et al.*, 2003; Möller *et al.*,

2009; Weber *et al.*, 2013). All the species (about 30) in the genus *Monophyllaea* are unifoliate; three different growth forms have evolved in *Streptocarpus*: caulescent species with a conventional shoot apical meristem, unifoliate species with the typical single phyllomorph, and rosulate species. In the last of these, a new phyllomorph develops on an older phyllomorph, and the entire body consists of multiple phyllomorphs.

Growth and differentiation of these plants express the activity of three meristems: the basal meristem of the larger cotyledon; the groove meristem, which is responsible for the development of a new phyllomorph and the inflorescence; and the petiolode meristem, which contributes to the growth of the midrib and petiolode (Jong and Burtt, 1975). Peculiar patterns of expression of *STM*, *PHAN/RS2/AS1* and *WUS* orthologues have been observed in these meristems (Gallois *et al.*, 2002; Harrison *et al.*, 2005; Mantegazza *et al.*, 2007, 2009).

8.5.5 Utricularia *(Bladderworts)*

Utricularia is quite a large genus, with ca. 230 species (Rutishauser, 2016a), most of which live fully immersed in water; but some are epiphytes on trees (e.g. *U. alpina*) or live on moist soil (e.g. *U. livida, U. sandersonii, U. longifolia*) (Taylor, 1989). Irrespective of their ecological preferences, all *Utricularia* species are provided with tiny bladders used as traps for small animal prey.

Most bladderworts produce two kinds of root-like or runner-like organs, respectively called rhizoids and stolons, both of which are usually interpreted as stem homologues, plus a range of leaf-like organs. Rather than 'true' leaf homologues, some authors have described the latter as fuzzy organs in which the 'developmental programs' of leaves and shoots are jointly expressed (Rutishauser and Isler, 2001; Rutishauser, 2016a).

In some species, e.g. *U. alpina* and *U. purpurea*, the branching of the stolon is extra-axillary, that is, the lateral branch has no sub-tending leaf; in other species, e.g. *U. dichotoma, U. longifolia* and

U. sandersonii, the subtending leaf points towards the distal end of the mother stolon while the axillary bud points in the opposite direction (Reut and Fineran, 2000).

Mechanisms specifying the identity of individual organs are unusually plastic: of the various primordia in a rosette, some will become leaves, others stolons, but their fate can be changed even at a very late stage of development. In some species, a stolon can eventually stop growing, completing its activity with the production of a terminal leaf. On the other hand, the tip of a nearly mature leaf can continue growing and turn into a new stolon.

Meristem identity, indeed, is not always irreversible. For example, conversion of root meristems to shoot meristems is known from other angiosperms and suggests some degree of homology between root and shoot. Examples have been reported, e.g. from *Nasturtium* and *Neottia* (Guédès, 1979).

8.5.6   Podostemaceae

The flowers of the Podostemaceae, aquatic plants of the tropics, do not differ much from those of the most closely related families, e.g. the Hypericaceae, but the vegetative structure of these plants deviates so strongly from the typical organization of flowering plants, in some genera at least, that it is often impossible to describe it in terms of roots, stems and leaves.

Of the basic architectural rules typical of the non-reproductive structures of angiosperms, none seems to be conserved in the Podostemaceae. This is also true of the Araceae Lemnoideae (see Section 10.1.2), but the unique organization of the latter is a strict correlate of miniaturization; the same might perhaps apply to some small Asian Podostemoideae, with shoots of about 1 cm or even less, but not to many other, much larger members of the family (Cook and Rutishauser, 2007). Among the oddities of the Podostemaceae (Cook and Rutishauser, 2007; Fujinami *et al.*, 2013; Rutishauser, 2016a) there are the exogenous (rather than endogenous) origin of daughter roots from mother roots, the lack of axillary branching, the endogenous (rather than exogenous) origin of lateral shoots (the flower buds especially), as well as the occurrence of epiphyllous flowers in some species, the strong flattening of the roots and the frequent loss of the root cap. In *Ledermanniella bowlingii*, epiphyllous shoots arise from the leaves. This whole description, however, presupposes that we can describe these plants using conventional morphological terms like root, shoot and leaf. In fact, it can be disputed if podostemads have stems and leaves directly comparable to the parts to which we apply these terms in the other angiosperms.

The Tristichoideae, the most basal clade of the Podostemaceae, are characteristic for the presence of a unique organ, the photosynthetic appendage known as the ramulus, which develops endogenously from the root tissue. The ramulus is provided with an apical meristem, forms scaly leaves and undergoes repeated sympodial branching from extra-axillary buds (Fujinami *et al.*, 2013). The ramulus has been alternatively interpreted as a short shoot, as a compound leaf or as a developmental mosaic combining both shoot and leaf features (Rutishauser and Huber, 1991; Rutishauser, 1997; Rutishauser and Isler, 2001).

In *Dalzellia zeylanica*, also of the Tristichoideae, there is no root and the flat shoot provided with scaly leaves on its margin and its upper surface adheres to the substrate like a lichen (Imaichi *et al.*, 2004; Fujinami and Imaichi, 2015). This structure is possibly the result of a congenital fusion between shoot axes of various orders (Imaichi *et al.*, 2004; Fujinami *et al.*, 2013; Fujinami and Imaichi, 2015; Rutishauser, 2016a).

Shoot apical meristems are cryptic and sometimes lacking in the Podostemoideae (Rutishauser *et al.*, 2003; Imaichi *et al.*, 2005; Koi *et al.*, 2005; Jäger-Zürn, 2007). The unusual horizontal body plan evolved in this family has been attributed to the loss of embryonic shoot and root meristems (Katayama *et al.*, 2011). Katayama *et al.* (2010) analysed in the root-borne shoots of selected species of podostemads the expression of genes known to be regulators of shoot meristems (*STM* and *WUS*) or to promote leaf identity (*ARP*) in model angiosperms. In *Terniopsis minor*, a representative of the basal tristichoid podostemads, an ordinary shoot meristem is still present and *STM* and *WUS* orthologues are expressed there; in *Hydrobryum japonicum*, which belongs to the more derived Podostemoideae, *STM* and *WUS* orthologues are initially expressed in the leaf primordium, but this expression disappears when the *ARP* orthologue is expressed in the distal part of the leaf (Katayama *et al.*, 2013).

9 Pheno-Evo-Devo

9.1 THE TEMPORAL DIMENSION OF THE
PHENOTYPE

Differences among plants are not limited to the morphological diversity of their parts and the spatial relationships among these, but extend to the temporal patterns and relationships of their differentiation, growth, maturation and eventual senescence and abscission, if any. If, as suggested in Chapter 1, we adopt a process view of development, rather than exclusively focusing on the morphological aspects of the resulting phenotypes, we disclose a whole, poorly explored chapter of developmental biology, the ontogeny of temporal phenotypes.

Evolvability, modularity, robustness and eventual evolution of temporal phenotypes are the subject of pheno-evo-devo, the virtually new chapter of evolutionary developmental biology which is the subject of the following pages.

Pheno-evo-devo spans from spatially and temporally restricted phenomena such as the abscission of individual parts of the flower to global aspects of a plant's life such as its annual, biennial or perennial habit (and the related choice between monocarpy and polycarpy; see Section 9.2.2), the periodization of a plant's life into more or less precisely defined phases, and the mechanisms ensuring e.g. the switch from a vegetative to a reproductive state. As a rule, these events are controlled by a combination of environmental cues and endogenous factors, integrated at the level of expression of a number of genes – no different from the interaction of genes and environment in controlling the shape of a leaf or the symmetry of a flower.

To some extent, morphological and temporal aspects of the phenotype cannot be considered separately, but I do not see any solid reason why the temporal aspects should not be addressed on their own, using descriptive and interpretive tools comparable to the weaponry available to morphology and morphogenetics.

In the following sections I will mostly focus on complex and obviously critical temporal phenotypes, i.e. those that characterize the whole life cycle of an individual plant, such as flowering and senescence, but a few examples will show how many other aspects have a place within the potential scope of pheno-evo-devo.

How long does it take for a leaf to grow from its first inception to maturity? The question is perhaps marginal in many instances and of very minor evolutionary relevance, but how can it pass unnoticed that a leaf of the fern *Osmunda cinnamomea* may take five years to complete development (Steeves and Sussex, 1989)? Or shall we ignore, from an evo-devo perspective, that the whole morphology of a shoot may depend on the joint effect of shorter or longer plastochrons combined with the time it takes for each leaf to grow to its full size? A number of plants have evolved peculiar schedules of leaf inception, growth and differentiation whose cellular and genetic background has not so far been investigated. An example is 'preformation' (Diggle, 1997), the initiation of organs a year or more prior to maturation and function, as observed in plants of arctic or alpine environments. In *Polygonum viviparum*, four years elapse between the inception of a leaf or inflorescence and its full maturity. As a consequence, up to five cohorts of primordia are borne simultaneously on an individual plant, each cohort initiated in a different year.

Plant physiologists have revealed the molecular basis of the mechanisms by which plants measure time, with reference to the alternation of day and night and to the change in the relative duration of the intervals of light and darkness in the yearly cycle of seasons. Plants, however, are not totally dependent on astronomical cues for regulating their clocks. An internal clock based on periodic pulses of gene expression has been discovered in a region of the *Arabidopsis* root tip termed the oscillation zone. This allows a regular positioning of lateral roots. This mechanism has the capacity to compensate for variation in temperature and other environmental conditions (Moreno-Risueno *et al.*, 2010, 2012).

9.2 THE LIFE SPAN OF PLANTS AND THEIR PARTS

9.2.1 *Age of Cells versus Age of Plants*

Longevity, or organismal life span, is largely determined by the period over which a plant's cells can function metabolically (Tomlinson and Huggett, 2012).

Diploid and autotetraploid individuals of the same species have different growth rhythms, and this may translate into different longevity, with a shift from annual to perennial habit. Müntzing (1936) described a perennial autotetraploid strain of *Zea mays*, contrasting with the annual habit of ordinary diploid maize.

In very long-lived monocots, such as the palms, growth is not dependent on a process like the secondary growth produced by the activity of bifacial cambium, as in non-monocots. By means of sustained primary growth, monocots retain living cells in their trunks throughout the individual plant's life. No conventional woody plant retains in its trunk differentiated and metabolically active cells for a similarly long time span, not even in the case of the oldest trees. The skeletal, woody cylinder of a *Pinus longaeva* may be up to 3000 years old, but the oldest living tissues in the tree are much younger: differentiated tissues can remain functional for about 45 years at most (Ewers and Schmid, 1981). Reliable age records for arborescent monocotyledons without secondary growth are few, but values of up to 720 years have been estimated for the palm *Livistona eastonii* (Hnatiuk, 1977) and up to 350 years for *Lodoicea sechellarum* (Savage and Ashton, 1983); among monocots other than palms, *Kingia australis* has been estimated to live up to 750 years (Lamont, 1980).

All meristematic cells, including those of the primary meristems (shoot and root tip) and those of the secondary meristems (vascular cambium), do not last long before dividing. The individual cells in the apical meristem of a 100-year-old pine have life spans of a little more than a year, even if the meristem, as continuous cell lineage, may have been there since the plant was just a seedling (Tomlinson and Huggett, 2012).

9.2.2 Senescence

Owing to the modular organization of plants, we must distinguish between the senescence and eventual abscission (see Section 9.9) or decay of individual modules such as leaves, floral parts or whole flowers, and the senescence and eventual death of the whole plant.

Leaf senescence is an active process that involves the increased expression of many hundreds of genes (Buchanan-Wollaston, 2007; Breeze *et al.*, 2008). This 'genetically controlled dismantling programme' (Jibran *et al.*, 2013, p. 547) is, as such, potentially prone to evolutionary change.

Environmental cues such as drought, waterlogging, high or low solar radiation, changing photoperiod, extreme temperatures, nutrient deficiencies and biotic stress such as pathogen attack are reported to induce leaf senescence in monocarpic plants, but also in iteroparous perennials. No mutation or treatment, including change in environmental condition, is known to allow the plant to escape leaf senescence completely (Jibran *et al.*, 2013).

In perennial plants, the senescence of individual leaves is tightly controlled, while the senescence of the whole plant, which can grow and reproduce for many years, does not seem to be regulated by development.

The regulation of plant senescence was studied in *Arabidopsis thaliana*. Comparisons of 16 accessions revealed natural variation in the relationship between leaf and plant senescence, and a significant correlation between plant senescence on the one side and the total number of leaves and siliques produced and the plant's bolting age on the other (Balazadeh *et al.*, 2008).

In evolutionary biology, senescence can be defined as a reduction in age-specific survival and fecundity that reflects the decline in the level of performance of different physiological functions in individuals of increasingly advanced age (Charlesworth, 1980). However, some long-lived plants do not exhibit any obvious sign of senescence. This is the case for *Dioscorea pyrenaica* (= *Borderea pyrenaica*), known to reach ages of more than 300 years, the longest life span recorded for a herbaceous plant (García and Antor, 1995a). A distinct scar is left on the tuber each year when the annual stem dies back in September (García and Antor, 1995a). This makes age determination of this dioecious plant very easy. Males start reproduction at 10–20 years of age, females at 15–35 (García and Antor, 1995b). García *et al.* (2011) found no evidence for senescence, not even in individuals aged 260 years. Growth and fecundity of both female and male individuals did not decrease at high ages; rather, survival and reproductive value increased with age.

Monocarpy, or semelparity, is characterized by a single massive reproductive episode, which is followed by the death of the organism. This life history is shared by annual and biennial plants, and also by some long-lived perennial plants, most famously bamboos (Seifriz, 1950; Janzen, 1976) and agaves. During vegetative development, senescence affects mostly the leaves and progresses, with growth, along the axis, or axes, of the plant. This may last weeks or months (annuals), one year (biennial) or many years (bamboo-type perennials). Things change with the transition to the reproductive phase: senescence now affects all organs of the plant and leads to its death. Most perennials, however, are polycarpic or iteroparous, that is, they reproduce at least twice during their lifetime. Characteristic of perennial polycarpic plants is their ability to dissociate the fate of their meristems: some of these become determinate and exhaust their growing potential with the production of inflorescences and flowers, but some indeterminate vegetative meristems are instead maintained. These meristems are apparently insensitive to growth-promoting signals until the reproductive season is over (Rohde and Bhalerao, 2007).

In these long-lived plants, ageing is thus not, or not obviously, dependent on the reproductive effort. In principle, we might expect that a plant's longevity is eventually limited by the steady accumulation of deleterious mutations in its cells.

It has been estimated that in the complete life cycle (say, zygote to zygote) of an annual plant such as maize, there are approximately 50 cell divisions. In perennials reaching an age of centuries, the cells of apical meristems are likely to go through a much higher number of divisions before their latest offspring are subject to floral induction. It is thus difficult to imagine that these cells could escape mutational changes, even if benefiting from DNA repair mechanisms or other internal solutions to buffer their genomes against the consequences of mutation (Walbot, 1996). It seems, however, that ageing is not directly related to the accumulation of somatic mutations in meristems (Thomas, 2003): no evidence of a steadily increasing number of mutations was found in bristlecone pines aged between 23 and 4713 years (Lanner and Connor, 2001).

Even in advanced age, apical and axillary meristems of perennial plants retain the capacity to produce new shoots; because of this, some physiologists suggest that the notion of senescence, if defined as a predictable, regulated process unavoidably leading to death, might not apply to the meristems (Munné-Bosch, 2008).

9.2.3   Annual versus Perennial Plants

According to the expected life span of the individual plants, three main strategies can be distinguished: annual, biennial and perennial. To a large extent, the different life expectancy is dependent on the length of the pre-reproductive phase and the possibility to escape ageing and death at the end of a reproductive season. Therefore, the three temporal strategies can be interpreted as combined effects of the control of the transition to flowering, on the one hand, and the control of senescence, on the other. As a consequence, transitions from one habit to the other will depend on changes in one of these mechanisms – flowering and senescence – or both. The whole picture, however, is a bit more complex.

As said above, plants that use the annual or the biennial reproductive strategies are monocarpic: in either case, meristems are allocated to flowering and the plant dies after seed production. Changes in the expression and activity of genes that control flowering time are involved in the switch between annual and biennial.

Most perennials are polycarpic, flowering over many years, with reproductive seasons interspersed by periods of vegetative growth. In these plants, timing of flowering can be regulated by very accurate controls, as in the synchronously flowering dipterocarps. These trees flower at irregular intervals separated by many years of vegetative growth (Ashton *et al.*, 1988). Mast years are typically the same for all species within a genus, but different species flower at different times of the year (Rohde and Bhalerao, 2007).

Many annuals are derived from perennial ancestors, including instances of multiple evolution of annuals within a genus (e.g. *Sidalcea*: Andreasen and Baldwin, 2001; *Nemesia*: Datson *et al.*, 2008). In *Euphorbia*, the annual life form has evolved several times from perennial ancestors, in different clades of subgenus *Esula* (Frajman and Schönswetter, 2011). However, in some genera, a reversal to the perennial strategy has also occurred (Tank and Olmstead, 2008). Some plant species, e.g. *Rorippa palustris*, include annual, biennial and perennial individuals.

Flowering time genes that are known to be involved in interspecific shifts between annual and perennial strategy are regulated by proteins in the SPL family of transcription factors. Functional studies in *Arabidopsis thaliana*, *Oryza sativa* and *Zea mays*, but also in *Aphanorrhegma patens*, suggest that the most common function of the *SPL* genes is in the regulation of vegetative phase change (developmental transition from juvenile to adult growth; Wu and Poethig, 2006; Schwarz *et al.*, 2008; Usami *et al.*, 2009) and reproductive phase change (vegetative to reproductive growth; Wang *et al.*, 2009a; Yamaguchi *et al.*, 2009; Preston and Hileman, 2010).

In the biennial strategy, flowering is delayed by a requirement for vernalization, the process by which prolonged exposure to cold renders plants competent to flower. Unlike photoperiod, which is perceived in leaves, cells of the shoot apical meristem directly sense cold and become vernalized. In many species, the vernalized state is mitotically stable, i.e. once shoot apical meristem cells have become vernalized they retain a memory of a prior cold exposure and the competence to flower. However, in the Brassicaceae and other plants, the vernalized state is not stable, and thus a transition to flowering promoted by the cold environmental conditions must take place during exposure to the latter (reviewed in Amasino, 2010).

9.3 PERIODIZATION

9.3.1 Phases of a Plant's Life

Plants and animals are multicellular organisms derived from different groups of unicellulars. Therefore, there is no a priori reason to expect that, in the two groups, multicellularity will be associated with strictly equivalent features of organization and broadly similar sequences of phases along the individual's development. Strictly speaking, universality of these features is not even granted within the metazoans: for example, the majority of sponges lack true tissues, and also in many sponges it is a matter of personal choice whether to describe some early, free-floating stages as embryos or as larvae. When it comes to plants, it would be sensible to use as sparingly or at least as critically as possible terms taken from animal developmental biology, but actual suggestions for a terminological revolution are beyond the scope of this book.

A useful way to approach the study of morphological and physiological changes along the life of an individual plant is to identify events that mark a more or less neat transition between two phases. A major divide is obviously the transition between the pre-reproductive vegetative phase and the following reproductive phase. However, in many plants the vegetative phase can be divided, in turn, into distinct temporal segments (usually two, juvenile and mature); the reproductive phase may be followed by a phase of senescence, eventually culminating in the plant's death, or by a new phase of vegetative growth, the first manifestation of a more or less long sequence of alternating reproductive and vegetative phases.

However, this is a very simplified and somewhat abstract account. In practice, we must acknowledge that stages exist in the mind of the biologist rather than in nature. Nevertheless, we can follow three practical criteria in determining where a recognizable phase of plant development begins, and where it ends. Phases are relatively stable, in that they cannot be easily reversed (Zimmerman *et al.*, 1985); they are generally associated with suites of morphological traits and physiological and biochemical properties that change in a coordinated fashion (Kerstetter and Poethig, 1998); and they are relatively discrete, that is, the traits that characterize one phase are generally replaced by those typical of the next phase, rather than added to them (Kerstetter and Poethig, 1998).

In practice, the evidence of phase transition is sometimes abrupt, sometimes smooth, especially in the transition from juvenile to adult vegetative phase. For example, changes in leaf shape may only affect the number of leaflets (e.g. in *Thalictrum flavum*) or the ratio of length to breadth (e.g. in *Polygonum cuspidatum*), but in other plants (e.g. *Campanula rotundifolia*) the lower leaves differ in shape completely from the upper leaves of the same plant. In the New Zealand tree species *Pseudopanax crassifolius*, as many as five different shapes appear in the sequence of seedling leaves before the juvenile leaves are formed (Gould, 1993). Often, especially in long-lived plants, the leaves produced at maturity are quite different from those of the juvenile phase, but the change goes through a virtually continuous series of intermediates. An example is *Hedera helix*, whose leaves change from the juvenile five-lobed form to the adult ovate shape (Fig. 9.1).

Age- or phase-correlated differences in leaf shape (heterophylly) are often more conspicuous when phase transition is accompanied by an abrupt change in external conditions, e.g. from submerged to aerial, as in the case of several aquatic plants. Here, the submerged leaves are often highly dissected and exhibit repressed laminar outgrowth compared with aerial leaves (Arber, 1920).

In *Ludwigia arcuata*, the submerged leaves are much narrower than the terrestrial leaves, similar to many other aquatic angiosperms,

FIGURE 9.1 Ivy (*Hedera helix*), a popular example of a heterophyllous plant. Left, juvenile leaves; right, leaves transitional to the shape of adult shoots.

e.g. *Hippuris vulgaris* and several *Ranunculus* species, but in the former species these diverging shapes are obtained in a different manner. Whereas in other species the difference in the leaf lamina can be largely traced to differences in the shape of the epidermal cells, the change of leaf shape in *L. arcuata* is due to changes in the numbers of epidermal cells aligned in transverse sections. The susceptibility of leaves to changes in shape above and below the water surface varies with the developmental stages: only the first three leaves respond to a change in environment, while older leaves do not (Kuwabara *et al.*, 2001).

In the aquatic plants of the genus *Callitriche*, section *Callitriche* s.s., there is a strong difference between the submerged leaves, with narrow laminae and separated by long plastochrons, and the leaves, mostly with broad ovate laminae, crowded in floating apical rosettes (Fig. 9.2). In this case, the widespread change from short to long plastochrons characterizing the transition from a basal rosette to

FIGURE 9.2 In this water-starwort (*Callitriche cophocarpa*), as in most species of the genus, the leaves forming the terminal, floating rosettes (inset) are very different from the submerged ones.

a sequence of caulinar leaves, as found e.g. in *Arabidopsis* or in many *Cardamine* species (Fig. 3.1), is reversed, in a sense, and it would be interesting to know how this unusual sequence of plastochrons has evolved here. In several *Callitriche* species, flowers are produced in both the axils of submerged leaves and in the apical rosette, but in *C. hamulata* flowers are restricted to the submerged segments; therefore, in this case the apical rosette documents, along the growing axis of the plant, a reversion to a vegetative phase following a reproductive phase.

In the amphibious *Rorippa aquatica* (= *Nasturtium lacustre*, *Neobeckia aquatica*), the environmental control of leaf dissection is due to the combined effect of immersion in water and temperature (Nakayama *et al.*, 2014): moving plants to lower or higher temperatures causes the production of leaves with the distal (older) half of the lamina reflecting the condition in which it was initially growing, while the proximal half of the leaf reflects the environmental conditions experienced after the change. This shows that in this species leaf morphology is continuously specified during development, and that these heterophyllous changes can occur abruptly.

At any given time, parts of the same individual can be in different physiological phases, that is, new leaves and even new vegetative shoots can develop while other parts of the plant are in the flowering phase (Li and Johnston, 2000). It would be interesting to know whether this form of functional modularity has been a specific target of selection or is a simple by-product of other structural (e.g. branching pattern, branch length) or functional (e.g. mobility and sequestration of florigenous factors) aspects of plant organization. In any case, it deserves to be investigated how much this expresses differences in the developmental plasticity of plants with different longevities, different branching patterns and different flowering schedules.

In some species there is a correlation between the length of the juvenile phase and the morphology of the adult plant, especially in the number of leaves produced. For example, in *Thlaspi arvense* late-flowering individuals produce over twice as many leaves as early-flowering individuals (McIntyre and Best, 1975). In *Lupinus*, there is a positive linear relationship between the number of leaves on the main axis and the date of flowering (Duthion *et al.*, 1994). A strong correlation between the number of days to first flower and the number of leaves in the rosette has been determined in ecotypes and mutants of *Arabidopsis thaliana* (Zagotta *et al.*, 1992).

In some woody plants (e.g. *Callistemon*), inflorescences continue growing after flowering and fruiting; however, their main axis reverts to a vegetative state, and may produce a new inflorescence at a later time (Endress, 2010b).

9.3.2 Heteroblasty

In perennial plants, age- or season-related differences extend to whole branches, involving differences in plastochron, in addition to heterophylly. According to Cronk (2009), heteroblasty is a particular kind of heterophylly depending more on plant age than on environmental condition. However, changes in the latter often occur simultaneously with developmental transitions, and thus the effects of extrinsic and intrinsic factors in determining heterophylly may overlap (Pabón-Mora and González, 2012).

The most obvious evidence of heteroblasty is in any case heterophylly, sometimes affecting the dissection and overall shape of the lamina, as in the aquatic plants mentioned above, or in *Fraxinus* and *Leea*, which show a transition from simple to compound leaves (Gerrath and Lacroix, 1997), or in leaf anatomy and morphogenesis, including difference between unifacial and bifacial laminae. In a few cases, heteroblasty translates into different leaf phyllotaxis, as in *Eucalyptus* (Goebel, 1900–1905; Cameron, 1970; Wiltshire *et al.*, 1998; Jaya *et al.*, 2010).

In *Arabidopsis*, heteroblasty is based partly on the developmental control of cell division by three independent genes, *PAUSED* (*PSD*), *SQUINT* (*SQT*) and *SQUAMOSA PROMOTER BINDING PROTEIN-LIKE 15* (*SPL15*), the last of which is controlled in turn by *miR156* (Usami *et al.*, 2009).

9.3.3   Regulating Phase Transitions

9.3.3.1   Juvenile to Adult to Reproductive

When and how is leaf identity (juvenile vs. adult) determined? The mechanism is not necessarily the same in all plants. In maize, where the juvenile versus adult leaf identity is determined when the leaf is > 700 µm long (Orkwiszewski and Poethig, 2000), factors external to the meristem are likely responsible for directing the fate of the hitherto undetermined organ. However, at least in *A. thaliana*, control factors already present in the meristem are certainly involved in specifying leaf kind. This was initially suggested by the discovery that a mutant allele of the *WUS* gene prevents the formation of juvenile leaves (Hamada *et al.*, 2000). Later genetic analyses in the same model plant have identified a large number of genes that converge to control the transition from juvenile to adult leaf and the switch of the meristem from vegetative growth to reproductive development (Poethig, 2009).

Two microRNAs, miR156 and miR172, are involved in the regulation of the juvenile-to-adult phase transition and in the transition to flowering (Wu *et al.*, 2009). The levels of expression of miR156 are high in early stages of development and decrease with plant age, while miR172 shows the opposite pattern. When miR156 levels decline, the SPL proteins increase, leading to the activation of miR172 and of the flowering genes and also inducing adult leaf features. The increasing expression of miR172 is accompanied by the progressive silencing of *AP2-like* genes, the expression of adult leaf traits and the induction of flowering. This antagonistic relationship between the two miRNAs involved in the regulation of heteroblastic development occurs also in species with dramatic changes in leaf shape, including *Eucalyptus* species (Hudson *et al.*, 2014), *Quercus acutissima* and *Hedera helix*, and in the transition from juvenile bipinnate compound leaves to adult phyllodes in *Acacia* species (Wang *et al.*, 2011).

Flowering is under the joint control of developmental and environmental factors. The latter include photoperiod, light quality and quantity, ambient growth temperature and vernalization (prolonged exposure to cold temperatures) (Schneitz and Balasubramanian, 2009). The flowering date of each plant species is quite strictly conserved and thus predictable from year to year.

In terms of their dependence on the photoperiod, three kinds of species are distinguished. In short-day plants, induction of flowering is dependent on a succession of short days; flowering of long-day plants requires instead a succession of long days; finally, in day-neutral plants, flowering is not controlled by day length (Hileman, 2012). *A. thaliana* is a facultative long-day plant. Eventually, however, *Arabidopsis* will also flower under short-day conditions (Schneitz and Balasubramanian, 2009).

The regulation of flowering is also dependent on precise temporal and, to some extent, spatial patterns of expression of a number of genes. Indeed, over 300 genes that can affect flowering time have been identified in *A. thaliana* (Bouche *et al.*, 2016).

In all flowering plants, external and internal stimuli eventually converge in the regulation of the main components of the florigen signal *FT* and its paralogue *TSF* (Andrés and Coupland, 2012; Song *et al.*, 2013).

Initially produced in the leaves, the FT and TSF proteins are carried to the meristems. Here these molecules interact with the transcription factor FD (Wigge *et al.*, 2005), forming transcriptional complexes (FT–FD and TSF–FD) which in turn activate the expression of the floral pathway integrators *SOC1* and *SPL*, and also of flower meristem identity genes. This chain of induction eventually culminates in flowering (Posé *et al.*, 2012). In *Arabidopsis*, once floral meristem identity genes are transcribed, the plant is irreversibly committed to flowering (Corbesier *et al.*, 2007), but in other species reversion to vegetative growth is observed, if exposure to the inductive photoperiod is not maintained (Tooke *et al.*, 2005).

LFY also plays a dual role in regulating floral meristem identity and floral organ patterning (Parcy *et al.*, 1998). Changes in the levels of expression of the *LFY* gene are an important element of the phase transition, but in *A. thaliana* these changes are modulated by photoperiod. In long days, soon after germination a rapid upregulation of *LFY* occurs, followed by early flowering. In short days, *LFY* expression increases gradually before flowering commences, and this occurs several weeks later than in long days (Blázquez *et al.*, 1997).

The floral transition is also controlled by the *FLC* genes, mainly acting as vernalization-controlled floral repressors, e.g. in *Arabidopsis* (Michaels and Amasino, 1999), *Brassica* (Tadege *et al.*, 2001; Schranz *et al.*, 2002) and *Beta vulgaris* (Reeves *et al.*, 2007). Natural variation in *FLC* gene activity has been found to be associated with variation in flowering time and differential vernalization response among ecotypes of *Arabidopsis* (Salomé *et al.*, 2011) and other Brassicaceae (Schranz *et al.*, 2002; Nah and Chen, 2010). Evolutionary changes in the regulation of *FLC* orthologues are involved in major changes in phenology, e.g. the evolution of the perennial life habit in *Arabis alpina* (Wang *et al.*, 2009b) and the regulation of floral bud dormancy as a response to cold, for example in *Citrus trifoliata* (Zhang *et al.*, 2009) and *Geum rupestre* (= *Taihangia rupestris*) (Du *et al.*, 2008).

The mechanism of action of the products of these genes is complex. A repressor complex of the two transcription factors, FLC and SVP (Michaels and Amasino, 1999; Sheldon *et al.*, 1999; Hartmann *et al.*, 2000; Li *et al.*, 2008), controls the integration of flowering signals. FLC represses *FT* expression in leaves and also directly represses the expression of *SOC1* and of the FT cofactor FD in shoot apical meristems (Searle *et al.*, 2006); as a consequence, the meristem becomes less sensitive to flowering signals (Liu *et al.*, 2009a).

In addition to the homologues of the floral meristem identity genes of *Arabidopsis*, meristem identity genes that are unique to grasses have been isolated. For example, in maize, floral meristem initiation is controlled by *INDETERMINATE SPIKELET1* (*IDS1*), an *AP2*-like gene (Della Pina *et al.*, 2014).

Genetic variation in vernalization in Brassicaceae is mostly due to alleles of the *FRIGIDA* (*FRI*) and *FLC* loci (Weigel, 2012, in *Arabis alpina*) (Koenig and Weigel, 2015). In *Thlaspi arvense*, early versus late flowering is controlled by a single locus (McIntyre and Best, 1978).

9.3.3.2 The Flowering Calendar and its Evolvability

Further series of floral temporal phenotypes include the duration of anthesis of the individual flower, the hours of the day or night at which it is open (the 'floral clock'), the overall duration of flowering of the flowers in an inflorescence, the relative timing of anthesis of the individual flowers within the inflorescence, the relative timing of male versus female maturity of hermaphrodite flowers, and the time of the flowering season in respect to the production of leaves. All these phenotypes are under developmental control, irrespective of the different and possibly large extent to which they are also subject to environmental influence; in the latter case, we must in any case consider that the mechanisms by which environmental inputs are translated into temporal phenotypes are themselves features of developmental control and, to some extent at least, the product of evolution – thus, legitimate subjects of evo-devo. Only a few examples are given here.

The length of anthesis may vary enormously within a family; in orchids, for example, one flower may last 9 months (*Grammatophyllum multiflorum*; Kerr, 1972), 1–3 months (*Paphiopedilum*), ca. 2 weeks (*Oncidium ornithorrhynchum*), 9 days (*Dendrobium crumenatum*), 1 week (*Bulbophyllum ecornutum*), 1 day (*Sobralia* spp.), or just 5 minutes (*Dendrobium appendiculatum*). The duration of the individual flower is usually reduced by fertilization, but is prolonged instead in some orchids (*Zygopetalum maculatum* [= *Z. mackaii*], *Z. crinitum*, *Lycaste virginalis* [= *L. skinneri*], *Anguloa uniflora*, *Dimorphorchis lowii* [= *Renanthera lowii*], *Listera ovata*) (Fitting, 1921). In other families, however, the temporal pattern of anthesis is conserved: according to data collected by Spencer Barrett and published by Primack (1985), all species in the Commelinaceae, Pontederiaceae, Malpighiaceae Turneroideae and Convolvulaceae have one-day flowers. Among the basally branching Winteraceae and Degeneriaceae, the flowers of four species last two days, those of another six species, 4–12 days (Thien, 1980). As a rule, in both dioecious and monoecious plants, female flowers last longer than male flowers.

As for the floral clock, within the Nymphaeaceae there has been a shift from diurnal to nocturnal flowering in *Victoria* and in several *Nymphaea* species of the tropics. These clades are highly nested within the family, and their phenological specialization is therefore young (Yoo *et al.*, 2005; Borsch *et al.*, 2008; Endress and Doyle, 2015).

In *Heliconia*, the anthesis of the whole inflorescence (up to 50 flowers) lasts over 40–200 days, but the individual flowers open for one day only, and only one flower in the whole inflorescence is open on a given day (Endress, 1994a).

In most plants, flowers develop later than the leaves borne on the same axis. Some species, e.g. *Galanthus nivalis*, *Crocus vernus*, *Primula officinalis*, *P. elatior* and *P. acaulis*, start leafing at the time of flowering, but the growth of leaves is only complete after the end of anthesis. In *Colchicum autumnale*, flowers develop before the leaves are visible (Troll, 1964).

Instead of producing a single flower, the individual floral meristem of certain angiosperms continues branching, eventually giving rise to a complex inflorescence. The best example for this process (floral reversion) is *Nymphaea prolifera*, where each eventually branching flower produces some perianth-like leaves before switching back to shoot apical meristem identity. This process is repeated up to three times, generating a branched structure comprising more than 100 sterile flowers eventually serving as vegetative propagules (Grob *et al.*, 2006).

Bisexual flowers are usually dichogamous, that is, male and female functions are separated in time. Among the angiosperms as a whole, protandry (male first) is more common than protogyny (female first) (Lloyd and Webb, 1986), but protogyny appears to be the ancestral condition in flowering plants (Endress, 2010c) and is nearly universal among the basal groups (ANA grade, magnoliids, Chloranthaceae, Ceratophyllaceae) with dichogamous flowers. Here, protandrous flowers have been reported only for two species of the Piperaceae (*Peperomia magnoliifolia*, *Piper xylosteoides*), plus a dubious record for *Magnolia delavayi*.

A reproductively inactive phase separating the female and male phases is widespread, e.g. in Annonaceae, Lauraceae and Aristolochiaceae (Endress, 2010c).

Synchronized dichogamy based on the day/night rhythm in flowers with anthesis lasting two days is known (at the individual and population level) in Eupomatiaceae, Annonaceae and Canellaceae (Endress, 2010c). A special kind of synchronized dichogamy is heterodichogamy. In this case, the population includes two morphs represented by individuals with reciprocal flowering behaviour. These morphs commonly occur in equal proportions, suggesting a simple genetic dimorphism. Heterodichogamy is known from only a few species. In *Persea americana*, the flowers of morph A individuals are functionally female in the morning and male in the afternoon, while morph B individuals are functionally male in the morning and female in the afternoon. In *Hernandia*, which is monoecious, heterodichogamy is based on unisexual flowers, but again with two morphs. In morph A plants, only female flowers are open in the morning and only male flowers in the afternoon; in morph B individuals, it is the other way round (Endress and Lorence, 2004).

A potential evolutionary pathway from synchronous dichogamy to heterodichogamy is suggested by *Eupomatia laurina* (Endress, 1984) and *Annona mucosa* (= *Rollinia jimenezi*) (Murray and Johnson, 1987), in which the individual flower is open for one day only and each tree tends to open a number of flowers every second day, with a flowerless day in between. At the population level, some individuals open flowers on the days on which the other individuals do not (Endress, 2010c).

9.4 ORDER OF PRODUCTION OF LATERAL ORGANS

9.4.1 *Synchronous and Asynchronous Leaf Whorls*

Lateral organs eventually arranged in a whorl are initiated either synchronously or asynchronously. A synchronous whorl consists of a single cycle of organs that are initiated simultaneously. Synchronous whorls may originate from independent leaf primordia, as in *Hippuris* and *Utricularia purpurea*, or from an annular bulge, as in *Ceratophyllum* (Rutishauser, 1999). As a rule, asynchronous leaf whorls are the result of a regular alternation between a series of extremely short internodes (as many as the lateral organs eventually forming the whorl) and a single long internode, separating one whorl from the next. Examples are found in a number of genera belonging to distantly related lineages, such as *Lilium, Peperomia, Polygala, Anagallis* and *Euphorbia* (Kwiatkowska, 1995, 1999; Rutishauser, 1999).

9.4.2 *Order of Production of Flower Parts*

Besides anthesis per se, flower biology is a cradle of temporal phenotypes. The following overview includes order of production of flower organs, opening sequence of flowers, ontogenetic changes in flower symmetry, order of production of flowers in the inflorescence, post-anthesis changes, and timing of abscission of flowers and their parts.

The order of inception of the floral organs is nearly universally fixed within a species; it is often shared by most or all members of a family, but not always, and often deviates, in characteristic ways, from the centripetal (acropetal) sequence we might expect to be the default condition (Erbar and Leins, 1997). Organ initiation is radially symmetrical and acropetal e.g. in most Winteraceae, Magnoliaceae, Ranunculaceae, mimosoid Fabaceae, Rosaceae, Hypericaceae, Rutaceae, Malvaceae, Caryophyllaceae and Solanaceae (Tucker, 1999). 'Chaotic' inception of the floral parts has occasionally been recorded, in *Achlys* (Endress, 1989), *Gleditsia* (Tucker, 1991) and *Calla* (Scribailo and Tomlinson, 1992).

There is no fixed relationship between the order of inception of floral organs and the resulting phyllotactic pattern (Endress, 2001c). In *Musa velutina*, for example, the sequence of flower initiation is usually correlated with the left-handedness of phyllotaxis, but the actual sequence is variable, even within the same inflorescence (Kirchoff, 2017).

Organs in the same whorl arise either sequentially or simultaneously. The direction of the evolutionary shift is mainly from spiral to whorl, but there are also cases where whorled flowers have turned into spiral flowers, e.g. Paeoniaceae and Theaceae (Ronse De Craene, 2016).

Even the inception of organs eventually arranged in a whorl (e.g. petals) is sometimes spiral rather than simultaneous (Erbar and Leins, 1997). In *Silene coeli-rosa* the two whorls of five stamens arise as a continuous helix. In the zygomorphic flowers of legumes, the first member (or members) of each pentamerous whorl form on the abaxial side, followed by the lateral ones, and finally by the member (or members) on the adaxial side of the flower (Tucker, 1984a, 1984b, 1987). This unidirectional order of floral organ initiation has also been documented in *Pinguicula* and *Utricularia* and in some representatives of the Scrophulariaceae, Solanaceae, Verbenaceae and Plantaginaceae (Tucker, 1999).

Centrifugal order in organ initiation has been documented in representatives of at least 32 families, including several Primulales, some Hamamelidaceae, the phytelephantoid palms and *Capparis*.

There are two different patterns of centrifugal development: intrazonal and interzonal (Rudall, 2010). In the first case, organs are initiated centrifugally within a single organ zone, e.g. the androecium. This is most obvious in many species with a high number of stamens, e.g. in some palms, in *Dillenia* and in *Populus*, where stamen primordia appear in a centrifugal sequence. However, the same sequence of inception occurs also in *Tradescantia* and *Arabidopsis*, both of which possess only two stamen whorls.

In other plants, an entire organ zone is initiated after the zone inside it. In this case (interzonal centrifugal development), the most common condition is that stamens precede petals. In *Lythrum salicaria*, the outer sepals (epicalyx) are initiated after the inner sepals, the inner stamen whorl after the gynoecium, and the petals after gynoecium and androecium (Sattler, 1972). In *Valeriana officinalis*, the sequence of organ inception is stamens, followed by petals, then sepals and finally carpels (Sattler, 1972).

Temporal intervals between the production of one organ and the next in the same whorl are often unequal, and in many instances allow for the formation of the primordia of some elements of the next whorl, before the former is completed. Idiosyncratic sequences have been reported for many plants. Erbar and Leins (1997) studied in detail a number of representatives of the Apiaceae and the Brassicaceae. A few examples follow, beginning with the Apiaceae.

In *Astrantia major*, floral development starts with the successive formation of three sepal–stamen common primordia (more about this in the next section), followed by petal initiation. Petals 4 and 5 originate nearly synchronously with the two remaining sepals. Only then is the initiation of the stamens completed, by the spiral inception of stamen primordia 4 and 5, followed by the nearly simultaneous inception of the two carpels. In *Foeniculum vulgare*, the five sepals, the five petals and the first stamen all develop simultaneously, followed by the remaining four stamen primordia. In *Levisticum officinale*, three stamens arise simultaneously with the members of the calyx and corolla, followed by stamen primordia 4 and 5. In both *Foeniculum* and *Levisticum*, the two carpels are formed simultaneously.

To describe floral organ inception in the Brassicaceae, we must take into account that there are six stamens: an external whorl of two 'transversal' stamens and an inner whorl of four elements. In *Iberis sempervirens*, the sequence is strictly acropetal: the four petals appear simultaneously, followed by the two transversal stamen primordia and, finally, by the four inner stamen primordia, whereas in *Fibigia clypeata* the four inner stamen primordia are formed before the two transversal ones, as in *Arabidopsis thaliana* (Bowman *et al.*, 1989; Hill and Lord, 1989; Smyth *et al.*, 1990). In *Alyssum saxatile* the first organs formed next to the sepals are the two transversal stamens, followed by the petal primordia and finally by the four inner stamen primordia. In *Capsella bursa-pastoris*, all six stamen primordia are initiated at the same time, followed by the petals. In *Brassica napus*, the two transversal stamens are formed first after the calyx; the four petals and the four inner stamens then arise simultaneously. In *Lepidium draba*, the four petals and the six stamens appear simultaneously.

Another example of opposite inception order of floral organs in plants of the same family is found in the Hamamelidaceae, where stamen initiation is centripetal in *Matudaea*, centrifugal in *Fothergilla* (Endress, 1976).

Stamen initiation is centrifugal in Nelumbonaceae, Aizoaceae, Cactaceae, Hypericaceae, Capparaceae, Cistaceae, Malvaceae and Theaceae, among others. It is bidirectional (or almost simultaneous) in *Caloncoba* and other Achariaceae and in the Fabaceae. In *Swartzia*, centrifugal and centripetal patterns were found in different species (Tucker, 2003). In Magnoliales, stamens are initiated centripetally, but at anthesis the inner ones often open earlier than the outer ones (Endress, 1994a).

9.4.3   Common Primordia

Flower organs often arise from common primordia which undergo fractionation (chorisis) at a later stage. Organs splitting off from the same primordium are often of the same kind (especially stamens, e.g. in Myrtaceae, Paeoniaceae, Hypericaceae, Cistaceae, Capparaceae, Resedaceae, Lythraceae and Lecythidaceae; Weberling, 1989), but frequent also are the common primordia out of which a petal and a stamen develop, as in the orchid genus *Bletia* and in the Primulaceae. A stamen–sepal common primordium occurs in *Astrantia major* (Weberling, 1989).

Common primordia are especially frequent and diverse in the monocots. The organs involved are tepals and stamens in the Liliaceae, petals and stamens in the Zingiberaceae, and sepals, petals and stamens in the Orchidaceae. Examples in the eudicots include petals and stamens in Primulaceae and Plumbaginaceae (Endress, 1994a; Leins and Erbar, 2008).

The two organs borne out of the same primordium do not grow and differentiate necessarily at the same pace: for example, in *Nandina domestica*, the petal is delayed in development with respect to the stamen with which it has shared the primordium (Feng and Lu, 1998).

9.4.4   Initiation and Opening of Flowers in the Inflorescence

Flowers are initiated in the inflorescence in strictly acropetal sequence, as a consequence of the strictly acropetal development of the branching shoot, in which each axillary bud is initiated together with the subtending leaf (Endress, 2010b). However, flowers do not necessarily develop to maturity and eventually open in the same sequence. The latter processes are often influenced by external conditions. In *Salix*, the flowers on the side of the inflorescence exposed to the sun may open first (Goebel, 1924). In large inflorescences with many flowers on a long axis, the opening of flowers does not begin with the lowermost units, but with some flowers above them; from this beginning, flowering will continue bidirectionally, towards the base and towards the tip of the inflorescence.

In the capitula of the Asteraceae, the sequence of initiation and eventual opening of the individual florets depends on the kind and arrangement of the florets in the inflorescence. In homogamous capitula, the progression is centripetal, while in heterogamous capitula, the pattern of initiation differs between flower types (Harris, 1995). The florets forming the central disc are the first ones to initiate, and their production proceeds in a centripetal manner. The initiation of marginal ray flowers starts later and usually proceeds centrifugally towards the margins of the inflorescence (Harris, 1991, 1995; Thomas *et al.*, 2009; Bello *et al.*, 2013), more rarely centripetally (Harris, 1995).

9.5 ONTOGENETIC CHANGES IN FLOWER SYMMETRY

Floral actinomorphy or zygomorphy is generally recognizable quite early, and is maintained from inception to full anthesis, but developmental changes in symmetry during flower development are far from rare. Three different schedules can be recognized: (1) the flower is actinomorphic in its early stages, but zygomorphic at maturity; (2) the flower is zygomorphic in the early stages, but actinomorphic at maturity; (3) flower symmetry changes more than once, most often from zygomorphy to actinomorphy and back to zygomorphy.

Among the monocots, the flowers of *Alpinia* become zygomorphic at the time the inner perianth whorl is initiated (Tucker, 1999).

Families containing at least some species in which flowers are zygomorphic at maturity but organ initiation is radial include Bignoniaceae, Lamiaceae, Polygalaceae, Lecythidaceae; a few examples are also found in the Fabaceae and Ranunculaceae.

Many strongly zygomorphic flowers, including representatives of the Orchidaceae, Zingiberaceae, Resedaceae, Fabaceae (subfamily Papilionoideae), Violaceae, Orobanchaceae and Plantaginaceae, manifest their mirror symmetry quite early (Tucker, 1999), but with differences in the time at which zygomorphy is first expressed (Armstrong and Douglas, 1989). In the Orobanchaceae Rhinantheae, zygomorphy is manifested by the unidirectional initiation of sepals, starting on the abaxial side, whereas in the Antirrhinoideae, zygomorphy is only visible at mid stage, after all parts are initiated, and is due to differential organ enlargement.

Among the taxa that express zygomorphy early in the initiation of floral parts, organs may be unidirectional from either the abaxial or the adaxial side, as in papilionoid legumes and in *Reseda*, respectively (Tucker, 1999).

In flowers with bilateral symmetry, zygomorphy is commonly less pronounced in early development than at maturity (Endress, 2006), especially in families with predominantly actinomorphic flowers, e. g. the Ranunculaceae (Endress, 1999). In legumes, because of the unidirectional initiation of flower organs, early zygomorphy is evident even in species with flowers actinomorphic at anthesis, such as *Cadia purpurea* (Tucker, 1999).

In *Viola odorata*, the simultaneous initiation of one median adaxial sepal and two abaxial sepals determines the onset of zygomorphy (Mayers and Lord, 1984). The two lateral sepals arise subsequently; when petals and stamens eventually appear in a unidirectional order from the abaxial side, zygomorphy becomes more pronounced (Tucker, 1999).

In the Brassicaceae, flowers are usually actinomorphic or disymmetric, but in *Iberis amara* a zygomorphic flower emerges when the petals begin to grow in a heterogeneous way (Busch and Zachgo, 2007) (see Section 9.11.4).

The opposite pattern of change (zygomorphy first, actinomorphy at maturity) is also widespread. Plants with flowers that are bilateral during development but become actinomorphic at maturity include *Buxus*, *Lythrum*, *Epilobium*, *Gaura*, *Fuchsia*, *Clarkia*, *Rhamnus*, *Asperula*, *Galium* and *Rubia*, all tetramerous (Tucker, 1999). Further examples of early zygomorphy in otherwise actinomorphic flowers are *Siparuna*, *Achlys*, *Trochodendron* and *Hypoxis*.

In many plants the growth of the abaxial half of the flower is initially delayed. There are examples in all major angiosperm clades (e.g. *Euptelea, Adoxa, Chrysosplenium* and *Veratrum*). The opposite condition, with the adaxial half of the flower delayed in early development, is found in *Acorus* and *Nymphaea*.

In the Apiaceae, the pentamerous symmetry of the other floral whorls drives a maturational change of the originally dimerous gynoecium towards zygomorphy. However, bilateral symmetry is sometimes transient. In *Steganotaenia* it is present in floral buds but vanishes during later flower and fruit development. In contrast, in *Polemanniopsis* it becomes more pronounced (Endress, 2012).

Three floral morphs (hermaphrodite, male and neuter flowers) coexist in the inflorescence of the mimosoid *Neptunia pubescens* (Tucker, 1999); all mature flowers are actinomorphic, but the male flowers become strongly zygomorphic during stamen differentiation.

Symmetry may change more than once during development. The flower of *Couroupita* is initially zygomorphic, then changes to actinomorphic, and finally back to zygomorphic (Endress, 1994a). In *Tiarella* and some *Chrysosplenium* species, zygomorphy is most pronounced in the early stages of flower development and again at the mature fruit stage but is least obvious at anthesis (Endress, 1999, 2006). In the gynoecium of *Catharanthus* and other Apocynaceae, a first developmental change from zygomorphy to disymmetry is followed by a further change from disymmetry to actinomorphy, in the apical part at least (Endress, 1999).

A good model for the study of the evolution of symmetry changes at anthesis would be the genus *Moringa*, the 13 species of which range from actinomorphic to highly zygomorphic flowers. In all species of this genus, traces of zygomorphy are visible at stages between petal organogenesis and anther differentiation. At later stages, zygomorphy is manifest by one petal being larger than the others and by the unidirectional maturation of the anthers; in many species, some staminodes may be missing. Late in development, the actinomorphic species show a trend towards increasing actinomorphy, whereas the zygomorphic features of early ontogeny are progressively accentuated throughout ontogeny in the zygomorphic species. Remarkable intraspecific variation was found in the flowers of the former, but not in those of the latter (Olson, 2003).

The flowers of some papilionoid Fabaceae become asymmetric at a late developmental stage as a result of the coiling or twisting of the style, sometimes also of the corolla. Ontogenetically late events also lead to asymmetry in enantiostylous flowers (see Section 2.7.3) and in the twisted beaks of the flowers of *Pedicularis* species (Tucker, 1999).

In *Senna*, asymmetry is expressed in the corolla from the early bud stage; in *S. aciphylla* and *S. wislizeni*, the outer stamens will also become asymmetric, but only transiently. In the related *Chamaecrista fasciculata*, the flower is asymmetric throughout organogenesis, beginning with the skewed shape of androecium and corolla, caused in the early stages by the precocious initiation of the left (or right) organs (Tucker, 1996, 1999). This form of asymmetric initiation is also present in *Schotia afra* (Tucker, 2001), but in this legume asymmetry disappears at later stages of development (Marazzi and Endress, 2008).

In several taxa with gynoecial asymmetry, in the form of enantiostyly, the style becomes deflected in late bud (*Wachendorfia paniculata, Dilatris corymbosa, Philydrum lanuginosum*) or only at anthesis (*Cyanella lutea, Monochoria australasica, Heteranthera, Solanum rostratum, Saintpaulia, Streptocarpus*) (Marazzi and Endress, 2008).

9.6 SEXUAL TEMPORAL PHENOTYPES

As long recognized (Darwin, 1877), two different developmental courses may similarly result in producing unisexual flowers; of these, accordingly, two types are recognized (Mitchell and Diggle, 2005). In type I, flowers are bisexual at inception and become unisexual at maturity, because only the androecium or the gynoecium is complete and functional. In type II, sex differentiation occurs when either stamens or carpels are initiated, to the exclusion of the others. As a consequence, in type I flowers, rudiments of the non-functional organ type are found at maturity; no vestigial sexual organs are found in type II flowers. Type II unisexual flowers are homoplasious and are produced by several distinct developmental pathways (Mitchell and Diggle, 2005).

In type I flowers, sexual organ abortion occurs, according to the species, either (1) before the initiation of stamen or carpel primordia or (2) shortly thereafter, or (3) just before the meiosis of the mega- or microspore mother cell(s), or even (4) after the latter event. Among the 292 species surveyed by Diggle *et al.* (2011), the loss of sexual organ function, in both male and female flowers, has been found to occur with equal frequency at each of those four stages, with no differences between monoecious and dioecious taxa. However, the abortion of the androecium or gynoecium occurs preferentially at the same stage in the female and male flowers of the same species.

Variation in the timing of organ abortion – both intra-individual and between individuals – has been reported for some species. In different individuals of *Epigaea repens*, at the time functionally female plants discontinue the development of stamens, the latter may not even have been initiated, or may have already grown, and differentiated until anthers were formed (although they will anyway remain devoid of pollen) (Clay and Ellstrand, 1981). It is not clear, however, how far these differences reflect genetic variation rather than environmental effects. Among the taxa surveyed by Diggle *et al.* (2011), inter- and intra-individual variation in the stage of abortion has been recorded both in the androecium and in the gynoecium. In any particular species, developmental variability in one type of organ is not associated with variability in the other, but in the whole species sample the overall pattern of variation is similar in functionally male and female plants. Differences in the stage at which organ abortion occurs have been recorded among taxa belonging to the same family, e.g. in the Cactaceae.

In different plants, different genes are involved in the control of the selective abortion of stamens or carpels in type I flowers; the limited evidence available to date suggests independent exaptation in the developmental processes leading to unisexual flowers of a number of pre-existing genes and pathways (Diggle *et al.*, 2011).

In the male flowers of *Zea mays*, the gynoecium is aborted, by programmed cell death, just after flower initiation, but in six different *TASSELSEED* (*TS*) mutants functional hermaphrodite flowers develop instead in the usually staminate terminal inflorescence. One of these genes (*TS2*) has been found in other grasses too, and also in other monocot families, but it is not restricted to the gynoecium and in some of the multiple locations where it is expressed it does not cause cell death at all (Malcomber and Kellogg, 2006).

In *Silene latifolia*, a male-specific pattern of expression of *CUC1*, *CUC2* and *STM* homologues was observed before the abortion of the gynoecium (Zluvova *et al.*, 2006). In the same plant, a *SUP* homologue is expressed in female flowers but not in male flowers (Kazama *et al.*, 2009). In *Arabidopsis*, this gene is probably involved in the development of the gynoecium, as it blocks the expression of B-class identity genes in the organs of the fourth whorl (Sakai *et al.*, 1995).

9.7 FLOWER COLOUR CHANGE AS A DEVELOPMENTAL PROCESS

Flower colour changes occur in at least 20% of angiosperm families. In a non-exhaustive survey (Weiss, 1995), on which the following account is based, change in flower colour was recorded for 393 species in 241 genera. Popular examples are found among the Boraginaceae, e.g. in *Pulmonaria* spp. (Fig. 9.3), and Verbenaceae, e.g. in *Lantana camara*.

Colour-changing taxa are known in approximately half of the families in the Lamiidae. This suggests a degree of phylogenetic inertia, but nevertheless multiple parallel evolution of this trait, and frequent reversals, are quite obvious: within one family, e.g. Solanaceae, flower colour changes are restricted to a number of (not necessarily closely related) genera, or even to some species within a genus. There are also examples of intraspecific variation. For example, in some

FIGURE 9.3   Flower colour change in lungwort (*Pulmonaria officinalis*).
A black and white version of this figure will appear in some formats. For the colour version, please refer to the plate section.

individual plants of *Lobularia maritima*, stamen filaments remain green throughout the life of the flower, but in others they turn from green to deep purple, and in some individuals of *Armeria maritima*, but not in others, the centre of the flower changes from whitish to pink.

To various degrees, floral colour changes may reflect the modularity of flower structure and development. Changes, indeed, can affect whole flowers (known in representatives of 48 families), but they are more often restricted to some flower organ only (examples in species of 61 families) and may affect either a part of a whorl, or an entire whorl, or several whorls or parts of whorls in combination. Closely related species, or even individual phenotypes within a species, can differ in the identity of the flower part(s) where colour change occurs. For example, in *Lantana camara*, this affects the entire corolla, but it is restricted to a central ring in *L. hirta* and *L. montevidense*. In the purple form of *Lobularia maritima*, colour change affects the entire flower, while in the white form it is restricted to the filaments.

9.8 POST-ANTHESIS CHANGES

Flower morphogenesis does not come to an end with full anthesis. Post-anthesis changes are very widespread, and commonly released by pollination.

Poorly conspicuous but very common changes are those affecting flower orientation, for example in some orchid species (van Doorn and Stead, 1997). A similar behaviour, depending on anatomical changes in the pedicel, is observed in many other plants. In *Cymbalaria muralis*, after the fertilization of the flower the pedicel becomes negatively phototropic, causing fruit to complete ripening in the darkness of cracks in walls and rocks (Weberling, 1989). In *Nymphaea* and *Vallisneria*, tight coiling of the pedicel following pollination of the flower pulls the latter under water.

In plants of the Boraginaceae such as *Myosotis* spp. and *Lappula squarrosa*, and in several species of the Asteraceae, Dilleniaceae and Malvaceae, sepals continue to grow after anthesis is over. Sepal growth is also observed in the Lamiaceae (Endress, 1994a), where the sepals take part in fruit development. In some Dipterocarpaceae, the fruits become zygomorphic by differential further growth of the sepals (Ashton, 2003).

In the Asteraceae, the persisting calyx is reduced to a ring of hair on the pappus that aids dispersal of the seed in the wind. Persistent calyces often enlarge in the fruit; the very peculiar case of *Physalis* is described in Section 7.7.1. In some Dilleniaceae, sepals become fleshy in the fruit.

In *Clerodendrum minahassae*, sepals become petaloid (thus, visually attractive) after anthesis (Rutishauser *et al.*, 2008).

9.9 RELATIVE TIME OF ABSCISSION

9.9.1 The Abscission of Leaves and its Evolvability

To the best of my knowledge, up to now the evolvability and evolutionary history of leaf abscission has not been the subject of targeted studies. Thus, in the absence of adequate understanding of the mechanisms underlying the arrival of the fittest, we are currently left with a wealth of descriptive information, mostly scattered in the taxonomic literature, which suggests a possibly intriguing diversity of evolutionary change. An easily accessible, although not necessarily exhaustive, source of information about the occurrence of deciduous leaves at the family level is Byng (2014). From a perusal of this work, we learn that leaf abscission has not been recorded in most monocot families but nevertheless occurs, more or less frequently, in Melanthiaceae, Iridaceae, Amaryllidaceae, Restionaceae and Poaceae. In other lineages, leaf abscission may occur in one taxon, but not in other closely related ones. For example, in the two living representatives of the Trochodendraceae, leaf abscission occurs in one (*Tetracentron sinense*) but not in the other (*Trochodendron aralioides*); in Menispermaceae, leaves are deciduous in *Calycocarpum* and in some *Menispermum*, but not in other taxa. As a rule, leaf abscission occurs in long-lived, woody plants, less frequently in herbaceous annuals. It might be rewarding to see if leaf abscission has evolved (or has been lost) within genera in which there has been transition from annual to perennial (or vice versa).

In some plants, the actual detachment of the leaf is long delayed in respect to the end of its physiological integration with the plant: in *Quercus* and *Fagus*, for example, leaves desiccate on the tree and fall only at the opening of the buds in the following spring. In our context, once more, it is a question not of the adaptive significance of this peculiar temporal phenotype, but of its evolvability. The circumstance that in this case only dead plant organs are involved should not obscure the fact that the temporal sequence of leaf desiccation and eventual detachment is in any case a (delayed) consequence of a physiological process previously deployed in living tissues.

9.9.2   *The Breaking Apart of Inflorescences*

The presence and precise spatial distribution of abscission zones determine the size and composition of the disseminules that will derive from an inflorescence following fertilization of the flowers and ripening of seeds. The size and nature of the disseminules range from the individual seed, in the case of plants with dehiscent fruits, to the individual fruit, to a group of fruits corresponding to a partial inflorescence, up to the whole infructescence, as in *Valerianella* and *Fedia* (Weberling, 1989).

The evolvability of abscission patterns is arguably highest in the Poaceae. One striking difference between grasses and their close outgroups is the occurrence of abscission zones in inflorescence axes other than those that directly bear the spikelet or floret: Joinvilleaceae and Flagellariaceae shed their seed as a fleshy fruit, the Ecdeiocoleaceae as a hard nut or a loculicidal capsule (Rudall *et al.*, 2005b). Among the grasses there is huge diversity in the size and morphological identity of the units through which seed dispersal occurs. These units range from the individual seed, as in *Sporobolus*, to the whole inflorescence, as in *Spinifex*. This diversity of dispersal units is a consequence of the formation of abscission zones in one or more of multiple available locations within the inflorescence (Doust *et al.*, 2014).

A first pattern consists of the whole inflorescence disarticulating to become the diaspore. In several cases, spikelets or florets separate following secondary abscission.

In a second group of grasses, diaspores consist of spikelets that remain attached to other structures, for example falling alone but still with their pedicel; otherwise, spikelets can be dispersed in groups, while joined by their fused pedicels; in still other instances, a segment of the rachis of the inflorescence is still attached to the falling spikelets. These various spikelet–branch combinations may remain unmodified, but sometimes some sterile branches are modified into feathery or spiny structures, as in *Cenchrus*.

In a third group, the spikelet, including the glumes, is detached as a whole.

Finally, in the fourth and most common pattern, parts of the spikelet disarticulate, thus releasing individual fertile florets, or florets attached to rachilla, or groups of florets.

From a human perspective, easy disarticulation of the inflorescence is a nuisance, and therefore races with superior grain retention have been favoured since the beginning of selection practices in agriculture. In the case of rice, plants with a flower pedicel that persists for longer result from a single base change in a regulatory element of the rice orthologue of a gene also associated with seed dispersal in *Arabidopsis*. But in the latter plant the gene regulates abscission at the level of the fruit (Roeder *et al.*, 2003) rather than the pedicel as in rice (Konishi *et al.*, 2006).

9.9.3 The Abscission of Flowers and Their Parts

In the angiosperms there is huge diversity in the timing of abscission of flowers and flower parts. Current knowledge of the developmental control of these events does not allow us yet to address the evolution of these temporal phenotypes from the perspective of the arrival of the fittest. Most studies of these phenomena only address their adaptive significance. To stimulate comparative and evo-devo research in this area, I offer here an overview of these temporal phenotypes (Table 9.1).

Most plants have a specific and predictable schedule of abscission of either the whole flower or the individual floral organs (modularity!). In a few plants, all flower organs are discarded simultaneously, either without withering, like the unfertilized flowers of *Liparis* and other orchids, and the flowers of *Mercurialis annua* and *Begonia* spp., or after withering, like the male flowers of the Cucurbitaceae. In some plants, male flowers fall before they shed their pollen, or right at the time this happens. For example, in aquatic plants such as *Elodea* and *Vallisneria*, the male flowers shed pollen after their abscission and eventual rise to the water surface. In *Mercurialis* spp., the abscission of the male flowers occurs as the anthers open and pollen is dispersed while these tiny flowers fall. More frequently, the different kinds of flower organs are abscised separately; in this case, the perianth can be discarded as a unit, either fresh (*Fuchsia*) or withered (*Oenothera, Epilobium*); otherwise, the calyx is discarded before the petals (Papaveraceae), together with petals (e.g. Ranunculaceae, Brassicaceae, *Impatiens*), or only

Table 9.1 *Patterns of abscission of inflorescences, flowers and flower parts, mainly after van Doorn and Stead (1997), with non-exhaustive examples.*

Abscission of whole inflorescences or individual flowers

Bisexual inflorescence	Oleaceae: *Olea europaea*; Acanthaceae: *Justicia brandegeeana* (= *Beloperone guttata*), *Pachystachys lutea*
Staminate inflorescence	Fagaceae: *Castanea, Fagus, Quercus*; Myricaceae: *Myrica*; Juglandaceae: *Juglans, Pterocarya*; Betulaceae: *Alnus, Betula, Carpinus, Corylus*; Salicaceae: *Populus, Salix*; Sapindaceae: *Aesculus*
Flower	Orchidaceae: *Cymbidium*; Iridaceae: *Diplarrena*; Asparagaceae: *Aloe, Arthropodium*; Arecaceae: *Phoenix*; Cannaceae: *Canna*; Marantaceae: *Maranta*; Proteaceae: *Macadamia*; Grossulariaceae: *Ribes*; Fabaceae: *Glycine*; Rosaceae: *Prunus*; Fagaceae: *Quercus*; Cucurbitaceae: *Cucumis*; Begoniaceae: *Begonia*; Euphorbiaceae: *Euphorbia*; Lythraceae: *Cuphea*; Onagraceae: *Fuchsia*; Anacardiaceae: *Mangifera*; Rutaceae: *Citrus*; Malvaceae: *Hibiscus, Theobroma*; Caryophyllaceae: *Dianthus*; Nyctaginaceae: *Mirabilis*; Cactaceae: *Schlumbergera*; Balsaminaceae: *Impatiens*; Primulaceae: *Anagallis*; Apocynaceae: *Asclepias*; Solanaceae: *Atropa*; Scrophulariaceae: *Verbascum*; Lamiaceae: *Clerodendrum, Salvia*; Bignoniaceae: *Campsis*; Caprifoliaceae: *Lonicera*

Abscission of sepals in plants in which petals are shed in a turgid state

Sepals (calyptra) often fall before petals	Papaveraceae
Sepals fall	Nymphaeaceae, some Ranunculaceae, Polygalaceae, some Onagraceae

Table 9.1 (*cont.*)

Sepals fall at about same time as petals	Balsaminaceae
Sepals of species showing petal abscission grow considerably, forming a balloon-like structure	Some Solanaceae: *Physalis, Nicandra*
Sepals remain	Saxifragaceae, Rosaceae, Linaceae, Geraniaceae, Lythraceae, Rutaceae, Cistaceae, several Primulaceae, several Ericaceae, Apocynaceae, Boraginaceae, Gesneriaceae, Lamiaceae, Orobanchaceae (in *Euphrasia, Pedicularis, Rhinanthus* etc. the sepals resume growing after pollination)

Abscission of sepals in plants in which petals are shed shortly after they wilt

Sepals fall	Berberidaceae, some Cucurbitaceae
Sepals fall or remain	Brassicaceae, Portulacaceae, several Solanaceae
Sepals fall together with petals, when there is a common abscission zone, or sepals fall after the petals when abscission zones are separate	Some Onagraceae
Sepals remain	Commelinaceae, Cannaceae, Violaceae, Hypericaceae, Rutaceae, Malvaceae, Polemoniaceae, Convolvulaceae, Verbenaceae, Apiaceae, most Asteraceae

Abscission of the androcoeum or its parts

Anthers fall; the filaments desiccate but remain	Some Geraniaceae
Stamens fall while fully turgid, before petal abscission	Some Papaveraceae, Balsaminaceae

Table 9.1 (cont.)

Stamens fall while turgid or somewhat wilted, with or shortly after the (almost) turgid petals	Some Papaveraceae, Rutaceae, Boraginaceae, Gesneriaceae, Scrophulariaceae, Lamiaceae, Acanthaceae
Stamens fall when wilted, with the wilted petals (tepals)	Some Liliaceae, Brassicaceae, Malvaceae, Portulacaceae, Polemoniaceae, Convolvulaceae, Solanaceae, Verbenaceae
Stamens fall following desiccation, but prior to tepal fall	Some Liliaceae
Stamens desiccate, then fall with or after the petals	Berberidaceae, Ranunculaceae, Violaceae, Onagraceae
Abscission of the style	
Style falls while fully turgid or somewhat wilted	Some Liliaceae and Iridaceae, some Rosaceae, some Rutaceae, some Solanaceae
Style falls following wilting, together with wilted petals and stamens	Malvaceae, Convolvulaceae
Style falls after desiccation	Some Liliaceae, Berberidaceae, Onagraceae, Boraginaceae

petals are discarded, as in *Fragaria, Pyrus, Rosa, Geranium* and *Linum* (Fitting, 1921; van Doorn and Stead, 1997).

In most plants of the basal ANA grade, as well as in *Hedyosmum* (the only living genus with a perianth in the Chloranthales), the tepals remain attached, but in the Austrobaileyales, the perianth is more or less caducous, with an abscission zone at the base of each tepal (Endress, 2008a).

In most plant families, the sepals remain attached and eventually desiccate. If not, sepals commonly fall after the petals, but the order of abscission is the opposite, e.g. in the Papaveraceae; in other plants, as in several Brassicaceae, sepals and petals fall at the same time. In still other plants, only the petals are subject to abscission,

FIGURE 9.4 An apple (*Malus pumila*) flower after fertilization, showing that petals have abscised.

while the sepals remain attached, as in many representatives of the Rosaceae (Fig. 9.4). In the Papaveraceae, the sepals, still united together at abscission, form a cap which is released when the flower opens (Fig. 9.5). A similar cap occurs in *Eucalyptus* spp.: depending on the species, this may be derived from sepals, petals, or both, and falls in any case upon flower opening. In a few plants, as in *Physalis* (see Section 7.7.1), the sepals resume growing after the flower is fertilized. The type of sepal abscission is often consistent within families, but in some families there are examples of both falling and persistent sepals.

In a number of families, petals are usually shed while fully turgid, but this is quite rare in monocots. Whether petals are still turgid or not at the time of abscission, this is generally uniform for plants of the same family. However, in the Iridaceae, Ericaceae and Primulaceae there are examples of abscission of turgid petals, but in other species of the same families petals wilt prior to abscission or do not fall at all. Shedding of petals accompanying fruit growth was observed in families in which petals are generally shed when still

FIGURE 9.5 In the common poppy (*Papaver rhoeas*), the abscission of the hairy sepals precedes the full expansion of petals. *A black and white version of this figure will appear in some formats. For the colour version, please refer to the plate section.*

turgid, e.g. in some Boraginaceae (e.g. *Symphytum, Pulmonaria*) and Lamiaceae, but also in species where petal wilting precedes shedding, as in several representatives of the Convolvulaceae, Malvaceae and Solanaceae. In some taxa, e.g. in several Liliaceae, Iridaceae, Cucurbitaceae, Gentianaceae, Hypericaceae and Campanulaceae, the petals desiccate, but remain attached to the developing fruit. In other species, the petals or the internal tepals remain attached to the developing fruit, while remaining fully turgid and showing additional growth, as in *Rumex* and in some monocots (*Eucomis comosa* [= *E. punctata*], *Paris quadrifolia, Veratrum nigrum, V. album*).

Stamens, if shed, may fall together with the petals, or separately. Generally, stamens that are not attached to the petals wilt and do not fall, irrespective of whether petals are subjected to abscission or wilting. In some flowers, stamens are shed when still fully turgid, in others when partially or completely wilted, in still others only after desiccation. In a few species, the anthers may also abscise from the filament, for example in *Geranium* spp., where the filaments subsequently desiccate and remain attached. More commonly, an abscission zone is found at the base of the filament.

In Magnoliales the stamens abscise early, when the anthers are still full of pollen. Similarly, the stamens of the submerged flowers of Ceratophyllaceae, which probably represent unistaminate flowers (Endress, 2004), detach from the plant and gain the water surface before releasing the pollen.

In many plants, the styles desiccate and remain attached to the fruit, as, for example, in most of the Linaceae, Polemoniaceae, Verbenaceae and Violaceae. In some families, however, the styles undergo abscission. At this time, styles can be still fully turgid, as in some Liliaceae, Rutaceae (*Citrus* spp.) and Rosaceae, or wilted, e.g. in *Eucalyptus* species. Styles fall otherwise only after desiccation, for example in Onagraceae and Boraginaceae. The abscission zone is usually located at the base of the style, or just above it, as in *Geum* and *Citrus*. Within the genus *Eucalyptus* the position of the style abscission zone is variable, and in some species no style abscission occurs.

9.10   A DEVELOPMENTAL HOURGLASS IN PLANTS?

In his masterwork on comparative (animal) embryology, von Baer (1828) established the principle that the embryos of the most diverse vertebrates are quite similar to each other during the early stages of development and progressively diverge over the next stages, gradually acquiring the characters that allow us to assign them unambiguously to the different classes first and then to the different orders, until they eventually show the first evidence of their belonging to a particular species. In other words, vertebrate embryos proceed from an early generalized state through more and more specific, diverging conditions. Later, Haeckel (1866) took the similarity between early embryos of different animals as a proof of common derivation from the same ancestors, in the context of his principle (the biogenetic 'law') according to which ontogeny recapitulates phylogeny.

Subsequent research has not only shown that recapitulation is at best just one of the possible ways in which a trace of phylogenetic history can still be read in the ontogeny of recent forms; perhaps even more interesting, it has also shown that similarity between embryos is greatest not at the very earliest stages, but somewhat later. Therefore, the ontogenetic pathways of different species (e.g. different vertebrates) begin at arbitrarily different points (as, for example, a bird egg and a human zygote) and converge towards the stage (the phylotypic stage; Sander, 1983) at which similarity is greatest, to diverge again towards their increasingly different shapes, as remarked by von Baer. This pattern of embryo morphology – first convergent, then divergent – has been dubbed the developmental hourglass, or egg-timer (Duboule, 1994).

In the last few years, a couple of studies have addressed the question, whether flowering plants also develop according to the hourglass model, that is, whether a phylotypic stage (a 'phytotype', we could say) is also recognizable in their embryonic development. The first approach to the problem was molecular; a revisitation at the level of morphology followed soon afterwards. In animals, a molecular approach was applied at approximately the same time by Kalinka *et al.* (2010).

Quint *et al.* (2012) used two complementary approaches to get evidence for a molecular embryonic hourglass in *Arabidopsis thaliana*. To understand this work, it might be useful to devote a few words to the morphological aspects of plant embryonic development.

In the embryogenesis of flowering plants, three major phases are recognized. The earliest phase is characterized by asymmetric cell divisions giving rise to a globular embryo, in which both the apical–basal axis and the radial axis are established, while the cells are still largely undifferentiated. In the intermediate phase, major organs and primordia are initiated; in the late phase, these will expand into the articulated, although still simple structure of the mature embryo, with hypocotyl, radicle and one or more cotyledons (De Smet *et al.*, 2010; Peris *et al.*, 2010).

There is considerable diversity in the pattern of cell divisions that gives rise to the globular embryo. This morphological diversity has a parallel in the early embryos of animals. Again as in animals, it is during the middle phase of embryogenesis that the basic body plan of the plant is laid down. At this time, the root and shoot meristems are specified. During this stage, the morphology of the embryo is quite uniform across the eudicots, regardless of any difference in previous segmentation pattern (Wardlaw, 1955; Mordhorst *et al.*, 1997). However, it is also during the middle phase that the morphology of monocot embryos deviates conspicuously from that of eudicot embryos (Mordhorst *et al.*, 1997; Chandler *et al.*, 2008). The most obvious differences are the differentiation of one cotyledon rather than two, the lateral rather than apical position where the shoot meristem is formed, and the consequent lack of alignment between the apical–basal axis established early in development and the new shoot–root axis. Moreover, at this stage monocot embryos are considerably more diverse than dicot embryos.

Looking for the possible presence of an embryonic hourglass in plants, Quint *et al.* (2012) compared two measures of molecular distance between *A. thaliana* and a number of plants for which suitable transcriptomic data were available. Two different measures of evolutionary distance were calculated: a transcriptome age index (TAI) and a transcriptome divergence index (TDI).

To calculate the TAI, Quint *et al.* (2012) assigned each *Arabidopsis* gene to its phylostratum (Domazet-Lošo *et al.*, 2007). This was accomplished by identifying the most distant phylogenetic node containing at least one species with a detectable homologue of that *Arabidopsis* gene. The lowest phylostratum included the *Arabidopsis* genes for which there is a homologue even in prokaryotes; to the following phylostrata were assigned the genes progressively restricted, respectively, to Eukaryota, Viridiplantae, Embryophyta, Tracheophyta, Magnoliophyta, eudicots, core eudicots, rosids, malvids, Brassicales and the genus *Arabidopsis*. The last phylostratum includes the youngest genes, with no homologue in any species other than *A. thaliana*. To calculate the TDI, the same authors determined, for the same genes, the sequence divergence between *A. thaliana* and its sister species *Arabidopsis lyrata* or another closely related member of the Brassicaceae.

Eventually, Quint *et al.* (2012) found that the transcriptomes of early plant embryonic stages are evolutionarily young (high TAI), those of the mid-embryogenic phase are older (low TAI), and those of later stages of embryogenesis are younger again. Likewise, they found that transcriptomes of early stages are rich in genes with recently diverged sequence (high TDI), those of the mid-embryogenic phase are more conserved (low TDI), while transcriptomes of later stages of embryogenesis are more divergent again. Both profiles pointed to the so-called torpedo stage of the embryo as the plant phylotypic stage, representing simultaneously the stage with the oldest as well as the most conserved transcriptome.

The middle period of plant embryogenesis was thus initially defined as the phylotypic stage based on molecular evidence (Quint *et al.*, 2012; Drost *et al.*, 2015). Cridge *et al.* (2016) raised the question of whether the same middle phase of plant embryogenesis can also be considered phylotypic based on morphology. The evidence is not strong; nevertheless, it is right at this phase that two major phylogenetic lines, the eudicots and the monocots, acquire major differentiating traits, approximately paralleling the differences between the germ-band versus pharyngula stages characteristic of arthropods and vertebrates in the hourglass pattern of animal embryonic development (Hall, 1998).

9.11 HETEROCHRONY

9.11.1 *Evolving Concepts of Heterochrony*

Heterochrony is evolutionary change in rates and timing of developmental processes (Klingenberg, 1998). Within this broad definition, a diversity of perspectives has been adopted, starting with the classic works of de Beer (1930, 1940), Gould (1977), Alberch *et al.* (1979) and McNamara (1986).

With a characteristic zoological bias, most of the traditional literature on the subject is framed in terms of *growth heterochrony*, i.e. evolutionary changes in the relationship between overall shape, as an expression of somatic development, and size, indicative of progression towards sexual maturity. However, many evolutionary changes affecting developmental processes are not changes in either form or size. For this reason, Velhagen (1997) and Smith (2001) suggested a different approach, which takes the name of *sequence heterochrony* and focuses on the variations in the temporal sequence of developmental events (e.g. the order of inception of different flower organs).

In terms of growth heterochrony, evolutionary changes may involve onset time, offset time and rate of somatic growth versus progression towards sexual maturity (Alberch *et al.*, 1979; Fink, 1982, 1988; Reilly, 1997). Accordingly, two main patterns are recognized (Box and Glover, 2010):

- Paedomorphosis: evolution of adult morphology reminiscent of an embryonic or juvenile ancestral form. Specifically, paedomorphosis can be due to a reduction in the rate of an ancestral developmental process (neoteny), or to the terminal truncation of ontogeny as a result of earlier cessation of an ancestral developmental process (progenesis).
- Peramorphosis: evolution of new adult morphology by terminal extension of the ancestral ontogeny.

In plants, the dissociation between vegetative and reproductive development may depend on separate genetic control of the timing of the transition from juvenile to adult foliage and of the onset of reproduction (Diggle, 1999), as demonstrated by Wiltshire *et al.* (1998) for some *Eucalyptus* species.

Sequence heterochrony relies on developmental modularity, a property that is itself subject to evolution. For example, flower organs of the same whorl may behave as a single module, but in some plants they are only loosely associated and may evolve independently. This is the case of the nectariferous petals of *Delphinium*, compared to ordinary petals (Guerrant, 1982). More significant is the evolutionary plasticity of the corolla (and the androecium) in *Bauhinia*, where the individual petals and stamens behave as separate homologues: within the genus there are species like *B. blakeana*, with five petals and three fertile stamens, alongside species like *B. divaricata*, with two petals and one functional stamen only (Wunderlin, 1983; Chen *et al.*, 2010).

The timing of first reproduction can be evolutionarily decoupled from the vegetative growth of the plant (Diggle, 1999). Reproduction can even occur before acquisition of adult vegetative characteristics in monoecious plants, including gymnosperms. In several species of spruce, pine and larch, female strobili are associated with the immature habit and male strobili with the mature habit (Greenwood, 1995). The reverse is true in *Cucurbita argyrosperma*, where the first staminate flowers appear in the axils of leaves transitional between juvenile and adult shape, whereas all ovulate flowers arise in association with adult leaves (Jones, 1992). In the latter species, changes in flowering time are associated with domestication. The cultivar *argyrosperma* begins producing flowers earlier than the wild subspecies: on average, it produces the first male flowers at node 12 and the first female flowers at node 30, while the wild progenitor does not produce male flowers until node 19 or female flowers until node 39.

9.11.2 Major Phenotypic Consequences of Heterochrony

The most obvious example of heterochrony is possibly the evolution of annual flowering plants from perennial ancestors, or vice versa. Less obvious examples are found virtually everywhere, for example in leaf and flower morphology. The type of leaf architecture, simple versus compound, can be related to heterochronic development. In three species of *Sorbus* studied by Merrill (1979), the different degree of leaf dissection correlates with the timing of initiation of the leaflets. Heterochronic changes affecting flower morphology are most diverse.

For example, heterochrony can be involved in the evolution of heterostyly (see Section 2.7.1), as in *Amsinckia*. This genus includes both homostylous and heterostylous species (Li and Johnston, 2010).

The growth curve for the pistil is identical in the homostylous *A. vernicosa* (typical form) and in the thrum flowers of the heterostylous var. *furcata*, while the growth curve of the stamen filament is identical in the homostylous species and in the pin flowers of the heterostylous species.

Gross morphological change as a consequence of heterochrony is one of the best examples of the non-linear character of the genotype → phenotype map. Suspicion of heterochrony is often reasonable, when molecular systematics suggests that a species (or group of species) long classified in an independent genus is phylogenetically deeply nested within a different (thus paraphyletic) genus. A developmental study can easily clarify the issue. Some nice examples from European orchids have been provided by Box *et al.* (2008). The flowers of the species traditionally classified in the genus *Nigritella* are easily distinguished from those ascribed to *Gymnadenia*, but this is largely due to the fact that in the latter the spurs and the lateral lobes of the corolla continue growing much longer than in the former. Similarly, flower characters differentiating the orchid known as *Coeloglossum viride* from those in *Dactylorhiza*, mainly deviating in the relative length of spur versus labellum, turn out to be simple heterochronic effects of differences in the growth rate of two parts of the flower.

Another example of large effects of heterochronic change on flower morphology is provided by two closely related species of *Delphinium*, of which one (*D. decorum*) is pollinated by bumblebees, as are most of the species in this genus, whereas the grossly divergent *D. nudicaule* is pollinated by hummingbirds. In the latter species, paedomorphic development of sepals and petals results in the mature flowers resembling flowers of *D. decorum* in the bud stage, except that the nectariferous petals of *D. nudicaule* are larger than those of its relative, as a consequence of peramorphic development (acceleration and hypermorphosis) (Guerrant, 1982).

Heterochronic changes were discovered by Kampny *et al.* (1993, 1994) in their comparative studies on the growth patterns responsible for the differences among three closely related genera of the Plantaginaceae, i.e. *Veronica, Veronicastrum* and *Pseudolysima-chion.* In *Veronica,* the size of the sepals increases more slowly in the early stages, but their growth continues after those of *Veronicastrum* have stopped. In *Veronica* again, the growth of the corolla lobes is retarded until the beginning of style formation, but eventually it speeds up compared to *Veronicastrum,* whereas the corolla tube of *Veronica* remains shorter than in the related genus, owing to later onset of growth and slower progress in elongating. In *Pseudolysimachion,* the developmental trajectories of most flower organs are intermediate between those of *Veronica* and *Veronicastrum.* Overall, different trajectories lead to similar end products; thus, closer similarity in the shapes of the mature organs is not necessarily indicative of closer relationship.

Heterochronic effects may add to differences in the morphological pattern of flower organ inception. In the Asteridae, Leins and Erbar (1997) distinguish two modes of development of a corolla tube. In late sympetaly the initially separated petals become joined at a later stage by fusion of their basal meristems; in early sympetaly, instead, petals are joined from the beginning of their development, and in extreme instances the petals are initiated on a common ring primordium (Erbar and Leins, 1996).

The developmental pathways leading to the very different capitula produced by three closely related genera of the subtribe Artemisiinae (Asteraceae) were studied by Ren and Guo (2015). The morphological differentiation of the discoid capitulum of *Stilpnolepis* from the radiate inflorescence of *Chrysanthemum* or the disciform one of *Ajania* is already manifest at an early stage. The development of the florets of the radiate capitula and the marginal ones of the disciform capitula lags behind that of the marginal disc florets throughout floral initiation and organogenesis. The primordia of their dorsal corolla lobes and all their stamens stop growing soon after initiation, whereas all florets in discoid capitula develop at a uniform rate and eventually mature as hermaphrodite actinomorphic flowers. Further differentiation giving rise, respectively, to radiate and disciform capitula occurs only at the time of floral maturation.

9.11.3   Heterochrony and the Chasmogamy-to-Cleistogamy Transition

According to Lord's (1981) count, no less than 287 plant species in 56 families produce cleistogamous flowers: these do not open at maturity and are self-pollinated, usually with good results in terms of seed production. In a majority of cases, plant species with cleistogamous flowers also produce normal, chasmogamous flowers, among which there can be cross-pollination.

More frequently, chasmogamous and cleistogamous flowers may occur on the same individual, and often on the same inflorescence, but in a few dozen species scattered among the Poaceae, Hydrocharitaceae, Juncaceae, Plantaginaceae, Acanthaceae, Gesneriaceae, Fabaceae, Malpighiaceae, Nyctaginaceae, Oxalidaceae, Rubiaceae and Linderniaceae the individual plant produces either cleistogamous or chasmogamous flowers only.

It has been suggested that cleistogamous flowers evolved from chasmogamous flowers as the result of heterochrony (Lord and Hill, 1987; Gould, 1988). The mature cleistogamous flower looks like a young bud of the chasmogamous flower of the same species, but cleistogamous flowers can evolve not only through paedomorphic (progenetic) development, but also through peramorphic development by acceleration and/or early onset of floral development (Li and Johnston, 2000).

Different heterochronic patterns are sometimes combined. Three different heterochronic processes – early onset of floral development (predisplacement), decreased growth rate of the whorls (except gynoecium width) and early offset time (progenesis) – have been mentioned as jointly contributing to the paedomorphic traits of the cleistogamous flowers of *Centaurea melitensis* (Porras and Muñoz, 2000). In *Viola odorata*, earlier maturation of the cleistogamous flowers is a consequence both of the smaller size of the floral primordium at its inception and of the faster developmental rate; eventually, cleistogamous flowers reach sexual maturity two weeks earlier than the chasmogamous flowers (Mayers and Lord, 1983). In *Lamium amplexicaule*, the cleistogamous flowers reach maturity about 10 days earlier than the chasmogamous flowers; in this species, this is a consequence of the accelerated floral development after microsporogenesis (Lord, 1979, 1982). In *Collomia grandiflora*, the maturity of the cleistogamous flowers is reached only two days before the maturity of the chasmogamous flowers; in this case, heterochrony is due to accelerated development at early developmental stages of the flower and to the earlier onset of microsporogenesis (Minter and Lord, 1983). In *Bromus catharticus* (= *B. unioloides*), the small anthers in the cleistogamous flowers were interpreted as the combined result of neotenous development of the anthers and their progenetic dehiscence (Langer and Wilson, 1965).

9.11.4   Heterochronic Genes

Among the first examples of mutants with heterochronic effects described in plants were *hairy-sheath-frayed1-0* in maize (Bertrand-Garcia and Freeling, 1991) and *early-flowering* in Arabidopsis (Zagotta *et al.*, 1992), but the first systematic investigation of heterochronic genes – to employ a term used in the same year by Ambros and Moss (1994) in a study on the temporal control of development in *Caenorhabditis elegans* – was the now classic paper on *Pisum sativum* by Wiltshire *et al.* (1994). In this species, nine heterochronic mutants were described, showing dramatic morphological changes caused by a diversity of heterochronic processes: neoteny, progenesis, acceleration, hypermorphosis. For example, plants with the recessive mutant allele *sn* produced only four leaves, all with four leaflets: their growth stopped before the transition to the adult vegetative phase, in which leaves with six leaflets are produced. Under short-day conditions, these tiny plants began to produce flowers in the axil of the first four-leaflet leaf. This early offset of vegetative development, coupled with earlier flowering and earlier senescence, is an example of progenesis.

Heterochronic effects are produced by mutations affecting the length of the plastochron, as in *plastochron1-1* (*pla1-1*) and *pla1-2*, two recessive alleles of a heterochronic gene of rice: in mutant plants, the plastochron is reduced to ca. one-half its length in the wild-type individuals (Itoh *et al.*, 1998). However, the onset of the reproductive phase and the total duration of the vegetative phase are not affected, with the result that in the vegetative phase the mutant produces nearly twice as many leaves as the wild type (Lord, 2001).

In *Arabidopsis thaliana*, the *HASTY* gene product promotes the juvenile state and inhibits the transition to the adult reproductive state; a loss-of-function mutant of this gene has heterochronic consequences, by accelerating the phase transition (Telfer and Poethig, 1998).

Addition of extra nodes is also released by mutant alleles of the *TEOPOD* loci (*tp1, tp2, tp3*) in *Zea mays*, at least in certain genetic backgrounds (Poethig, 1988). As a result of the production of additional juvenile leaves, the ears are formed in the axils of later leaves compared with plants with wild-type alleles (Diggle, 1999).

Heterochronic genes may also change the architecture of the inflorescence. In *Antirrhinum majus*, a change in the timing or site of expression of *FLO* leads to a change of inflorescence type. For example, when the activation of *FLO* is delayed, a compound cyme (thyrse) was produced where a single flower would normally have been found (Coen *et al.*, 1990; Coen and Nugent, 1994).

The evolution of zygomorphic flowers is often the result of heterochronic change. This has been studied in depth in the case of *Iberis amara*. In the Brassicaceae, zygomorphy likely evolved via a shift of the expression of *TCP1* to late stages of flower development (Busch and Zachgo, 2007, 2009). In *Arabidopsis*, with actinomorphic flowers, *TCP1* is expressed early and only transiently in the adaxial domain of flower meristems. At the time the floral organs are initiated, *TCP1* expression has been already shut down. Temporal and spatial patterns of expression of the *TCP1* homologue in *Iberis amara* are different. This expression is strong, at later developmental stages, in the two adaxial petals, causing a reduced rate of cell proliferation; as a consequence, these organs remain shorter than the two abaxial petals, in which *TCP1* is not expressed (Cubas, 2004; Busch and Zachgo, 2007).

9.11.5 Heterochrony and Plant Phylogeny

Recognizing temporal patterns as a distinct class of phenotypes, as suggested in this chapter, is perhaps less unconventional than many researchers would imagine. In fact, temporal phenotypes – specifically, heterochronic patterns – may contain phylogenetic signals, and have actually been used in phylogenetic reconstruction (Minelli, 2015a; see also Minelli, 2015b for zoological examples).

Cauliflory, the production of flowers from the main woody axis of the plant rather than from younger branches or new shoots, can be regarded as a consequence of the heterochronic separation of vegetative and reproductive growth. Cauliflory is limited to quite a small number of taxa scattered among several distantly related families, including Moraceae, Malvaceae, Myrtaceae and Fabaceae. This distribution is a sign of multiple independent evolution of this trait. At a lower taxonomic level, however, cauliflory is an apomorphy of the clade *Zygia*, in the Fabaceae (Grimes, 1999). The effects on flower morphology of heterochronies such as anticipation or retardation of a whole whorl with respect to another, or of a single organ (e.g. a sepal or a petal) with respect to the other elements of its whorl, were demonstrated by Prenner and Klitgaard (2008) in a comparison of the developmental schedule of the unusual flower of *Duparquetia orchidacea* with those of other legumes (*Cercis canadensis, Petalostylis labicheoides, Labichea lanceolata, Ormocarpum sennoides* subsp. *hispidum* and *Tamarindus indica*).

10 Evolutionary Trends

10.1 EVOLUTIONARY TRENDS IN PLANT BODY COMPLEXITY

10.1.1 Number of Parts versus Divergence within a Series

It is a widespread opinion, although one rarely formulated in explicit terms, that, as a rule at least, recently evolved developmental processes responsible for the production of serial structures are far from strict: as a consequence, the number of elements in a series would be initially indeterminate and possibly large. Additionally, in the absence of specific control, the shape of these serially arranged parts would be largely the same throughout the series, but also subject to remarkable individual variation. With the subsequent evolution of more precise control mechanisms, with more rigorously localized gene expression, the number of these elements would be subjected to reduction and stabilization, perhaps accompanied by divergent specialization involving individual elements or localized sets of a few elements within the series. This is indeed the trend that dominates the evolution of flower organs, as documented in Chapter 7.

Similar trends have often been described in animals, with respect to number and specialization of body segments, vertebrae, teeth and other serially repeated features or body parts. In zoology and in animal palaeontology, this macroevolutionary trend is often discussed under the name of Williston's rule (or even Williston's law), based on a quite liberal interpretation of Williston's (1914) remarks on the evolution of skull bones in vertebrates. Irrespective of any possible historical abuse in attaching Williston's name to this principle (Minelli, 2003), it is important to say that this is at best an empirical generalization that applies to a number of instances, but is full of interesting exceptions, including clear examples of reversal of the expected trend. A couple of these can be taken from animals, before we return to flowering plants. The plesiomorphic number of vertebrae – whatever this was in early teleost fishes and in early reptiles, respectively – was certainly lower than the highest vertebral counts in eel relatives (ca. 750 in *Nemichthys*) and in snakes (more than 300 in some typhlopids); and the very high number of leg pairs in some geophilomorph centipedes (up to 191 in *Gonibregmatus plurimipes*) is the result of an exceptionally strong and evolutionarily late increase, compared to the 15 leg pairs reconstructed as the basal number in the Chilopoda (reviewed in Minelli, 2003).

To illustrate with a historical example how often the many-to-few trend in the number of serially arranged parts has been reversed in plant evolution, I will show how scattered in current classifications are the genera that Linnaeus grouped in the same class (Polyandria) because of their large number of stamens; more precisely, I will consider only the 36 genera classified by Linnaeus (1754) in the subclass Monogyna (with one pistil). Some of these genera do indeed belong to the basal ANA grade or to the lower eudicots, but the majority belong to the eudicots, within which many taxa have independently (re)evolved high numbers of stamens. Here is the full list of these Linnaean genera, redistributed according to modern views:

Nymphaeales: Nymphaeaceae (*Nymphaea*)
Ranunculales: Papaveraceae (*Argemone, Bocconia, Chelidonium, Papaver, Sanguinaria*); Berberidaceae (*Podophyllum*); Ranunculaceae (*Actaea*)
Dilleniales: Dilleniaceae (*Delima*, in Linnaeus as *Tetracera*)
Fabales: Fabaceae (*Mimosa; Saraca*, as *Sarracca*)
Oxalidales: Elaeocarpaceae (*Elaeocarpus, Sloanea*)
Malpighiales: Ochnaceae (*Ochna*); Chrysobalanaceae (*Chrysobalanus*); Calophyllaceae (*Calophyllum, Mammea, Mesua*); Clusiaceae (*Clusia; Garcinia*, as *Cambogia*)
Myrtales: Myrtaceae (*Plinia*)
Malvales: Muntingiaceae (*Muntingia*); Malvaceae (*Bombax, Tilia, Microcos, Corchorus*); Bixaceae (*Bixa*); Cistaceae (*Cistus*); Dipterocarpaceae (*Vateria*)

Brassicales: Capparaceae (*Capparis* incl. *Breynia, Morisona*)
Caryophyllales: Caryophyllaceae (*Dianthus*, as *Caryophyllus*)
Cornales: Loasaceae (*Mentzelia*)
Ericales: Marcgraviaceae (*Marcgravia*); Theaceae (*Camellia*, as *Thea*)

10.1.2 Regressive Morphological Trends

In the broad context of the evolutionary history of the land plants, the reduction in size and complexity of the gametophyte, eventually represented by a pollen grain, or an ovule with the accompanying embryonic sac, is arguably the most dramatic example of a regressive trend, but most of this history was already complete before the emergence of the angiosperms. In this section I will thus briefly focus on other examples of morphological simplification whose whole history falls within that of the angiosperms; most of these represent recent, even if morphologically conspicuous, transitions limited to smaller clades, from the individual species to the family.

Two main syndromes can be distinguished. The first is found in plants that have abandoned autotrophy, completely or to a large extent at least; the second embraces instead the anatomical correlates of miniaturization.

Two sets of angiosperms have abandoned autotrophy: those parasitic on other angiosperms and those that fully depend on their association with fungal mycelia. Among the first there are hemiparasitic plants such as *Melampyrum* and *Striga*, which are to some extent autotrophic and do not differ much from a conventional, fully autotrophic plant, and holoparasitic plants, such as *Orobanche* and *Rafflesia*, fully committed to heterotrophy. The latter habit is accompanied by a more or less extensive deconstruction of the vegetative architecture (in *Cuscuta*, among others, even the cotyledons are suppressed), whereas flowers and fruits are generally similar to those of their non-parasitic relatives, or even more conspicuous and complex, as in the case of *Rafflesia*. Adopting a parasitic lifestyle is thus one of the circumstances that most clearly exemplifies the constructional modularity of the plant body.

There are approximately 4500 species of parasitic plants (Heide-Jorgensen, 2013). These belong to ca. 30 families: Aristolochiaceae (Piperales), Lauraceae (Laurales), Cynomoriaceae (Saxifragales), Krameriaceae (Zygophyllales), Apodanthaceae (Cucurbitales), Rafflesiaceae (Malpighiales), Cytinaceae (Malvales), Balanophoraceae, Santalaceae, Loranthaceae and another 15 or more families (Santalales), Mitrastemonaceae (Ericales), Boraginaceae (Boraginales), Convolvulaceae (Solanales) and Orobanchaceae (Lamiales). This distribution suggests a minimum of 12 independent evolutionary events by which parasitic plants have emerged (Westwood *et al.*, 2010). Remarkably, virtually all parasitic plants belong to the core eudicots, and it is particularly striking that there are no monocots among them. This contrasts with the phylogenetic distribution of the little more than 500 species of mycoheterotrophic plants known to date: slightly fewer than half of these belong to the Orchidaceae, many others to different monocot families (Burmanniaceae, Thismiaceae, Triuridaceae, Corsiaceae, Iridaceae, Orchidaceae, Petrosaviaceae), and only a very small number have evolved within eudicot families (Ericaceae, Polygalaceae and Gentianaceae) (Merckx, 2013). The lack of endosperm in orchid seeds has possibly contributed to the success of obligate mycoheterotrophy in a number of representatives of this family, but a number of additional factors must probably be taken into account to explain the different evolvability of both parasitism and mycoheterotrophy in different angiosperm lineages.

In other plants, a strong simplification of the entire body is a correlate of miniaturization. The best example is provided by the Lemnoideae (Fig. 10.1), a small (37 species in 5 genera) but ecologically successful group of Araceae whose vegetative parts are reduced to a floating scale or even to a minuscule blob of green matter, less than one millimetre across; roots are retained in most species, but not in *Wolffia* and *Wolffiella*. This extreme morphological simplification is accompanied by an extreme reduction of the flowers (in *Wolffia*, these are unisexual and consist of a single stamen or a single carpel,

FIGURE 10.1 Duckweeds provide the most extreme examples of miniaturization and morphological deconstruction among the flowering plants. Here, common duckweed (*Lemna minor*).

respectively) and by the loss of vessels. In the angiosperms as a whole, this is an extremely rare condition, limited to *Amborella* and *Ceratophyllum*, were it not for its prevalence in a number of families of the Alismatales, including the sea grasses (Posidoniaceae, Zosteraceae, Hydrocharitaceae and Cymodoceaceae), all of which grow in sea water. It can be speculated that in this clade vessels have been lost with the colonization of salty waters. Within the Araceae, besides the Lemnoideae only the Orontioideae lack vessels.

The morphological trends accompanying miniaturization deserve reconsidering from an evo-devo perspective. Overall reduction in size affects different traits in different ways – changes being more obvious sometimes in the size of individual organs (leaves, flowers or specific flower parts), sometimes in their number; the temporal phenotype can also be affected. Miniaturization is in some cases a consequence of growing in dramatically impoverished environmental conditions (Hungerformen). Arber (1920) described the extreme phenotype of a plant of the genus *Littorella* flowering in a dry year, in which the stamen filament, of normal length, was longer than the rest of the plant. In other cases diminutive phenotypes have become fixed, as in the case of *Cardamine cubita* from the South Island, New Zealand (Heenan *et al.*, 2013). This miniaturized species is of special interest because it is closely related to the model plant *C. hirsuta*, from which it differs in a unique suite of reduced vegetative and floral characters. In *C. cubita*, cotyledons are persistent. None of the observed individuals, including one in fruit, produced more than three or four leaves. The overall size of the flower is extraordinarily reduced and is one of the smallest among the Brassicaceae. The whole flower is only 0.8–0.9 mm in diameter, the sepals 0.6–0.7 mm long and the anthers ca. 0.14 mm long. Miniaturization has also brought about a reduction in the number of some flower parts: petals are zero rather than four, stamens two rather than six. The two usual carpels have been retained, but each carpel bears only one ovule, rather than many. It is fair to say that the reduction in the number of stamens in the tiny *C. cubita* could easily be predicted from the instability of this trait observed in two related species, in both of which the number of stamens is variable and is more often four than six, i.e. *C. hirsuta* (Matsuhashi *et al.*, 2012) and *C. lacustris* (Garnock-Jones and Johnson, 1987; Heenan, 2002). Reduction to four or even two stamens has been otherwise reported, in the Brassicaceae, for several species of *Lepidium* (Lee *et al.*, 2002; de Lange *et al.*, 2013).

IO.2 PARALLELISM AND CONVERGENCE

Convergence and parallelism should not be regarded as synonyms, as suggested by many authors (e.g. Wiley, 1981; Williams and Ebach, 2007). In principle, there is parallel evolution only when the developmental processes directly responsible for the shared trait are homologous; otherwise, we should describe the relationship between similar phenotypes as convergence (Powell, 2007). In practice, however, the evidence required for a correct application of this distinction in any particular case is seldom available. Moreover, to refer to 'the generators directly causally responsible for the homoplastic event' (Powell, 2007, p. 567) implies the essentialist view of animal construction so obvious in Owen's (1843) famous definition of homologue ('the same organ in different animals under every variety of form and function') and still present, while perhaps not so obvious, in the traditional all-or-nothing views of homology criticized in Section 2.6.

Between the late 1970s and the early 1980s, just as systematists working on several animal groups, from fishes to midges, were rapidly accepting cladistic concepts and methods, a parallel development in angiosperm systematics was extensively rejected based on the argument that parallelism and convergence are all too frequent in plants, to such an extent that a search for unique synapomorphies is too often doomed to failure. These objections, however, were overturned as soon as plant systematists began to search for molecular, rather than morphological, synapomorphies, although homoplasy does occur at the level of sequences, too. Eventually, the successful revision of angiosperm systematics based on molecular methods has helped to reveal that the amount of homoplasy actually occurring at the morphological level is even larger than anticipated. Endress (2011) indirectly provided an excellent sample of the taxonomic distributions of convergent major traits, in a review paper in which he listed a number of 'innovations' that turn out to be largely characteristic of a larger or smaller lineage, but also show some 'occasional occurrence' in other clades. From his list I extract the following examples:

- Pentamerous flower: predominant in core eudicots but also sporadically occurring in Ranunculales, Sabiaceae and a few monocots
- Zygomorphic flower: independently evolved in a few major clades such as Orchidaceae, Zingiberales, Fabaceae, Polygalaceae, Lamiales, Dipsacales, but also in many Asterales and occasionally elsewhere
- Perianth differentiated into sepals and petals: predominant in core eudicots, Commelinales and Zingiberales, following independent evolution in each of the three clades; but also sporadically found elsewhere, e.g. among the Alismatales and Ranunculales

- Sympetaly: characteristic of asterids, especially euasterids, but sporadically present also in other core eudicots (e.g. Crassulaceae, Malvaceae), monocots (e.g. some Burmanniaceae) and basal eudicots (e.g. some Menispermaceae)
- Fusion of stamens and petals in the same sectors: common in monocots, dominant in euasterids; but also in a few Ranunculales and in the Sabiaceae
- Pollinaria: a key innovation independently evolved in Orchidaceae and Apocynaceae
- Syncarpy: a key innovation in two clades, core eudicots and monocots; occasionally present in basal angiosperms (e.g. Nymphaeaceae, Annonaceae, Piperaceae and Aristolochiaceae) and basal eudicots (e.g. Papaveraceae, Buxaceae)

Some features, relatively labile in basal angiosperms, exhibit a clear trend towards stabilization in more derived groups. For example, floral phyllotaxis is stabilized as whorled, rather than spiral, both in monocots and in core eudicots; the number of organs in a whorl is stabilized as three in monocots and five in core eudicots. However, other features are quite uniform in basal angiosperms and become more diverse in derived clades. Flowers in most basal angiosperms are actinomorphic, whereas zygomorphy and asymmetry characterize a number of derived clades, including large and successful families, e.g. Orchidaceae and Fabaceae, as well as Lamiaceae and other lamiids. Fused floral organs are infrequent in basal angiosperms, but occur frequently in derived clades; sympetaly, in particular, enables the evolution of a diversity of floral shapes (Endress, 2011).

As long as morphological comparisons are intended as a means for reconstructing phylogeny, homoplasy (in its several forms: reversal, convergence, parallelism) is nothing but a nuisance. However, parallelism, convergence and reversal are part of evolutionary history no less than the emergence and conservation of derived traits. The notorious pervasiveness of homoplasy in flowering plants has even prompted a plant morphologist to say that morphology is the *science of convergences* rather than the science of homologies (Weber, 2003).

Similarities due to causes other than common ancestry can result from a diversity of processes (Washburn *et al.*, 2016) including genetic drift and a prolonged exposure to similar selective forces, but this is not the whole story. A broader, evo-devo perspective is required, in particular, to understand the strongly uneven occurrence of these phenomena across the phylogenetic tree, and to explain how whole structural modules have been extensively conserved even in cases where most of the body architecture has been involved in a dramatic history of homoplastic change. In other words, this large, and in many circles unpopular, chapter of plant morphology must be explored in terms of modularity and evolvability, but an evo-devo re-examination of parallelism and convergence is, to a large extent, a matter for future research. The short account given below includes a number of examples of potentially interesting cases and a brief mention of the few studies that have hitherto been devoted to some of them.

Major examples of convergence among distantly related lineages of flowering plants are offered by specialized floral structures, such as the independently evolved keel zygomorphic flowers of the Fabaceae and Polygalaceae (see Sections 7.4.3 and 10.2.2), and the similarly synorganized flowers of the Orchidaceae and Apocynaceae Asclepioideae (see Sections 7.6.1.3 and 10.2.2). However, much more frequently, convergence is found in vegetative structures, while flowers, on the whole, are more conservative. As expected, convergence is much more frequent in relatively simple organs, such as the leaves, but it is more impressive when it affects the whole habit of the plant. However, from an evo-devo point of view, we must always take into account that the convergent evolutionary transitions affecting gross aspects of the phenotype are not necessarily more complex, or less probable, than those by which 'simple' leaf morphologies have convergently evolved.

In the end, morphology is only one of the perspectives from which we can describe the occurrence of convergence and parallelism in plant evolution. On the one hand, convergent morphological evolution is subtended by changes in the underlying developmental processes, and changes in the latter do not necessarily map in a simple way onto changes in morphology; on the other, morphology has more or less direct and more or less important functional correlates that may have played a strong role in the fixation of a convergent trait. An obvious example is dioecy, which is thought to have evolved not less than 100 times to account for the 160 plant families in which dioecious species are present (Charlesworth and Guttman, 1999). Quite likely, the actual number of transitions to dioecy is even larger, because this condition has probably evolved multiple times independently within one family, as in the cases of the Arecaceae, Rosaceae and Euphorbiaceae (Mitchell and Diggle, 2005).

10.2.1 Convergence in Habitus and Vegetative Organs

Textbook examples of convergence are the succulent stems so typical of the Cactaceae but also evolved independently, e.g. in several taxa of the Euphorbiaceae, Asteraceae and Apocynaceae. Another remarkable growth form that has evolved multiple times from herbaceous ancestors is the rosette shrub. This habit evolved independently at least three times in the Asteraceae alone, with *Dendrosenecio* in the high mountains of tropical East Africa, *Espeletia* in the northern Andes and *Argyroxiphium* on the Hawaiian Islands, but also occurs in some genera of the Campanulaceae Lobelioideae, Caprifoliaceae and Bromeliaceae (Givnish, 2010).

A unique growth form, vaguely resembling a palm with a bare, unbranched stem bearing a crown of very large leaves, is represented by the giant lobeliads of the tropical mountains in East Africa and Hawaii. According to Antonelli (2009), all giant lobeliads derived from a single evolutionary transition, but the opposite scenario of a convergent evolution of this life form is much more likely (Givnish, 2010).

The latter example indicates that a sound phylogeny is required, to decide whether the transition to an unusual phenotype has occurred only once, within a lineage, or multiple times in parallel, or convergently. The risk of wrong interpretations is particularly high when closely related taxa are involved. These are likely to share similar evolvability towards the same phenotype, but this does not mean that they form a monophyletic group. Recent phylogenetic investigations are revealing a number of examples of multiple parallel transitions within one genus, evolutionary pathways that would deserve closer study from an evo-devo perspective. An example is the genus *Sparganium*, within which floating-leaved species have arisen multiple times from emergent ancestors (Sulman *et al.*, 2013). More attractive is the case of *Streptocarpus*, because of the very unusual phenotypes that have evolved within this genus. As mentioned in Section 8.5.4, three main body architectures are found in these plants. There are caulescent species, with a normal shoot apical meristem and a conventional body structure; unifoliate species, with a single leaf-like organ, corresponding to one of the two cotyledons, growing to up to 0.75 m and bearing the inflorescence; and rosulate species, in which the morphology of the unifoliate kind is iterated in the form of leaves arising from leaves. Phylogenetic studies have revealed that both the unifoliate and the rosulate growth form have evolved several times, and there have also been some reversals (Müller and Cronk, 2001).

Tendrils are among the best examples of convergence in plant morphology, having derived, in different families, from very different organs: stipules in the Smilacaceae, leaflets in the Fabaceae and the Bignoniaceae Bignonieae, shoots in the Cucurbitaceae, inflorescences in the Passifloraceae and Vitaceae. One of the reasons why the true identity of tendrils is often difficult to interpret is that they may or may not bear leaves and buds.

A controversial case is represented by the tendrils of *Passiflora* species. According to Shah and Dave (1970), these tendrils derive from accessory axillary buds and are homologous to a flower (the central flower of a dichasium or the first flower of a monochasium; see Section 5.10.1). A different interpretation is suggested, however, by Prenner (2014), based on his studies of flower ontogeny in *P. lobata*, which support the view that the tendril represents the sterile terminus of the inflorescence.

In the grapevine and its relatives, i.e. in the majority of the Vitaceae, there are tendrils arranged in characteristic positions, i.e. opposite typical leaves, but not at every node; instead, in several species the alternation between nodes with a tendril and nodes without is very precise (Bell, 2008).

The tendril of the Vitaceae has long been considered homologous to an inflorescence, an interpretation supported by the occasional occurrence of structures intermediate between a typical tendril and an inflorescence (Boss and Thomas, 2002). Eventually, the studies of Calonje et al. (2004) provided molecular support to the hypothesis that the tendrils of Vitis are modified reproductive organs. In the grapevine, the joint expression of the putative Vitis homologues of FUL and AP1 is not limited to inflorescence and flower meristems, but is also found throughout tendril development. This is not due to an unrestricted expression of these genes throughout the whole plant, as neither of these genes is expressed in vegetative organs such as leaves or roots. On the other hand, FUL and AP1 expression in the developing tendrils is not dependent on the onset of the flowering transition, but is found also in the tendrils formed by very young plants. There are obvious similarities between the tendrils of Passifloraceae and Vitaceae, but Prenner (2014) cautioned that they are not necessarily homologous, and even floated the possibility that not all tendrils are homologous among all species of a large genus such as Passiflora.

Expression of AP1 orthologues was also detected in the tendrils and the inflorescences (but not in the leaves) in several species of the Vitaceae other than Vitis vinifera, further supporting the interpretation of the tendril as a structure with a remarkable degree of homology with the inflorescence (Zhang et al., 2015).

Different is the origin of the leaf tendrils of legumes, interpreted as abaxialized leaflets or segments of the leaf rachis (Tattersall et al., 2005). In Pisum sativum, the development of the tendril is controlled by the transcription factor TENDRIL-LESS (TL) (Hofer et al., 2009); in loss-of-function mutants of TL, tendrils are transformed into leaflets.

TL orthologues have been identified in tendrilled legumes including *Vicia* spp., *Lens culinaris* and *Lathyrus odoratus*, but not in the non-tendrilled *Medicago truncatula* (Hofer *et al.*, 2009).

Tendrils are also common, and diverse, within the Bignoniaceae Bignonieae. Sousa-Baena *et al.* (2014b) identified the main steps in the evolution of these structures. The plesiomorphic condition was reconstructed within the tribe as a tendril-less, bi-ternate leaf; at the base of the group known as the 'core Bignonieae', a new leaf type evolved, with two undivided leaflets and a tendril divided into three branches, each of which is regarded as a homologue to a secondary foliolule of the terminal leaflet. This reduction of the meristematic capacity of the leaf primordium probably resulted from a lesser duration of the *KNOX1* expression in leaf primordia, possibly with accompanying changes in *WOX1/3* activity (Nakata *et al.*, 2012). Reversal to tendril-less leaves is observed in 10 species. Two of them derive from ancestors provided with the trifid-tendrilled leaves described in the previous lines, the others from ancestors bearing two- to three-foliolate leaves ending in a single unbranched tendril. In *Adenocalymma*, the terminal unbranched tendril is replaced by a pinnate leaflet. The genetic control of tendrilled leaf development was studied in three species (*Amphilophium buccinatorium*, *Dolichandra unguiscati* and *Bignonia callistegioides*). Transcripts of both STM and LFY/FLO were found in leaf primordia, associated with regions from which either leaflets or tendril branches originate (Sousa-Baena *et al.*, 2014a).

Convergence in leaf shape and arrangement is extremely widespread and includes also examples among less common leaf forms, such as the terete and ensiform leaves that have evolved repeatedly in monocots (Rudall and Buzgo, 2002).

Convergence, however, is not frequently mentioned in discussions about phyllotactic patterns, possibly because their usual mathematical simplicity suggests that the regular arrangement of lateral organs can be explained by a kind of mechanical necessity (cf. Hofmeister's rule; Section 5.7) or, in more abstract terms, described in what a mathematician would call the simplest and thus the most elegant way. However, this is an aspect that may deserve closer scrutiny, as suggested, for example, by the following list of alternative ways by which vascular plants may produce whorls with four or more leaves (based on Rutishauser, 1999). New shoots may start with dimerous or trimerous whorls, followed by whorls with higher numbers of leaves, caused, among others:

- by continuous increase in size of the apical meristem (e.g. *Equisetum*, *Hippuris*)
- in taxa with decussate phyllotaxis, by the failure of every second internode to elongate, thus collapsing two successive leaf pairs to form a tetramerous whorl (e.g. *Silene stellata*)
- by replacement of stipules by leaves (e.g. *Galium* and allies)
- by shift of basal leaflets of compound leaves to around the node (e.g. *Limnophila*, probably also *Ceratophyllum*)
- by spiral inception of all leaves, but with elongation limited to the kth internode, thus forming k-merous whorls (e.g. *Lilium*, *Polygonatum*)
- by adding supernumerary leaves between a first series of helically arranged leaves (*Acacia baueri* and *A. verticillata*); this will eventually result in chaotic phyllotaxis, as recently illustrated by Rutishauser (2016b) in a number of *Acacia* species and in *Cananga odorata* and other representatives of the Annonaceae
- by additional leaves of a whorl arising on an annular bulge surrounding the node of a previously formed, whorl-founding leaf (*Hydrothrix*)

10.2.2   *Convergence in Flower Structure*

Several examples of parallel evolution of similar flower morphologies have been mentioned in Chapter 7. Particularly striking, and thus worth mentioning here again, is the similarity between the papilionate flowers independently evolved in the Polygalaceae (Forest *et al.*, 2007; Bello *et al.*, 2009) and in the Fabaceae (Westerkamp and Weber, 1999; Prenner, 2004; Bello *et al.*, 2012). In the former family, this specialized flower type is restricted to the tribe Polygaleae, whereas different papilionate morphologies have evolved in parallel within the legume family (Pennington *et al.*, 2000).

In the plant body, flowers are indeed *the* hotspot for the evolution of novelties. Some of these, like spurs and nectaries, are best expressed in mature flowers and are thus more easily targeted by selection; less conspicuous, but nonetheless important in canalizing the final phenotype, are ring meristems, an innovation in the shape of the floral apical meristem, which takes the form of a ridge whose centre is either depressed or occupied by the gynoecial primordium. The initiation of organ primordia on a ring meristem, rather than from small individual meristems, is frequently accompanied by an increase in the number of flower organs. Ring meristems have evolved multiple times, with examples of their occurrence scattered among, e.g., Capparaceae, Bixaceae, Fabaceae, Malvaceae and Dilleniaceae. Flowers with ring meristems are not necessarily actinomorphic. In those in which radial symmetry is eventually obtained at maturity, organ inception is often centripetal or centrifugal, but it is unidirectional in some representatives of the Rhizophoraceae (Juncosa, 1988). Among the monocots, mature flowers with organs arising from a ring meristem are eventually zygomorphic (Orchidaceae, Zingiberaceae, Costaceae) or asymmetric (Cannaceae and Marantaceae) (Tucker, 1999).

Circular outgrowths or marginal extensions of the receptacle have arisen many times. In some clades, these novel structures grow to a very large size and become as conspicuous as the perianth. Examples include the hypanthium of *Tropaeolum* and *Punica*, the corona or paracorolla of *Narcissus* (Fig. 10.2) and the corona of *Passiflora* (Fig. 10.3) (Ronse De Craene and Brockington, 2013).

Coronas, usually petaloid, are found indeed in a number of distantly related genera (I mentioned in Section 7.6.2 their occurrence in the Amaryllidaceae and Velloziaceae among the monocots, and in the Apocynaceae among the core eudicots) and have different origins (Endress and Matthews, 2006a; Ronse De Craene, 2010; Sajo *et al.*, 2010; Ronse De Craene and Brockington, 2013). The coronas of the Velloziaceae and those of the Passifloraceae (as mentioned) are part of the hypanthium; those of *Tulbaghia* are instead an appendage of tepals; those of the Sapindaceae and Tamaricaceae, an appendage

FIGURE 10.2 Flower of a large-cupped daffodil (*Narcissus*) with conspicuous corona. *A black and white version of this figure will appear in some formats. For the colour version, please refer to the plate section.*

FIGURE 10.3 Flower of purple passionflower (*Passiflora incarnata*). The receptacle of the flower is enlarged by a corona formed from numerous long filaments issued from the hypanthium, loosely surrounding the gynoecium – a good example of evolutionary innovation. Courtesy of Ioannis Schinezos.

FIGURE 10.4 Flowers of a hybrid cultivar of columbine (*Aquilegia*) showing the five long spurs, an unmistakeable innovation of this genus of the Ranunculaceae. *A black and white version of this figure will appear in some formats. For the colour version, please refer to the plate section.*

of petals; those of the Apocynaceae, an appendage of stamens. Staminodes, sometimes (*Napoleonaea*) together with modified and fused petals, contribute to the complex coronas of the Lecythidaceae (Endress, 1994a; Ronse De Craene, 2011).

In several lineages, hollow outgrowths, often hooked at the tip, are present on one, two or more perianth organs. These spurs have evolved especially in eudicots, in corollas more often than in calyces. Outside the eudicots, spurs are found only in some orchids and in a few Liliaceae (e.g. *Tricyrtis*) (Soltis *et al.*, 2005). Most spurred flowers, such as *Delphinium*, some Plantaginaceae and *Heterotoma*, have only one spur. Two spurs are characteristic of *Lamprocapnos spectabilis*. Examples of genera with four or five spurs per flower are *Aquilegia* (Hodges, 1997a, 1997b) (Fig. 10.4), *Epimedium* and *Halenia* (von Hagen and Kadereit, 2002).

Floral nectaries are diverse morphologically and histologically (reviewed in Endress, 2011). In basal angiosperms, nectaries are located on various floral organs, such as tepals, stamens or carpels, and are not elaborate. In monocots, nectaries are often found on the flanks of carpels, less frequently on tepals (Pandanales, Orchidaceae, Liliales) or other floral organs (some Iridaceae). In basal eudicots, nectaries occur most frequently on staminodes or carpels. Disc nectaries have evolved in core eudicots in the area between androecium and gynoecium, and other forms of nectaries have evolved many times in various groups; characteristic are the hair nectaries of the Malvales and Dipsacales. Loss of nectaries, sometimes followed by the evolution of novel structures of similar kind, but in a different location, has occurred many times in the history of flowering plants (Endress, 2011).

The most striking level of convergence in flower structure is arguably represented by the occurrence of similar, complex forms of synorganization involving essentially the same organs, such as between the androecia and gynoecia in two plant lines as distantly related as ascleps and orchids (see Sections 7.6.2 and 7.6.4).

The most striking aspect of this example of convergence in synorganization is that it is based on flowers with different symmetry: actinomorphy in the Apocynaceae, zygomorphy in the Orchidaceae. Endress (2015) reconstructs in the following way the putative sequences in which the many innovations eventually culminating in the uniquely synorganized flowers of these two plant clades have likely evolved:

Apocynaceae: (1) fixation of pentamery; (2) fixation of syncarpy, at least at the base of the gynoecium; (3) fixation of a single whorl for each category of flower organs; (4) sympetaly; (5) postgenital fusion of carpel tips and anthers with the gynoecium; (6) ontogenetic remoulding of the style head, with transition from the anatomical dimery to the morphological and functional pentamery; (7) production of 4 or 2 pollinia per stamen; (8) production of 5 pollinaria per flower.

Orchidaceae: (1) fixation of trimery with two perianth whorls and two androecial whorls; (2) congenital fusion of all tepals; (3) differentiation of the lip; (4) reduction of the number of stamens from 6 to 2 or 3;

(5) complete syncarpy; (6) reduction of the number of stamens to 1; (7) production of 4 or 2 pollinia per stamen; (8) production of 1 or 2 pollinaria per flower (Endress, 2015).

I close this short section with one example of convergence in fruit morphology. Different combinations of floral parts and accessory structures are responsible for the independent evolution of wing-like appendages (important for fruit dispersal) in some genera of the Juglandaceae. Wings correspond to a trilobed bract in *Engelhardia* and *Oreomunnea*, but are derived from the fusion of bracteoles and lateral sepals in *Platycarya*, from bracteoles in *Pterocarya*, from bracts and bracteoles fused together in *Cyclocarya* (Manos and Stone, 2001).

10.2.3 Brassicaceae: a Cradle of Convergence

From an evolutionary point of view, convergence between distantly related lineages is more impressive than convergence among close relatives, but the latter is likely to occur much more frequently, because close relatives are likely to share similar evolvability. Convergent or parallel evolution of similar traits in multiple representatives of a family, or even of a single genus, may be difficult to uncover before the whole group has been subjected to a sound phylogenetic analysis and the distribution of the different character states, for any trait potentially affected by homoplasy, has been accurately mapped onto the resulting tree. Eventually, a phylo-evo-devo analysis may reveal the existence of clades that are particularly prone to homoplasy. For example, this is the case in the Brassicaceae, with some 3700 species currently classified in more than 300 genera, in which convergent evolution has occurred, according to Bailey *et al.* (2006), in nearly every morphological feature used to define tribes and genera. Here are a few examples of tricky characters and problematic (usually polyphyletic) taxa.

Homoplasy affects almost all aspects of the morphology of the Brassicaceae, but especially the fruit, whose shape and structure have been extensively used for delimiting genera. Remarkable are the siliques longitudinally divided into two halves through a septum corresponding to the shortest diameter of the fruit's elliptical cross-section, a feature that evolved independently in 25 of the 44 tribes recognized to date (Franzke et al., 2011).

Shrubs evolved independently, in this family largely dominated by herbaceous species, in at least 12 distantly related tribes; lianas are found in four distinct clades. Gamosepaly is known in 10 genera of five tribes; divided petals evolved independently in *Schizopetalon, Ornithocarpa, Dryopetalon, Berteroa* and *Draba verna* (Franzke et al., 2011). The species long classified in the genus *Arabis* belong to at least two distinct lineages, with extensive morphological convergence (Koch et al., 2001; Koch and Al-Shehbaz, 2002; Mitchell-Olds et al., 2005), one closely related to *Arabidopsis*, the other to *Draba*.

10.3   EVOLUTIONARY REVERSAL

In the small family Chloranthaceae (75 species in four genera), flowers are unisexual, except in *Sarcandra*, where a bisexual (hermaphrodite) reproductive unit has secondarily evolved. In principle, this is just one among a huge number of histories of evolutionary reversal that we can trace along the branches of the phylogenetic tree of the angiosperms. However, the case of *Sarcandra* deserves special mention, because this transition to a reproductive unit in which male and female organs are united could be described, technically, as a reinvention of the angiosperm flower!

Similar to parallelism and convergence, evolutionary reversal is an aspect of phenotypic evolution that cannot be satisfactorily explained by framing the question purely in terms of adaptation: the necessary complement is an adequate understanding of the system's evolvability. As in other sections of this book, I am dealing here with a topic about which large amounts of circumstantial evidence are available, but targeted studies from an evo-devo perspective have been very limited to date.

A first level of analysis is to estimate asymmetries in rates of change among character states. This question was addressed by Geeta *et al.* (2012) in their study of leaf complexity in the flowering plants. Seven data sets were considered, the largest of which included a sample of 560 angiosperm species scattered across the phylogenetic tree, while the other six – all derived from the former – were restricted to representatives of lower clades of various size and taxonomic rank (Sapindales, Apiales, Papaveraceae, Fabaceae, *Lepidium*, *Solanum*). Within the largest data set, simple leaves – more frequent than dissected or compound leaves among the living angiosperms, and reconstructed as the ancestral character state in the whole clade – are for the most part retained. The multiple transitions to dissected or compound leaves have been followed by an even larger number of reversals to the simple leaf; moreover, loss of leaf complexity has been, on average, quicker than change in the opposite direction. The same trends were observed within the large genus *Solanum*, but in the other clades the transitions from simple to dissected or compound leaves have been more frequent than the opposite.

The following examples of evolutionary reversal are taken from flower morphology. As mentioned in Section 2.3, zygomorphy has been repeatedly lost in most of the major zygomorphic lineages of flowering plants (Donoghue *et al.*, 1998; Citerne *et al.*, 2006; Zhou *et al.*, 2008; Pang *et al.*, 2010; Zhang *et al.*, 2010). Multiple independent transitions from bilateral to radial symmetry have even occurred within one family, e.g. in the Malpighiaceae, with *Psychopterys*, *Sphedamnocarpus* and *Lasiocarpus* (Zhang *et al.*, 2010, 2013b). Within the Lamiales, in which this reversal has occurred a number of times (Smith *et al.*, 2004), the transition to actinomorphy is more frequent in the clades in which the zygomorphic flower had retained the fifth stamen, although as a sterile staminode (Baum, 1998; Endress, 1998; Smith *et al.*, 2004).

Reversal to actinomorphy is thus a convergent trend (Washburn *et al.*, 2016); its underlying mechanisms are far from uniform, but are likely to involve in any case changes in the regulation and expression of *CYC*-like genes. This brings us back to the most popular among the old discoveries that we regard as iconic antecedents of plant evo-devo, i.e. to a naturally occurring mutant phenotype of *Linaria vulgaris* described more than 250 years ago by Linnaeus under the name of *peloria*, in which the symmetry of the flower is changed from bilateral to radial. In peloric *Linaria*, the homologue of the *CYC* gene is extensively methylated and transcriptionally silent. Demethylation of the *Linaria CYC* homologue occurs occasionally during somatic development; this causes restoration of gene expression and reversal to the zygomorphic phenotype (Cubas *et al.*, 1999). The epigenetic nature of the peloria phenotype is peculiar to *Linaria vulgaris*, but *CYC*-like genes are widely involved in the evolutionary transitions from zygomorphic to actinomorphic flowers, at least in the Lamiales and allies (Luo *et al.*, 1996; Donoghue *et al.*, 1998; Coen, 1999; Cubas *et al.*, 1999; Endress, 1999; Ree and Donoghue, 1999; Citerne *et al.*, 2000).

In many reversals from zygomorphy to actinomorphy, flower merosity has been conserved. This is the case in species of the Gesneriaceae with radially symmetrical flowers, all with five petals and five stamens, like their zygomorphic relatives (Endress, 1998, 2001a; Citerne *et al.*, 2000). Five petals are also recognizable in the peloric forms frequently occurring in nature in *Streptocarpus*, but in this genus peloric flowers with four petals have been occasionally reported (Bateson and Bateson, 1891).

In several members of the Lamiales, 4-merous actinomorphic flowers have derived from ancestors with 5-merous zygomorphic flowers, as in *Plantago*, *Veronica* and *Buddleja*. This has been described as resulting from the 'dorsalization' of two adaxial and two abaxial petals (Theißen, 2000), or in terms of complete reduction of the median adaxial floral sector (Endress, 2012).

An opposite trend, with increase from the 5-merous flower of the zygomorphic ancestors to the 6-merous flower of the actinomorphic descendants, has also been recorded in a few Lamiales, e.g. in *Sibthorpia* (Endress, 1994a).

Multiple mechanisms have been hypothesized by which bilateral flower symmetry might be lost, depending on changed regulation and expression of *CYC*-like genes (Preston *et al.*, 2011a; Hileman, 2014a). One possibility is the complete loss of *CYC*-like gene expression in flowers, because of either gene loss or changes in their regulation. But regulatory evolution may have resulted instead in expanded expression of *CYC*-like genes across the dorsoventral axis of developing flowers. Finally, the evolution of radial symmetry from *CYC*-dependent bilateral symmetry could arise through mechanisms independent of the functional or regulatory evolution of *CYC*-like genes.

Citerne *et al.* (2006) compared the expression patterns of two paralogous *CYC*-like genes in two legumes, one with typical zygomorphic flowers, *Lupinus nanus*, the other with actinomorphic flowers, *Cadia purpurea*. In *Lupinus*, the adaxial expression was identical for the two paralogues, but in *C. purpurea* the two *CYC* copies were differentially expressed during late stages of organ development, when the *CYC1A* homologue was only weakly expressed in adaxial petals, whereas the expression domain of the *CYC1B* homologue expanded to all petals, presumably causing their uniform adaxialization.

The role of *CYC*-like genes in the reversion to actinomorphy was also studied in the Gesneriaceae (Smith *et al.*, 2004). Expression patterns in three species are now available for comparison. *Chirita heterotricha* has zygomorphic flowers in which the adaxial petals are smaller than the remaining petals and the only two fertile stamens are both formed in an abaxial position, whereas the lateral stamens develop into sterile filamentous structures and the adaxial stamen is fully aborted; *Oreocharis benthamii* has a weakly zygomorphic corolla and an adaxial staminode; *O. leiophylla* (= *Bournea leiophylla*) has actinomorphic flowers with five petal lobes and five stamens of equal size and shape. Four *CYC2* genes were isolated from *Chirita*, two of which (*ChCYC1C* and *ChCYC1D*) are expressed in the adaxial petals and adaxial staminodes, but the expression domain of *ChCYC1C* extends into the lateral stamens: according to Gao *et al.* (2008), this could be indicative of a role in stamen abortion. It seems that, relative to *CYC* expression in *Antirrhinum*, a spatial *CYC2* expression shift resulted in altered flower symmetry in *Chirita*. In *Oreocharis benthamii*, *ObCYC1* is expressed in adaxial regions of the flower, but this expression decreases at later stages of development (Du and Wang, 2008). In *O. leiophylla*, *OlCYC1* is expressed in the early adaxial stamen and petals, but expression ceases before anthesis: this has been indicated as responsible for the almost fully actinomorphic flowers of this species (Zhou *et al.*, 2008).

In the Plantaginaceae, different changes in the expression of *CYC*-like genes or in genes downstream of them are involved in the reversal to flower actinomorphy. Available comparisons contrast the zygomorphic *Digitalis purpurea* with the actinomorphic *Aragoa abietina* and *Plantago major*. Of the two (at least) expected *CYC*-like paralogues, only one has been found in *Plantago lanceolata* and *P. major*, suggesting a recent gene loss event (Hileman and Baum, 2003; Preston *et al.*, 2009; Reardon *et al.*, 2009), compared with the genomes of two close relatives of *Plantago*: two *CYC*-like genes are present in *Veronica serpyllifolia* and three in *Digitalis purpurea*. Preston *et al.* (2011b) demonstrated that, in addition to a dorsally expressed *CYC*-like gene, *Plantago* species have also lost the downstream targets *RAD* and *DIV*. But this is not true of *Aragoa*. The single *CYC*-like gene found in *P. major* is expressed across all regions of the flower, similar to the expression of its orthologue in the closely related *Veronica serpyllifolia*.

Uniform and overlapping patterns of expression of both *CYC*-like paralogues are found instead in two Malpighiaceae with actinomorphic flowers: in *Psychopterys*, both genes are expressed across the dorsoventral flower axis, whereas in *Sphedamnocarpus* neither gene is expressed in flowers (Zhang *et al.*, 2013b).

Somehow parallel to the reversal of flowers from bilateral to radial symmetry is the reversal from zygomorphy to disymmetry in the Papaveraceae Fumarioideae. The disymmetric floral ground plan, also present in *Hypecoum*, is more easily observed in *Lamprocapnos* and related genera because of the presence of two spurs in one of the two symmetry planes. Reduction of one of the spurs (as in *Corydalis*;

FIGURE 10.5 *Corydalis cava,* zygomorphic flowers with one spur.

Fig. 10.5) causes transition to zygomorphy. From the latter condition, flowers can revert to disymmetry either by re-evolving a second spur, as exceptionally observed in *Corydalis solida* or *Capnoides sempervirens,* or by losing their spur, as in *Corydalis cheilanthifolia* (Endress, 1999).

Reversal to homostyly has been reconstructed in some lineages in which heterostyly had previously evolved. In *Amsinckia* (see Section 2.7.1), the homostylous *A. vernicosa* has evolved from a heterostylous ancestor comparable to *A. furcata,* whereas the homostylous *A. tessellata* var. *gloriosa* has evolved from a heterostylous ancestor comparable to *A. douglasiana* (Barrett, 2010).

10.4 SALTATIONAL EVOLUTION

'A key problem in understanding the evolution of morphology is that it is impossible to predict from observations of a final shape the patterns of growth that produced it' (Canales *et al.,* 2010, p. 27). The genotype → phenotype map is complex, largely unpredictable and, in any case, very far from linear. Small genetic changes may translate into dramatic phenotypic differences, and large genetic rearrangement may have minor consequences at the phenotypic level. In this scenario, saltational evolution is not forbidden.

The most elementary examples of saltational changes are provided by events by which the number of elements in a floral whorl is multiplied, either by splitting of each primordium into two equal organs, or by replacement of each individual primary primordium with a fascicle of secondary primordia, from each of which a complete organ will develop. According to Nuraliev *et al.* (2014), saltationary events are responsible for the explosive polymery of *Schefflera* flowers. These plants exhibit by far the most extreme floral polymery within the Araliaceae: 19–43 stamens and 15–33 carpels are found in a flower of *Schefflera subintegra*, 60–172 stamens and 60–138 carpels in *Schefflera pueckleri*. Nuraliev *et al.* (2014) suggest a saltational multiplication of floral elements accompanied by a loss of individuality of sepals in the calyx and petals in the corolla, followed by further polymerization of stamens and carpels, possibly dependent on mutations in *CLV*-like genes.

Bateman and DiMichele (1994, 2002; see also Rudall and Bateman, 2002, 2003) hypothesized that strongly deviant phenotypes – Goldschmidt's (1940) 'hopeful monsters' – are steadily generated owing to mutations in genes that control morphogenesis, thanks to the modularity of the developing plants, which provides opportunity for positional shifts of body parts (heterotopy) and changes in the temporal schedule of morphogenetic events (heterochrony).

Indeed, as remarked by Theißen (2006), this happens, for example, when changes in the expression domains of floral homeotic genes translate into homeotic transformations of floral organs, as seen in the transformation of sepals into carpelloid organs, or of petals into staminoid organs (Bradley *et al.*, 1993). Theißen (2009) has explicitly defended saltational evolution as a concept necessary to describe a number of key innovations and changes in plant architecture that are unlikely to have evolved as the cumulative result of a large number of individually small modifications. Before the advent of evo-devo, the very idea of saltational evolution as an explanation for macroevolutionary transitions (Goldschmidt, 1940) was strictly banned as heretical. However, an appreciation of the non-linear character of the genotype → phenotype map is enough for us to realize how major phenotypic changes can be accomplished in a leap. For example, a single-gene mutation was probably responsible for the evolution of the bilaterally symmetrical orchid flowers from an ancestor with radially symmetrical ones (Theißen, 2009).

To be sure, the arrival of a hopeful monster (that is, its generation through a sustainable sequence of developmental events) does not guarantee its eventual survival, especially in local environments where a well-established wild type is abundant. But Theißen and colleagues (Hintz *et al.*, 2006; Nutt *et al.*, 2006; Theißen, 2009) have at last successfully launched a floral homeotic mutant of *Capsella bursa-pastoris* as a model system to investigate the performance of a hopeful monster in the wild. In this mutant, known as *stamenoid petals* (*spe*), petals are completely replaced by stamens, while all other floral organs are as in the wild type. This is not one of the usual homeotic mutants isolated in the lab; instead, it has been found in natural habitats, where it has probably existed for a number of generations, thus demonstrating a fitness comparable to that of the wild type.

11 Looking Ahead

Focusing on individual aspects of a plant's phenotype, for example the symmetry of flowers, the complexity of the leaf margin or the number of stamens, is usually a sensible way to begin an experimental study in plant developmental biology, often supported by a good degree of modularity that allows us to ignore interferences from neighbouring structures. However, this strategy should not put a limit to the questions we ask, especially when development is considered from an evolutionary perspective. To be sure, pleiotropy belongs in the vocabulary of all biology students and multiple phenotypic effects of the same 'developmental genes' are frequently registered, as often mentioned in the previous pages. However, genes or gene cascades with possibly systemic effects are generally overlooked, except perhaps when we are confronted with strongly miniaturized organisms, either animals or plants, in which most of the usual anatomical complexity has been swept away.

However, genes with systemic developmental effects are probably the rule, although their effects are mostly masked by other, locally dominant factors. Genes with systemic effects are easily accommodated within a process view of development based on the modulation of basic uniform dynamics. If so, there is no need (better, no reason) to invoke co-option to explain the recurrence of similar patterning over different body axes. Co-option in the patterning, for example, of the proximodistal axis of a leaf of pattern-controlling genes already involved in patterning of another axis (e.g. in the phyllotactic distribution of the leaves along the stem) would imply, in fact, the previous existence of something like an unpatterned leaf. In the alternative interpretation defended here in Section 8.4, the default condition is for patterning to affect all existing axes in a systemic way, and to evolve with them, until additional local markers cause

335

FIGURE 11.1 A part of a compound leaf of meadowsweet (*Filipendula ulmaria*), with anisophyllous leaflets – large and small – alternating along the leaf axis.

divergent patterns to evolve along the different axes (cf. Minelli, 2000). Eventually, as Johann Wolfgang von Goethe, Agnes Arber and many other botanists have remarked, the distance between shoot and leaf is not necessarily as great as elementary plant morphology accounts would suggest (Fig. 11.1).

Systemic effects, I dare to say, are obvious in the forms of paramorphism discussed in Section 8.4 (e.g. the correspondence between phyllotactic pattern and pairing of leaflets in decussate compound leaves), although actual experimental proof is badly needed. But I would suggest that systemic effects are much more widespread and can take different forms. I did not mention these in the previous chapters, which are based on the best available evidence – thus I have left them for these last pages, devoted to final reflections and some tentative suggestions for a future research agenda.

To articulate the point, I suggest we use *Taraxacum* as a model. Let's ignore the idiosyncrasies of its ovules and the problems in providing an adequate taxonomic treatment of the hundreds of

FIGURE 11.2 Dandelion
(*Taraxacum* sp.), leaf.

apomictic clones that specialists are able to recognize in this genus. Let's focus instead on the leaf, which is more or less deeply lobed (Fig. 11.2), but always with the tips of the lateral lobes pointing basipetally rather than acropetally – a quite unusual feature, common however to many other genera of the Asteraceae with a ligulate capitulum (Fig. 11.3). Is there any relationship between the 'forced' polarity of the leaf lobes and the crowding of the ligulate florets in the capitulum?

Leaf developmental polarity is indeed a topic that deserves further study (Tsukaya, 2014). Within one family (Papaveraceae) there are genera (those of the Chelidonioideae, and *Argemone* and the *Papaver* clade in the Papaveroideae) in which the leaflets of the compound leaves differentiate basipetally, as usual, but also genera (*Romneya*, the *Papaver* clade, *Eschscholzia* and *Hunnemannia*) in which the progression is acropetal. In some species of *Hypecoum* and

FIGURE 11.3 Dandelion (*Taraxacum* sp.), capitulum. *A black and white version of this figure will appear in some formats. For the colour version, please refer to the plate section.*

the *Papaver* clade there is even a third pattern, with divergent progression, partly acropetal, partly basipetal (Gleissberg, 1998a, 1998b; Gleissberg and Kadereit, 1999).

Expanding and articulating in full detail a research agenda for plant evo-devo is an open book to which a number of researchers are contributing, as amply documented in the previous chapters. Progress will be achieved by adding new entries to the list of model species or by systematically expanding the currently available database on the spatial and temporal expression patterns of 'developmental genes' – valuable factual contributions to the progress of the discipline, but this is not enough. If plant evo-devo (and evo-devo generally) is really a trading place where different biological disciplines can meet (see Section 1.1), it will be profitable and legitimate also to get the contributions of a number of research traditions ranging from comparative morphology to phylogenetics, from comparative physiology to population genetics. Unavoidably, but also luckily, the dialogue will require not only a critical re-examination of fundamental concepts, but also abandoning obsolete notions and introducing new points of view, some of which may elicit grudging reactions from some, or many, of us. No harm. Fresh perspectives will stimulate new research, new discussion and eventually progress.

Here and there, in this book, I have not hesitated to float unconventional ideas on topics such as homology, temporal phenotypes and even the nature of development. I do not dare to add here specific suggestions on how to articulate a future research agenda on any of these topics. I prefer to close this book by mentioning two aspects that have not been seriously addressed to date from an evo-devo perspective; at least the second of these questions is very unconventional.

1 Parasitic and miniaturized plants. In many parasitic plants, e.g. *Cuscuta*, important architectural traits such as phyllotaxis and leaf structure have been lost, more or less completely. Still more devastating have been the architectural effects of miniaturization in *Wolffia* and, to some extent, even in less extreme forms of duckweeds such as *Lemna*. Many architectural traits have also been strongly modified in a number of succulents, with the loss of foliage leaves. Thus the question: what is left in any of these plants of the genes and gene expression cascades that in 'normal' plants control the production and patterning of the features that are not present any more in the body of these plants? Is there any common pattern, at the genetic level, in these multiple histories pertaining to distantly related lineages and resulting in such diversity of unusual phenotypes?

2 Plant galls. Plant galls have very seldom been considered as products of developmental processes, something they seriously deserve. In galls as 'extended phenotypes' we find examples of combinatorial homology, of phenotypic plasticity, of paramorphism. I only mention here the combinatorial homology manifested in the axillary buds of *Marcetia taxifolia* whose developmental fate is deviated by a galling midge to become pistil-like galls, like a miniaturized version of the flower that might otherwise have grown in the same site (Ferreira and Isaias, 2014).

Enough of this incursion by a zoologist into plant evo-devo. I leave, inviting botanists to reciprocate.

References

Aagaard, J. E., Olmstead, R. G., Willis, J. H. & Phillips, P. C. (2005). Duplication of floral regulatory genes in Lamiales. *American Journal of Botany*, 92: 1284–1293.

Aida, M., Ishida, T., Fukaki, H. *et al.* (1997). Genes involved in organ separation in *Arabidopsis*: an analysis of the cup-shaped cotyledon mutant. *Plant Cell*, 9: 841–857.

Aida, M., Ishida, T. & Tasaka, M. (1999). Shoot apical meristem and cotyledon formation during *Arabidopsis* embryogenesis: interaction among the *CUP-SHAPED COTYLEDON* and *SHOOT MERISTEMLESS* genes. *Development*, 126: 1563–1570.

Aida, M. & Tasaka, M. (2006). Morphogenesis and patterning at the organ boundaries in the higher plant shoot apex. *Plant Molecular Biology*, 60: 915–928.

Ainsworth, C., Crossley, S., Buchanan-Wollaston, V., Thangavelu, M. & Parker, J. (1995). Male and female flowers of the dioecious plant sorrel show different patterns of MADS box gene expression. *Plant Cell*, 7: 1583–1598.

Airoldi, C. A., Bergonzi, S. & Davies, B. (2010). Single amino acid change alters the ability to specify male or female organ identity. *Proceedings of the National Academy of Sciences of the United States of America*, 107, 18898–18902.

Airoldi, C. A. & Davies, B. (2012). Gene duplication and the evolution of plant MADS-box transcription factors. *Journal of Genetics and Genomics*, 39: 157–165.

Al-Shehbaz, I. A., Beilstein, M. A. & Kellogg, E. A. (2006). Systematics and phylogeny of the Brassicaceae (Cruciferae): an overview. *Plant Systematics and Evolution*, 259: 89–120.

Alados, C. L., Escos, J., Emlen, J. M. & Freeman, D. C. (1999). Characterization of branch complexity by fractal analyses. *International Journal of Plant Sciences*, 160: S147–S155.

Alapetite, E., Baker, W. J. & Nadot, S. (2014). Evolution of stamen number in Ptychospermatinae (Arecaceae): insights from a new molecular phylogeny of the subtribe. *Molecular Phylogenetics and Evolution*, 76: 227–240.

Alberch, P. (1991). From genes to phenotype: dynamical systems and evolvability. *Genetica*, 84: 5–11.

Alberch, P., Gould, S. J., Oster, G. F. & Wake, D. B. (1979). Size and shape in ontogeny and phylogeny. *Paleobiology*, 5: 296–317.

Albert, V. A., Gustafsson, M. H. G. & Di Laurenzio, L. (1998). Ontogenetic systematics, molecular developmental genetics, and the angiosperm petal. In *Molecular Systematics of Plants II: DNA Sequencing*, eds. D. E. Soltis, P. S. Soltis & J. A. Doyle. Boston, MA: Kluwer, pp. 349–374.

Albert, V. A., Oppenheimer, D. & Lindqvist, C. (2002). Pleiotropy, redundancy and the evolution of flowers. *Trends in Plant Science*, 7: 297–301.

Almeida, A. M. R., Yockteng, R., Otoni, W. C. & Specht, C. D. (2015a). Positive selection on the K domain of the AGAMOUS protein in the Zingiberales suggests a mechanism for the evolution of androecial morphology. *Evo-Devo*, 6: 7.

Almeida, A. M. R., Yockteng, R. & Specht, C. D. (2015b). Evolution of petaloidy in the Zingiberales: an assessment of the relationship between ultrastructure and gene expression patterns. *Developmental Dynamics*, 244: 1121–1132.

Alonso-Cantabrana, H., Ripoll, J. J., Ochando, I. *et al.* (2007). Common regulatory networks in leaf and fruit patterning revealed by mutations in the *Arabidopsis* ASYMMETRIC LEAVES1 gene. *Development*, 134: 2663–2671.

Alvarez, J. & Smyth, D.R. (1999). CRABS CLAW and SPATULA, two *Arabidopsis* genes that control carpel development in parallel with *AGAMOUS*. *Development*, 126: 2377–2386.

Alvarez-Buylla, E. R., Ambrose, B. A., Flores-Sandoval, E. *et al.* (2010). B-function expression in the flower center underlies the homeotic phenotype of *Lacandonia schismatica* (Triuridaceae). *Plant Cell*, 22: 3543–3559.

Amasino, R. (2010). Seasonal and developmental timing of flowering. *The Plant Journal*, 61: 1001–1013.

Ambros, V. & Moss, E. G. (1994). Heterochronic genes and the temporal control of *C. elegans* development. *Trends in Genetics*, 10: 123–127.

Ambrose, B. A., Espinosa-Matías S., Vázquez-Santana S. *et al.* (2006). Comparative floral developmental series of the Mexican triurids support a euanthial interpretation for the unusual floral structures of *Lacandonia schismatica* (Lacandoniaceae). *American Journal of Botany*, 93: 15–35.

Ambrose, B. A. & Ferrándiz, C. (2013). Development and the evolution of plant form. *Annual Plant Reviews*, 45: 277–320.

Ambrose, B. A., Lerner, D. R., Ciceri, P. *et al.* (2000). Molecular and genetic analyses of the *silky1* gene reveal conservation in floral organ specification between eudicots and monocots. *Molecular Cell*, 5: 569–579.

Andreasen, K. & Baldwin, B. G. (2001). Unequal evolutionary rates between annual and perennial lineages of checker mallows (*Sidalcea*, Malvaceae): evidence from 18S–26S rDNA internal and external transcribed spacers. *Molecular Biology and Evolution*, 18: 936–944.

Andrés, F. & Coupland, G. (2012). The genetic basis of flowering responses to seasonal cues. *Nature Reviews Genetics*, 13: 627–639.

Andriankaja, M., Dhondt, S., De Bodt, S. *et al.* (2012). Exit from proliferation during leaf development in *Arabidopsis thaliana*: a not-so-gradual process. *Developmental Cell*, 22: 64–78.

Angenent, G. C. & Colombo, L. (1996). Molecular control of ovule development. *Trends in Plant Science*, 1: 228–232.

Antonelli, A. (2009). Have giant lobelias evolved several times independently? Life form shifts and historical biogeography of the cosmopolitan and highly diverse subfamily Lobelioideae (Campanulaceae). *BMC Biology*, 7: 82.

Antonius, K. & Ahokas, H. (1996). Flow cytometric determination of polyploidy level in spontaneous clones of strawberries. *Hereditas*, 124: 285.

APG IV (2016). An update of the Angiosperm Phylogeny Group classification for the orders and families of flowering plants: APG IV. *Botanical Journal of the Linnean Society*, 181: 1–20.

Appel, O. & Al-Shehbaz, I. A. (2003). Cruciferae. In *The Families and Genera of Vascular Plants, Vol. 5*, eds. K. Kubitzki & C. Bayer. Berlin: Springer, pp. 75–174.

Arabidopsis Genome Initiative (2000). Analysis of the genome sequence of the flowering plant *Arabidopsis thaliana*. *Nature*, 408: 796–815.

Arber, A. (1918). The phyllode theory of the monocotyledonous leaf, with special reference to anatomical evidence. *Annals of Botany*, 32: 465–501.

Arber, A. (1920). *Water Plants: A Study of Aquatic Angiosperms*. Cambridge: Cambridge University Press.

Arber, A. (1921). The leaf structure of the Iridaceae, considered in relation to the phyllode theory. *Annals of Botany*, 35: 301–336.

Arber, A. (1928). Studies in the Gramineae. V. 1). On *Luziola* and *Dactylis*. 2). On *Lygeum* and *Nardus*. *Annals of Botany*, 42: 391–407.

Arber, A. (1941). The interpretation of leaf and root in the angiosperms. *Biological Review*, 16: 81–105.

Arber, A. (1950). *The Natural Philosophy of Plant Form*. Cambridge: Cambridge University Press.

Armbruster, W. S., Debevec, E. M. & Willson, M. F. (2002). Evolution of syncarpy in angiosperms: theoretical and phylogenetic analyses of the effects of carpel fusion on offspring quantity and quality. *Journal of Evolutionary Biology*, 15: 657–672.

Armstrong, J. & Douglas, A. W. (1989). The ontogenetic basis for corolla aestivation in Scrophulariaceae. *Bulletin of the Torrey Botanical Club*, 116: 378–389.

Arnaud, N. & Laufs, P. (2013). Plant development: brassinosteroids go out of bounds. *Current Biology*, 23, 152–154.

Ashton, P. S. (2003). Dipterocarpaceae. In *The Families and Genera of Vascular Plants, Vol. 5*, eds. K. Kubitzki & C. Bayer. Berlin: Springer, pp. 182–197.

Ashton, P. S., Givnish, T. J. & Appanah, S. (1988). Staggered flowering in the Dipterocarpaceae: new insights into floral induction and the evolution of mast fruiting in the aseasonal tropics. *American Naturalist*, 132: 44–66.

Autran, D., Jonak, C., Belcram, K. *et al.* (2002). Cell numbers and leaf development in *Arabidopsis*: a functional analysis of the *STRUWWELPETER* gene. *EMBO Journal*, 21: 6036–6049.

Bachmann, K. & Gailing, O. (2003). The genetic dissection of the stepwise evolution of morphological characters. In *Deep Morphology: Toward a Renaissance of Morphology in Plant Systematics*, eds. T. F. Stuessy, V. Mayer & E. Hörandl. Königstein: Koeltz, pp. 35–62.

Bahadur, B., Reddy, N. P., Rao, M. M. & Farooqui, S. M. (1984). Corolla handedness in Oxalidaceae, Linaceae and Plumbaginaceae. *Journal of the Indian Botanical Society*, 63: 408–411.

Bailey, C. D., Koch, M. A., Mayer, M. *et al.* (2006). Toward a global phylogeny of the Brassicaceae. *Molecular Biology and Evolution*, 23: 2142–2160.

Bainbridge, K., Guyomarc'h, S., Bayer, E. *et al.* (2008). Auxin influx carriers stabilize phyllotactic patterning. *Genes and Development*, 22: 810–823.

Baker, C. C., Sieber, P., Wellmer, F. & Meyerowitz, E. M. (2005). The early extra petals1 mutant uncovers a role for microRNA miR164c in regulating petal number in *Arabidopsis*. *Current Biology*, 15: 303–315.

Balazadeh, S., Parlitz, S., Mueller-Roeber, B. & Meyer, R. C. (2008). Natural developmental variation in leaf and plant senescence in *Arabidopsis thaliana*. *Plant Biology*, 10: 136–147.

Baldwin, J. M. (1896). A new factor in evolution. *American Naturalist*, 30: 441–451, 536–553.

Banks, J. A., Nishiyama, T., Hasebe, M. *et al.* (2011). The *Selaginella* genome identifies genetic changes associated with the evolution of vascular plants. *Science*, 332: 960–963.

Barclay, I. R. (1975). High frequencies of haploid production in wheat (*Triticum aestivum*) by chromosome elimination. *Nature*, 256: 410–411.

Barker, M. S., Kane, N. C., Matvienko, M. *et al.* (2008). Multiple paleopolyploidizations during the evolution of the Compositae reveal parallel patterns of duplicate gene retention after millions of years. *Molecular Biology and Evolution*, 25: 2445–2455.

Barkoulas, M., Galinha, C., Grigg, S. P. & Tsiantis, M. (2007). From genes to shape: regulatory interactions in leaf development. *Current Opinion in Plant Biology*, 10: 660–666.

Barkoulas, M., Hay, A., Kougioumoutzi, E. *et al.* (2008). A developmental framework for dissected leaf formation in the *Arabidopsis* relative *Cardamine hirsuta*. *Nature Genetics*, 40: 1136–1141.

Barow, M. & Meister, A. (2003). Endopolyploidy in seed plants is differently correlated to systematics, organ, life strategy and genome size. *Plant Cell and Environment*, 26: 571–584.

Barrett, R. D. H. & Schluter, D. (2008). Adaptation from standing genetic variation. *Trends in Ecology and Evolution*, 23: 38–44.

Barrett, S. C. H. (ed.) (1992). *Evolution and Function of Heterostyly*. Berlin: Springer.

Barrett, S. C. H. (2002). The evolution of plant sexual diversity. *Nature Reviews Genetics*, 3: 274–283.

Barrett, S. C. H. (2010). Darwin's legacy: The forms, function and sexual diversity of flowers. *Philosophical Transactions of the Royal Society B*, 365: 351–368.

Barrett, S. C. H., Jesson, L. K. & Baker, A. M. (2000). The evolution of stylar polymorphisms in plants. *Annals of Botany*, 85 (Suppl. A): 253–265.

Barth S., Geier, T., Eimert, K. *et al.* (2009). *KNOX* overexpression in transgenic *Kohleria* (Gesneriaceae) prolongs the activity of proximal leaf blastozones and drastically alters segment fate. *Planta*, 230: 1081–1091.

Barthélémy, D. & Caraglio, Y. (2007). Plant architecture: a dynamic, multilevel and comprehensive approach to plant form, structure and ontogeny. *Annals of Botany*, 99: 375–407.

Bartholmes, C., Hidalgo, O., Gleissberg, S. (2012). Evolution of the *YABBY* gene family with emphasis on the basal eudicot *Eschscholzia californica* (Papaveraceae). *Plant Biology*, 14: 11–23.

Bartlett, M. E. & Specht, C. D. (2010). Evidence for the involvement of *GLOBOSA*-like gene duplications and expression divergence in the evolution of floral morphology in the Zingiberales. *New Phytologist*, 187: 521–541.

Bartlett, M. E. & Specht, C. D. (2011). Changes in expression pattern of the *TEOSINTE BRANCHED1*-like genes in the Zingiberales provide a mechanism for evolutionary shifts in symmetry across the order. *American Journal of Botany*, 98: 227–243.

Barton, M. K. (2010). Twenty years on: the inner workings of the shoot apical meristem, a developmental dynamo. *Developmental Biology*, 341: 95–113.

Bateman, R. M. & DiMichele, W. A. (1994). Saltational evolution of form in vascular plants: a neoGoldschmidtian synthesis. In *Shape and Form in Plants and Fungi*, eds. D. S. Ingram & A. Hudson. London: Academic Press, pp. 63–102.

Bateman, R. M. & DiMichele, W. A. (2002). Generating and-filtering-major phenotypic novelties: NeoGoldschmidtian saltation revisited. In *Developmental Genetics and Plant Evolution*, eds. Q. C. B. Cronk, R. M. Bateman & J. A. Hawkins. London: Taylor & Francis, pp. 109–159.

Bateman, R. M., Hilton, J. & Rudall, P. J. (2006). Morphological and molecular phylogenetic context of the angiosperms: contrasting the 'top-down' and 'bottom-up' approaches to inferring the likely characteristics of the first flowers. *Journal of Experimental Botany*, 57: 3471–3503.

Bateman, R. M., Hilton, J. & Rudall, P. J. (2011). Spatial separation and developmental divergence of male and female reproductive units in gymnosperms, and their relevance to the origin of the angiosperm flower. In *Flowers on the Tree of Life*, eds. L. Wanntorp & L. P. Ronse De Craene. Cambridge: Cambridge University Press, pp. 8–48.

Bateson, W. & Bateson, A. (1891). On variations in the floral symmetry of certain plants having irregular corollas. *Journal of the Linnean Society, Botany*, 28: 386–424.

Baum, D. A. (1998). The evolution of plant development. *Current Opinion in Plant Biology*, 1: 79–86.

Baum, D. A. & Donoghue, M. J. (2002). Transference of function, heterotopy and the evolution of plant development. In *Developmental Genetics and Plant Evolution*, eds. Q. C. B. Cronk, R. M. Bateman & J. A. Hawkins. London: Taylor & Francis, pp. 52–69.

Baum, D. A. & Hileman, L. C. (2006). A developmental genetic model for the origin of the flower. In *Flowering and its Manipulation*, ed. C. Ainsworth. Sheffield: Blackwell, pp. 3–27.

Beaulieu, J. M. & Donoghue, M. J. (2013). Fruit evolution and diversification in campanulid angiosperms. *Evolution*, 67: 3132–3144.

Beilstein, M. A., Al-Shehbaz, I. A. & Kellogg, E. A. (2006). Brassicaceae phylogeny and trichome evolution. *American Journal of Botany*, 93: 607–619.

Beilstein, M. A., Nagalingum, N. S., Clements, M. D. et al. (2010). Dated molecular phylogenies indicate a Miocene origin for *Arabidopsis thaliana*. *Proceedings of the National Academy of Sciences of the United States of America*, 107: 18724–18728.

Bell, A. (2008). *Plant Form: An Illustrated Guide to Flowering Plant Morphology*, new edition. Portland, OR; London: Timber Press.

Bell, E. M., Lin, W., Husbands, A. Y. et al. (2012). *Arabidopsis* LATERAL ORGAN BOUNDARIES negatively regulates brassinosteroid accumulation to limit growth in organ boundaries. *Proceedings of the National Academy of Sciences of the United States of America*, 109: 21146–21151.

Bellini, C., Pacurar, D. I. & Perrone, I. (2014). Adventitious roots and lateral roots: similarities and differences? *Annual Review of Plant Biology*, 65: 639–666.

Bello, M. A., Álvarez, I., Torices, R. & Fuertes-Aguilar, J. (2013). Floral development and evolution of capitulum structure in *Anacyclus* (Anthemideae, Asteraceae). *Annals of Botany*, 112: 1597–1612.

Bello, M. A., Bruneau, A., Forest, F. & Hawkins, J. A. (2009). Elusive relationships within order Fabales: phylogenetic analyses using matK and rbcL sequence data. *Systematic Botany*, 34: 102–114.

Bello, M. A., Rudall, P. J. & Hawkins, J. A. (2012). Combined phylogenetic analyses reveal interfamilial relationships and patterns of floral evolution in the eudicot order Fabales. *Cladistics*, 28: 393–421.

Benková, E., Michniewicz, M., Sauer, E. *et al.* (2003). Local, efflux-dependent auxin gradients as a common module for plant organ formation. *Cell*, 115: 591–602.

Benlloch, R., Berbel, A., Serrano-Mislata, A. & Madueno, F. (2007). Floral initiation and inflorescence architecture: a comparative view. *Annals of Botany*, 100: 659–676.

Bennett, M. D. & Leitch, I. J. (2012). *Plant DNA C-values Database* (release 6.0, December 2012). http://data.kew.org/cvalues (accessed September 2017).

Berbel, A., Navarro, C. & Ferrándiz, C. *et al.* (2001). Analysis of *PEAM4*, the pea *AP1* functional homologue, supports a model for *AP1*-like genes controlling both floral meristem and floral organ identity in different plant species. *The Plant Journal*, 25: 441–451.

Berger, B. A., Thompson, V., Lim, A., Ricigliano, V. & Howarth, D.G. (2016). Elaboration of bilateral symmetry across *Knautia macedonica* capitula related to changes in ventral petal expression of *CYCLOIDEA*-like genes. *EvoDevo*, 7: 8.

Berger, Y., Harpaz-Saad, S., Brand, A. *et al.* (2009). The NAC-domain transcription factor GOBLET specifies leaflet boundaries in compound tomato leaves. *Development*, 136: 823–832.

Bergthorsson, U., Adams, K. L., Thomason, B. & Palmer, J. D. (2003). Widespread horizontal transfer of mitochondrial genes in flowering plants. *Nature*, 424: 197–201.

Bergthorsson, U., Richardson, A. O., Young, G. J., Goertzen, L. R. & Palmer, J. D. (2004). Massive horizontal transfer of mitochondrial genes from diverse land plant donors to the basal angiosperm *Amborella*. *Proceedings of the National Academy of Sciences of the United States of America*, 101: 17747–17752.

Bertrand-Garcia, R. & Freeling, M. (1991). *Hairy-sheath-frayed1-0*: a systemic, heterochronic mutant of maize that specifies slow developmental stage transitions. *American Journal of Botany*, 78: 747–765.

Bharathan, G., Goliber, T. E., Moore, C. *et al.* (2002). Homologies in leaf form inferred from *KNOXI* gene expression during development. *Science*, 296: 1858–1860.

Bissell, E. K. & Diggle, P. K. (2008). Floral morphology in *Nicotiana*: architectural and temporal effects on phenotypic integration. *International Journal of Plant Sciences*, 169: 225–240.

Blaser, J. L. (1954). The morphology of the flower and inflorescence of *Mitchella repens*. *American Journal of Botany*, 41: 533–539.

Blázquez, M. A., Ferrándiz, C., Madueno, F. & Parcy, F. (2006). How floral meristems are built. *Plant Molecular Biology*, 60: 855–870.

Blázquez, M. A., Soowal, L. N., Lee, I. & Weigel, D. (1997). *LEAFY* expression and flower initiation in *Arabidopsis*. *Development*, 124: 3835–3844.

Blázquez, M. A. & Weigel, D. (2000). Integration of floral inductive signals in *Arabidopsis*. *Nature*, 404: 889–892.

Blein, T., Hasson, A. & Laufs, P. (2010). Leaf development: what it needs to be complex. *Current Opinion in Plant Biology*, 13: 75–82.

Blein, T., Pulido, A., Vialette-Guiraud, A. *et al.* (2008). A conserved molecular framework for compound leaf development. *Science*, 322: 1835–1839.

Bliss, B. J., Wanke, S., Barakat, A. *et al.* (2013). Characterization of the basal angiosperm *Aristolochia fimbriata*: a potential experimental system for genetic studies. *BMC Plant Biology*, 13: 13.

Bohs, L., Weese, T., Myers, N. *et al.* (2007). Zygomorphy and heteranthery in *Solanum* in a phylogenetic context. *Acta Horticulturae*, 745: 201–224.

Borsch, T., Löhne, C. & Wiersema, J. (2008). Phylogeny and evolutionary patterns in Nymphaeales: integrating genes, genomes and morphology. *Taxon*, 57: 1052–1081.

Boss, P. K. & Thomas, M. R. (2002). Association of dwarfism and floral induction with a grape 'green revolution' mutation. *Nature*, 416: 847–850.

Bouche, F., Lobet, G., Tocquin, P. & Perilleux, C. (2016). FLOR-ID: an interactive database of flowering-time gene networks in *Arabidopsis thaliana*. *Nucleic Acids Research*, 44: D1167–D1171.

Boudaoud, A. (2010). An introduction to the mechanics of morphogenesis for plant biologists. *Trends in Plant Science*, 15: 353–360.

Bowers, J. E., Chapman, B. A., Rong, J. & Paterson, A. H. (2003). Unravelling angiosperm genome evolution by phylogenetic analysis of chromosomal duplication events. *Nature*, 422: 433–438.

Bowman, J. L. (1997). Evolutionary conservation of angiosperm flower development at the molecular and genetic levels. *Journal of Biosciences*, 22: 515–527.

Bowman, J. L., Alvarez, J., Weigel, D., Meyerowitz, E. M. & Smyth, D. R. (1993). Control of flower development in *Arabidopsis thaliana* by *APETALA1* and interacting genes. *Development*, 119: 721–743.

Bowman, J. L., Bruggemann, H., Lee, J.-Y. & Mummenhoff, K. (1999). Evolutionary changes in floral structure within *Lepidium* L. (Brassicaceae). *International Journal of Plant Sciences*, 160: 917–929.

Bowman, J. L., Smyth, D. R. & Meyerowitz, E. M. (1989). Genes directing flower development in *Arabidopsis*. *Plant Cell*, 1: 37–52.

Bown, D. (2000). *Aroids: Plants of the Arum Family*, 2nd edn. Portland, OR: Timber Press.

Box, M. S., Bateman, R. M., Glover, B. J. & Rudall, P. J. (2008). Floral ontogenetic evidence of repeated speciation via paedomorphosis in subtribe Orchidinae (Orchidaceae). *Botanical Journal of the Linnean Society*, 157: 429–454.

Box, M. S. & Glover, B. J. (2010). A plant developmentalist's guide to paedomorphosis: reintroducing a classic concept to a new generation. *Trends in Plant Science*, 15: 241–246.

Bradley, D., Carpenter, R., Copsey, L. *et al.* (1996). Control of inflorescence architecture in *Antirrhinum*. *Nature*, 379: 791–797.

Bradley, D., Carpenter, R., Sommer, H., Hartley, N. & Coen, E. (1993). Complementary floral homeotic phenotypes result from opposite orientations of a transposon at the *plena* locus of *Antirrhinum*. *Cell*, 72: 85–95.

Bradley, D., Ratcliffe, O., Vincent, C., Carpenter, R. & Coen, E. (1997). Inflorescence commitment and architecture in *Arabidopsis*. *Science*, 275: 80–83.

Braybrook, S. A. & Kuhlemeier, C. (2010). How a plant builds leaves. *Plant Cell*, 22: 1006–1018.

Breeze, E., Harrison, E., Page, T. *et al.* (2008). Transcriptional regulation of plant senescence: from functional genomics to systems biology. *Plant Biology*, 10 Suppl 1: 99–109.

Breuil-Broyer S., Trehin, C., Morel, P. *et al.* (2016). Analysis of the *Arabidopsis superman* allelic series and the interactions with other genes demonstrate developmental robustness and joint specification of male–female boundary, flower meristem termination and carpel compartmentalization. *Annals of Botany*, 117: 905–923.

Brigandt, I. & Love, A. C. (2010). Evolutionary novelty and the evo-devo synthesis: field notes. *Evolutionary Biology*, 37: 93–99.

Brigandt, I. & Love, A. C. (2012). Conceptualizing evolutionary novelty: moving beyond definitional debates. *Journal of Experimental Zoology Part B: Molecular and Developmental Evolution*, 318: 417–427.

Broadley, M. R., White, P. J., Hammond, J. P. *et al.* (2008). Evidence of neutral transcriptome evolution in plants. *New Phytologist*, 180: 587–593.

Brockington, S. F., Roolse, A., Randall, J. *et al.* (2009). Phylogeny of the Caryophyllales sensu lato: revisiting hypotheses on pollination biology and perianth differentiation in the core Caryophyllales. *International Journal of Plant Sciences*, 170: 627–643.

Brockington, S. F., Rudall, P. J., Frohlich, M. W. *et al.* (2011). 'Living stones' reveal alternative petal identity programs within the core eudicots. *The Plant Journal*, 69: 193–203.

Brody, A. & Morita, S. I. (2000). A positive association between oviposition and fruit set: female choice or manipulation? *Oecologia*, 124: 418–425.

Broholm, S. K., Tähtiharju, S., Laitinen, R. A. *et al.* (2008). A TCP domain transcription factor controls flower type specification along the radial axis of the *Gerbera* (Asteraceae) inflorescence. *Proceedings of the National Academy of Sciences of the United States of America*, 105: 9117–9122.

Broholm, S. K., Teeri, T. H. & Elomaa, P. (2014). Molecular control of inflorescence development in Asteraceae. *Advances in Botanical Research*, 27: 297–334.

Brookfield, J. F. Y. (2009). Evolution and evolvability: Celebrating Darwin 200. *Biology Letters*, 5: 44–46.

Brunetti, R., Gissi, C., Pennati, R. *et al.* (2015). Morphological evidence that the molecularly determined *Ciona intestinalis* type A and type B are different species: *Ciona robusta* and *Ciona intestinalis*. *Journal of Zoological Systematics and Evolutionary Research*, 53: 186–193.

Buchanan-Wollaston, V. (2007). Senescence in plants. In *Encyclopedia of Life Sciences*. Chichester: Wiley.

Buchholz, J. T. (1946). Volumetric studies of seeds, endosperm, and embryos in *Pinus ponderosa* during embryonic differentiation. *Botanical Gazette*, 108: 232–244.

Budd, G. E. (1999). Does evolution in body patterning genes drive morphological change or vice versa? *BioEssays*, 21: 326–332.

Burgeff, C., Liljegren, S. J., Tapia-Lopez, R., Yanosky, M. F. & Alvarez-Buylla, E. R. (2002). MADS-box gene expression in lateral primordia, meristems and differentiated tissues of *Arabidopsis thaliana* roots. *Planta*, 214: 365–372.

Bürglin, T. R. (2005). Homeodomain proteins. In *Encyclopedia of Molecular Cell Biology and Molecular Medicine*, ed. R. A. Meyers. Weinheim: Wiley-VCH, pp. 179–222.

Busch, A., Horn, S., Mühlhausen, A., Mummenhoff, K. & Zachgo, S. (2012). Corolla monosymmetry: evolution of a morphological novelty in the Brassicaceae family. *Molecular Biology and Evolution*, 29: 1241–1254.

Busch, A. & Zachgo, S. (2007). Control of corolla monosymmetry in the Brassicaceae *Iberis amara*. *Proceedings of the National Academy of Sciences of the United States of America*, 104: 16714–16719.

Busch, A. & Zachgo, S. (2009). Flower symmetry evolution: towards understanding the abominable mystery of angiosperm radiation. *BioEssays*, 31: 1181–1190.

Buzgo, M. & Endress, P. K. (2000). Floral structure and development of Acoraceae and its systematic relationships with basal angiosperms. *International Journal of Plant Sciences*, 161: 23–41.

Buzgo, M., Soltis, P. S. & Soltis, D. S. (2004). Floral developmental morphology of *Amborella trichopoda* (Amborellaceae). *International Journal of Plant Sciences*, 165: 925–947.

Byng, J. W. (2014). *The Flowering Plants Handbook: A Practical Guide to Families and Genera of the World*. Hertford: Plant Gateway.

Byrne, M. (2012). Making leaves. *Current Opinion in Plant Biology*, 15: 24–30.

Byrne, M. E. (2006). Shoot meristem function and leaf polarity: the role of class III HD-ZIP genes. *PLoS Genetics* 2 (6): e89.

Byrne, M. E., Barley, R., Curtis, M. *et al.* (2000). *ASYMMETRIC LEAVES1* mediates leaf patterning and stem cell function in *Arabidopsis*. *Nature*, 408: 967–971.

Caddick, L. R., Rudall, P. J. & Wilkin, P. (2000). Floral morphology and development in Dioscoreales. *Feddes Repertorium*, 111: 189–230.

Cai, H., Liu, X., Vanneste, K. *et al.* (2014). The genome sequence of the orchid *Phalaenopsis equestris*. *Nature Genetics*, 47: 65–72.

Callos, J. D. & Medford, J. I. (1994). Organ positions and pattern formation in the shoot apex. *The Plant Journal*, 6: 1–7.

Calonje, M., Cubas, P., Martinez-Zapater, J. M. & Carmona, M. J. (2004). Floral meristem identity genes are expressed during tendril development in grapevine. *Plant Physiology*, 135: 1491–1501.

Cameron, R. J. (1970). Light intensity and growth of *Eucalyptus* seedlings. I. Ontogenetic variation in *E. fastigiata*. *Australian Journal of Botany*, 18: 29–43.

Canales, C., Barkoulas, M., Galinha, C. & Tsiantis, M. (2010). Weeds of change: *Cardamine hirsuta* as a new model system for studying dissected leaf development. *Journal of Plant Research*, 123, 25–33.

Cantino, P. D., Doyle, J. A., Graham, S. W. *et al.* (2007). Towards a phylogenetic nomenclature of Tracheophyta. *Taxon*, 56: 822–846.

Carles, C. C., Choffnes-Inada, D., Reville, K., Lertpiriyapong, K. & Fletcher, J. C. (2005). *ULTRAPETALA1* encodes a SAND domain putative transcriptional regulator that controls shoot and floral meristem activity in *Arabidopsis*. *Development*, 132: 897–911.

Carles, C. C. & Fletcher, J. C. (2003). Shoot apical meristem maintenance: the art of a dynamic balance. *Trends in Plant Science*, 8: 394–401.

Carlquist, S. (1969). Toward acceptable evolutionary interpretations of floral anatomy. *Phytomorphology*, 19: 332–362.

Carlquist, S. (1980). *Hawaii, a Natural History*, 2nd edn. Lawai, Kauai, HI: Pacific Tropical Botanical Garden.

Carlsbecker, A., Tandre, K., Johanson, U., Englund, M. & Engstrom, P. (2004). The MADS-box gene *DAL1* is a potential mediator of the juvenile-to-adult transition in Norway spruce (*Picea abies*). *The Plant Journal*, 40: 546–557.

Carlson, J. E., Leebens-Mack, J. H., Wall, P. K. *et al.* (2006). EST database for early flower development in California poppy (*Eschscholzia californica* Cham., Papaveraceae) tags over 6,000 genes from a basal eudicot. *Plant Molecular Biology*, 62: 351–369.

Carlson, S. E., Howarth, D. G. & Donoghue, M. J. (2011). Diversification of *CYCLOIDEA*-like genes in Dipsacaceae (Dipsacales): implications for the evolution of capitulum inflorescences. *BMC Evolutionary Biology*, 11: 325.

Carpenter, R. & Coen, E. S. (1990). Floral homeotic mutations produced by transposon-mutagenesis in *Antirrhinum majus*. *Genes and Development*, 4: 1483–1493.

Carraro, N., Peaucelle, A., Laufs, P. & Traas, J. (2006). Cell differentiation and organ initiation at the shoot apical meristem. *Plant Molecular Biology*, 60: 811–826.

Caruso, C., Rigato, E. & Minelli, A. (2012). Finalism and adaptationism in contemporary biological literature. *Atti dell'Istituto Veneto di Scienze Lettere ed Arti, Classe di Scienze Fisiche, Matematiche e Naturali*, 170: 69–76.

Castel, R., Kusters, E. & Koes, R. (2010). Inflorescence development in petunia: through the maze of botanical terminology. *Journal of Experimental Botany*, 61: 2235–2246.

Causier, B., Castillo, R., Xue, Y., Schwarz-Sommer, Z. & Davies, B. (2010b). Tracing the evolution of the floral homeotic B- and C-function genes through genome synteny. *Molecular Biology and Evolution*, 27: 2651–2664.

Causier, B., Castillo, R., Zhou, J. *et al.* (2005). Evolution in action: following function in duplicated floral homeotic genes. *Current Biology*, 15: 1508–1512.

Causier, B., Schwarz-Sommer, Z. & Davies, B. (2010a). Floral organ identity: 20 years of ABCs. *Seminars in Cell and Developmental Biology*, 21: 73–79.

Cavalier-Smith, T., Chao, E. E., Snell, E. A. *et al.* (2014). Multigene eukaryote phylogeny reveals the likely protozoan ancestors of opisthokonts (animals, fungi, choanozoans) and Amoebozoa. *Molecular Phylogenetics and Evolution*, 81: 71–85.

Cevik, V., Ryder, C. D., Popovich, A. *et al.* (2010). A *FRUITFULL*-like gene is associated with genetic variation for fruit flesh firmness in apple (*Malus domestica* Borkh.). *Tree Genetics and Genomes*, 6: 271–279.

Chae, E., Tan, Q. K. G., Hill, T. A. & Irish, V. F. (2008). An *Arabidopsis* F-box protein acts as a transcriptional co-factor to regulate floral development. *Development*, 135: 1235–1245.

Champagne, C. & Sinha, N. (2004). Compound leaves: equal to the sum of their parts? *Development*, 131: 4401–4412.

Champagne, C. E., Goliber, T. E., Wojchiechowski, M. F. *et al.* (2007). Compound leaf development and evolution in the legumes. *Plant Cell*, 19: 3369–3378.

Chanderbali, A. S., Albert, V. A., Leebens-Mack, J. *et al.* (2009). Transcriptional signatures of ancient floral developmental genetics in avocado (*Persea americana*; Lauraceae). *Proceedings of the National Academy of Sciences of the United States of America*, 106: 8929–8934.

Chanderbali, A. S., Berger, B. A., Howarth, D. G., Soltis, D. E. & Soltis, P. S. (2017). Evolution of floral diversity: genomics, genes and gamma. *Philosophical Transactions of the Royal Society B*, 372: 20150509.

Chandler, J., Nardmann, J. & Werr, W. (2008). Plant development revolves around axes. *Trends in Plant Science*, 13: 78–84.

Chapman, M. A., Tang, S., Draeger, D. *et al.* (2012). Genetic analysis of floral symmetry in Van Gogh's sunflowers reveals independent recruitment of *CYCLOIDEA* genes in the Asteraceae. *PLoS Genetics*, 8: e1002628.

Charlesworth, B. (1980). *Evolution in Age-structured Populations*. Cambridge: Cambridge University Press.

Charlesworth, D. & Guttman, D. S. (1999). The evolution of dioecy and plant sex chromosome systems. In *Sex Determination in Plants*, ed. C. C. Ainsworth. Oxford: Bios Scientific, pp. 25–49.

Chen, C., Xu, Y., Zeng, M. & Huang, H. (2001). Genetic control by *Arabidopsis* genes *LEUNIG* and *FILAMENTOUS FLOWER* in gynoecium fusion. *Journal of Plant Research*, 114: 465–469.

Chen, C. B., Wang, S. P. & Huang, H. (2000). LEUNIG has multiple functions in gynoecium development in *Arabidopsis*. *Genesis*, 26: 42–54.

Chen, J. J., Janssen, B. J., Williams, A. & Sinha, N. (1997). A gene fusion at a homeobox locus: alterations in leaf shape and implications for morphological evolution. *Plant Cell*, 9: 1289–1304.

Chen, T. C., Zhang, D. X., Larsen, K. & Larsen, S. S. (2010). *Bauhinia* Linnaeus. In *Flora of China, 10*, eds. Z. Y. Wu, P. H. Raven & D. Y. Hong. Beijing: Science Press; St. Louis, MO: Missouri Botanical Garden Press, pp. 6–21.

Chen, X. (2012). Small RNAs in development: insights from plants. *Current Opinion in Genetics and Development*, 22: 361–367.

Chitwood, D. H., Headland, L. R., Ranjan, A. *et al.* (2012). Leaf asymmetry as a developmental constraint imposed by auxin-dependent phyllotactic patterning. *Plant Cell*, 24: 1–10.

Cho, E. & Zambryski, P. C. (2011). *ORGAN BOUNDARY1* defines a gene expressed at the junction between the shoot apical meristem and lateral organs.

Proceedings of the National Academy of Sciences of the United States of America, 108: 2154–2159.

Cho, J. W., Park, S. C., Shin, E. A. *et al.* (2004). Cyclin D1 and p22^{ack1} play opposite roles in plant growth and development. *Biochemical and Biophysical Research Communications*, 324: 52–57.

Chuang, C. F., Running, M. P., Williams, R. W. & Meyerowitz, E. M. (1999). The *PERIANTHIA* gene encodes a bZIP protein involved in the determination of floral organ number in *Arabidopsis thaliana*. *Genes and Development*, 13: 334–344.

Chuck G., Meeley, R. &, Hake, S. (2008). Floral meristem initiation and meristem cell fate are regulated by the maize *AP2* genes *ids1* and *sid1*. *Development*, 135: 3013–3019.

Chung, Y. Y., Kim, S. R., Kang, H. G. *et al.* (1995). Characterization of two rice MADS box genes homologous to *GLOBOSA*. *Plant Science*, 109: 45–56.

Citerne, H., Jabbour, F., Nadot, S. & Damerval, C. (2010). The evolution of floral symmetry. *Advances in Botanical Research*, 54: 85–137.

Citerne, H. L., Möller, M. & Cronk, Q. C. B. (2000). Diversity of *cycloidea*-like genes in Gesneriaceae in relation to floral symmetry. *Annals of Botany*, 86: 167–176.

Citerne, H. L., Pennington, R. T. & Cronk, Q. C. (2006). An apparent reversal in floral symmetry in the legume *Cadia* is a homeotic transformation. *Proceedings of the National Academy of Sciences of the United States of America*, 103: 12017–12020.

Citerne, H. L., Reyes, E., Le Guilloux, M. *et al.* (2017). Characterization of *CYCLOIDEA*-like genes in Proteaceae, a basal eudicot family with multiple shifts in floral symmetry. *Annals of Botany*, 119: 367–378.

Clark, S. E. (2001). Meristems: start your signaling. *Current Opinion in Plant Biology*, 4: 28–32.

Clarke, E. (2011). Plant individuality: a solution to the demographer's dilemma, *Biology and Philosophy*, 27: 321–361.

Classen-Bockhoff, R. (1992). Florale Differenzierung in komplex organisierten Asteraceenköpfen. *Flora*, 186: 1–22.

Classen-Bockhoff, R. (1996). Functional units beyond the level of the capitulum and cypsela in Compositae. In *Compositae: Biology and Utilization*, ed. D. J. N. Hind. Kew: Royal Botanic Gardens, pp. 129–160.

Classen-Bockhoff, R. (2001). Plant morphology: the historic concepts of Wilhelm Troll, Walter Zimmermann and Agnes Arber. *Annals of Botany*, 88: 1153–1172.

Clausen, R. E. & Mann, M. C. (1924). Inheritance of *Nicotiana tabacum*. V. The occurrence of haploid plants in interspecific progenies. *Proceedings of the National Academy of Sciences of the United States of America*, 10: 121–124.

Clausing, G. & Renner, S. S. (2001). Evolution of growth in epiphytic Dissochaeteae (Melastomataceae). *Organisms Diversity and Evolution*, 1: 45–60.

Clay, K. & Ellstrand, N. (1981). Stylar polymorphism in *Epigaea repens*, a dioecious species. *Bulletin of the Torrey Botanical Club*, 108: 305–310.

Clifford, H. T. (1987). Spikelet and floral morphology. In *Grass Systematics and Evolution*, eds. T. R. Soderstrom, K. W. Hilu, C. S. Campbell & M. E. Barkworth. Washington, DC: Smithsonian Institution Press, pp. 21–30.

Clune, J., Mouret, J. B. & Lipson, H. (2013). The evolutionary origins of modularity. *Proceedings of the Royal Society B*, 280: 20122863.

Coen, E. (1999). *The Art of Genes. How Organisms Make Themselves.* Oxford: Oxford University Press.

Coen, E. S., Doyle, S., Romero, J. M. *et al.* (1991). Homeotic genes controlling flower development in *Antirrhinum*. *Development, Supplement*, 1: 149–155.

Coen, E. S. & Meyerowitz, E. M. (1991). The war of the whorls: genetic interactions controlling flower development. *Nature*, 353: 31–37.

Coen, E. S. & Nugent, J. M. (1994). The evolution of flowers and inflorescences. *Development*, 1994 (supplement): 107–116.

Coen, E., Rolland-Lagan, A.-G., Matthews, M., Bangham, J.A. & Prusinkiewicz, P. (2004). The genetics of geometry. *Proceedings of the National Academy of Sciences of the United States of America*, 101: 4728–4735.

Coen, E. S., Romero, J. M., Doyle, S. *et al.* (1990). *Floricaula*: a homeotic gene required for flower development in *Antirrhinum majus*. *Cell*, 63: 1311–1322.

Cohen, J. I. (2014). A phylogenetic analysis of morphological and molecular characters of Boraginaceae: evolutionary relationships, taxonomy, and patterns of character evolution. *Cladistics*, 30: 139–169.

Cole, M., Nolte, C. & Werr, W. (2006). Nuclear import of the transcription factor SHOOT MERISTEMLESS depends on heterodimerization with BLH proteins expressed in discrete sub-domains of the shoot apical meristem of *Arabidopsis thaliana*. *Nucleic Acids Research*, 34: 1281–1292.

Colombo, L., Battaglia, R. & Kater, M. M. (2008). *Arabidopsis* ovule development and its evolutionary conservation. *Trends in Plant Science*, 13: 444–450.

Colombo, L., Franken, J., Koetje, E. *et al.* (1995). The petunia MADS box gene *FBP11* determines ovule identity. *Plant Cell*, 7: 1859–1868.

Conner, J. K. (2012). Quantitative genetic approaches to evolutionary constraint: how useful? *Evolution*, 66: 3313–3320.

Conti, L. & Bradley, D. (2007). *TERMINAL FLOWER1* is a mobile signal controlling *Arabidopsis* architecture. *Plant Cell*, 19: 767–778.

Cook, C. D. K. & Rutishauser, R. (2007). Podostemaceae. In *The Families and Genera of Vascular Plants, Vol. 9*, ed. K. Kubitzki. Berlin: Springer, pp. 304–344.

Cooke, T. J., Poli, D. & Cohen, J. D. (2003). Did auxin play a crucial role in the evolution of novel body plans during the Late Silurian-Early Devonian radiation of land plants? In *The Evolution of Plant Physiology: From Whole Plants to Ecosystems*, eds. A. R. Hemsley & I. Poole. London: Linnean Society of London, pp. 85–107.

Corbesier, L., Vincent, C., Jang, S. H. *et al.* (2007). FT protein movement contributes to long-distance signaling in floral induction of *Arabidopsis*. *Science*, 316: 1030–1033.

Corley, S. B., Carpenter, R., Copsey, L. & Coen, E. (2005). Floral asymmetry involves an interplay between TCP and MYB transcription factors in *Antirrhinum*. *Proceedings of the National Academy of Sciences of the United States of America*, 102: 5068–5073.

Coudert, Y., Périn, C., Courtois, B., Khong, N. G. & Gantet, P. (2010). Genetic control of root development in rice, the model cereal. *Trends in Plant Science*, 15: 219–226.

Couvreur, T. L. P., Franzke, A., Al-Shehbaz, I. A. *et al.* (2010). Molecular phylogenetics, temporal diversification and principles of evolution in the mustard family (Brassicaceae). *Molecular Biology and Evolution*, 27: 55–71.

Crane, P. R. (1985). Phylogenetic analysis of seed plants and the origin of angiosperms. *Annals of the Missouri Botanical Garden*, 72: 716–793.

Crane, P. R., Friis, E. M. & Pedersen, K. R. (1995). The origin and early diversification of angiosperms. *Nature*, 374: 27–33.

Crane, P. R. & Kenrick, P. (1997). Diverted development of reproductive organs: a source of morphological innovation in land plants. *Plant Systematics and Evolution*, 206: 161–174.

Cremer, F., Lönnig, W. E., Saedler, H. & Huijser, P. (2001). The delayed terminal flower phenotype is caused by a conditional mutation in the *CENTRORADIALIS* gene of snapdragon. *Plant Physiology*, 126: 1031–1041.

Cridge, A. G., Dearden, P. K. & Brownfield, L. R. (2016). Convergent occurrence of the developmental hourglass in plant and animal embryogenesis? *Annals of Botany*, 117: 833–843.

Cronk, Q. C. B. (2001). Plant evolution and development in a post-genomic context. *Nature Reviews Genetics*, 2: 607–619.

Cronk, Q. C. B. (2009). *The Molecular Organography of Plants*. Oxford: Oxford University Press.

Cronquist, A. (1968). *The Evolution and Classification of Flowering Plants*. Boston, MA: Houghton Mifflin.

Cronquist, A. (1981). *An Integrated System of Classification of Flowering Plants*. New York, NY: Columbia University Press.

Cubas, P. (2004). Floral zygomorphy, the recurring evolution of a successful trait. *Bioessays*, 26: 1175–1184.

Cubas, P., Vincent, C. & Coen, E. (1999). An epigenetic mutation responsible for natural variation in floral symmetry. *Nature*, 401: 157–161.

Cui, R., Han, J., Zhao, S. *et al.* (2010). Functional conservation and diversification of class E floral homeotic genes in rice (*Oryza sativa*). *The Plant Journal*, 61: 767–781.

Czapek, A. (1898). Die inverse Orientierung der Blätter von *Alstroemeria*. *Flora*, 85: 418–430.

Dahmann, C., Oates, A. C. & Brand, M. (2011). Boundary formation and maintenance in tissue development. *Nature Reviews Genetics*, 12: 43–55.

Dai, M., Zhao, Y., Ma, O. *et al.* (2007). The rice *YABBY1* gene is involved in the feedback regulation of gibberellin metabolism. *Plant Physiology*, 144: 121–133.

Damerval, C., Citerne, H., Le Guilloux, M. *et al.* (2013). Asymmetric morphogenetic cues along the transverse plane: shift from disymmetry to zygomorphy in the flower of Fumarioideae. *American Journal of Botany*, 100: 391–402.

Damerval, C., Le Guilloux, M., Jager, M. & Charon, C. (2007). Diversity and evolution of *CYCLOIDEA*-like TCP genes in relation to flower development in Papaveraceae. *Plant Physiology*, 143: 759–772.

Damerval, C. & Nadot, S. (2007). Evolution of perianth and stamen characteristics with respect to floral symmetry in Ranunculales. *Annals of Botany*, 100: 631–640.

Daniell, H., Lin, C. S., Yu, M. & Chang, W. J. (2016). Chloroplast genomes: diversity, evolution, and applications in genetic engineering. *BMC Genome Biology*, 17(1): 1–29.

Danyluk, J., Kane, N. A., Breton, G. *et al.* (2003). TaVRT-1, a putative transcription factor associated with vegetative to reproductive transition in cereals. *Plant Physiology*, 132: 1849–1860.

D'Arcy, W. (1991). The Solanaceae since 1976, with a review of its biogeography. In *Solanaceae III: Taxonomy, Chemistry, Evolution*, eds. J. G. Hawkes, R. W. Lester, M. Nee & R. N. Estrada. Kew: Royal Botanic Gardens; London: Linnean Society of London, pp. 75–137.

Darwin, C. (1859). *On the Origin of Species by Natural Selection*. London: J. Murray.

Darwin, C. (1877). *The Different Forms of Flowers on Plants of the Same Species*. London: Murray.

Datson, P. M., Murray, B. G. & Steiner, K. E. (2008). Climate and the evolution of annual/perennial life-histories in *Nemesia* (Scrophulariaceae). *Plant Systematics and Evolution*, 270: 39–57.

Davies, B., Motte, P., Keck, E. *et al.* (1999). *PLENA* and *FARINELLI*: redundancy and regulatory interactions between two *Antirrhinum* MADS-box factors controlling flower development. *EMBO Journal*, 18: 4023–4034.

Davis, C. & Wurdack, K. (2004). Host-to-parasite gene transfer in flowering plants: phylogenetic evidence from Malpighiales. *Science*, 305: 676–678.

Davis, C. C. & Anderson, W. R. (2010). A complete generic phylogeny of Malpighiaceae inferred from nucleotide sequence data and morphology. *American Journal of Botany*, 97: 2031–2048.

Davis, C. C. & Xi, Z. (2015). Horizontal gene transfer in parasitic plants. *Current Opinion in Plant Biology*, 26: 14–19.

Dawkins, R. (1976). *The Selfish Gene*. Oxford: Oxford UniversityPress.

de Beer, G. R. (1930). *Embryology and Evolution*. Oxford: Clarendon Press.

de Beer, G. R. (1940). *Embryos and Ancestors*. Oxford: Clarendon Press.

de Bruijn, S., Angenent, G. C. & Kaufmann, K. (2012). Plant 'evo-devo' goes genomic: from candidate genes to regulatory networks. *Trends in Plant Science*, 17: 441–447.

de Lange, P. J., Heenan, P. B., Houliston, G. J., Rolfe, J. R. & Mitchell, A. D. (2013). New *Lepidium* (Brassicaceae) from New Zealand. *PhytoKeys*, 24: 1–147.

de Martino, G., Pan, I., Emmanuel, E., Levy, A. & Irish, V. F. (2006). Functional analyses of two tomato *APETALA3* genes demonstrate diversification in their roles in regulating floral development. *Plant Cell*, 18: 1833–1845.

De Smet, I., Lau, S., Mayer, U. & Jürgens, G. (2010). Embryogenesis: the humble beginnings of plant life. *The Plant Journal*, 61: 959–970.

de Vries, H. (1904). *Species and Varieties: Their Origin by Mutation*. Chicago, IL: Open Court.

Della Pina, S., Souer, E. & Koes, R. (2014). Arguments in the evo-devo debate: say it with flowers! *Journal of Experimental Botany*, 65: 2231–2242.

DeMason, D. A. & Schmidt, R. J. (2001). Roles of the *Uni* gene in shoot and leaf development of pea (*Pisum sativum*): phenotypic characterization and leaf development in the *uni* and *uni-tac* mutants. *International Journal of Plant Sciences*, 162: 1033–1051.

Dengler, N. G. (1999). Anisophylly and dorsiventral shoot symmetry. *International Journal of Plant Sciences*, 160: S67–S80.

Dengler, N. G., Dengler, R. E. & Kaplan, D. R. (1982). The mechanism of plication in palm leaves: histogenic observations on the pinnate leaf of *Chrysalidocarpus lutescens. Canadian Journal of Botany*, 60: 2976–2980.

Derelle, R., Lopez, P., Le Guyader, H. & Manuel, M. (2007). Homeodomain proteins belong to the ancestral molecular toolkit of eukaryotes. *Evolution and Development*, 9: 212–219.

Deroin, T. (2007). Floral vascular pattern of the endemic Malagasy genus *Fenerivia* Diels (Annonaceae). *Adansonia*, 29: 7–12.

Dewitte, W., Scofield, S., Alcasabas, A. A. *et al.* (2007). *Arabidopsis* CYCD3 D-type cyclins link cell proliferation and endocycles and are rate-limiting for cytokinin responses. *Proceedings of the National Academy of Sciences of the United States of America*, 104: 14537–14542.

Di Giacomo, E., Sestili, F., Iannelli, M. A. *et al.* (2008). Characterization of *KNOX* genes in *Medicago truncatula*. *Plant Molecular Biology*, 67: 135–150.

Dickinson, T. A. (1978). Epiphylly in angiosperms. *Botanical Review*, 44: 181–232.

Diggle, P. K. (1991). Labile sex expression in andromonoecious *Solanum hirtum*: sources of variation in mature floral structure. *Canadian Journal of Botany*, 69: 2033–2043.

Diggle, P. K. (1993). Developmental plasticity, genetic variation and the evolution of andromonoecy. *American Journal of Botany*, 80: 967–973.

Diggle, P. K. (1994). The expression of andromonoecy in *Solanum hirtum*: phenotypic plasticity and ontogenetic contingency. *American Journal of Botany*, 81: 1354–1365.

Diggle, P. K. (1995). Architectural effects and the interpretation of patterns of fruit and seed development. *Annual Review of Ecology and Systematics*, 26: 531–552.

Diggle, P. K. (1997). Extreme preformation in an alpine *Polygonum viviparum*: an architectural and developmental analysis. *American Journal of Botany*, 84: 154–169.

Diggle, P. K. (1999). Heteroblasty and the evolution of flowering phenologies. *International Journal of Plant Sciences*, 160: S123–S134.

Diggle, P. K. (2003). Architectural effects on floral form and function: a review. In *Deep Morphology: Toward a Renaissance of Morphology in Plant Systematics*, eds. T. Stuessy, E. Hörandl & V. Mayer. Königstein: Koeltz, pp. 63–80.

Diggle, P. K. (2014). Modularity and intra-floral integration in metameric organisms: plants are more than the sum of their parts. *Philosophical Transactions of the Royal Society B*, 369: 20130253.

Diggle, P. K., Di Stilio, V. S., Gschwend, A. R. *et al.* (2011). Multiple developmental processes underlie sex differentiation in angiosperms. *Trends in Genetics*, 27: 368–376.

Dilcher, D. L. & Crane, P. R. (1984). *Archaeanthus*: an early angiosperm from the Cenomanian of the western interior of North America. *Annals of the Missouri Botanical Garden*, 71: 351–383.

Dinneny, J. R., Weigel, D. & Yanofsky, M. F. (2005). A genetic framework for fruit patterning in *Arabidopsis thaliana*. *Development*, 132: 4687–4696.

Dinneny, J. R., Yadegari, R., Fischer, R. L., Yanofsky, M. F. & Weigel, D. (2004). The role of JAGGED in shaping lateral organs. *Development*, 131: 1101–1110.

Dinneny, J. R. & Yanofsky, M. F. (2005). Drawing lines and borders: how the dehiscent fruit of *Arabidopsis* is patterned. *BioEssays*, 27: 42–49.

Ditta, G., Pinyopich, A., Robles, P., Pelaz, S. & Yanofsky, M. F. (2004). The SEP4 gene of *Arabidopsis thaliana* functions in floral organ and meristem identity. *Current Biology*, 14: 1935–1940.

Dodds, P. N. & Rathjen, J. P. (2010). Plant immunity: towards an integrated view of plant-pathogen interactions. *Nature Reviews Genetics*, 11: 539–548.

Doebley, J., Stec, A. & Hubbard, L. (1997). The evolution of apical dominance in maize. *Nature*, 386: 485–488.

Domazet-Lošo, T., Brajković, J. & Tautz, D. (2007). A phylostratigraphy approach to uncover the genomic history of major adaptations in metazoan lineages. *Trends in Genetics*, 23, 533–539.

Domoney, C., Duc, G., Ellis, T. H. N. *et al.* (2006). Genetic and genomic analysis of legume flowers and seeds. *Current Opinion in Plant Biology*, 9: 133–141.

Donnelly, P. M., Bonetta, D., Tsukaya, H., Dengler, R. & Dengler, N. G. (1999). Cell cycling and cell enlargement in developing leaves of *Arabidopsis*. *Developmental Biology*, 215: 407–419.

Donoghue, M. J. (1992). Homology. In *Keywords in Evolutionary Biology*, eds. E. F. Keller & E. A. Lloyd. Cambridge, MA: Harvard University Press, pp. 170–179.

Donoghue, M. J. & Ree, R. H. (2000). Homoplasy and developmental constraint: a model and an example from plants. *American Zoologist*, 40: 759–769.

Donoghue, M. J., Ree, R. H. & Baum D. A. (1998). Phylogeny and the evolution of flower symmetry in the Asteridae. *Trends in Plant Science*, 3: 311–317.

Doust, A. N. (2001). The developmental basis of floral variation in *Drimys winteri* (Winteraceae). *International Journal of Plant Sciences*, 162: 697–717.

Doust, A. N., Mauro-Herrera, M., Francis, A. D. & Shand, L. C. (2014). Morphological diversity and genetic regulation of inflorescence abscission zones in grasses. *American Journal of Botany*, 101: 1759–1769.

Doyle, J. A. (2008). Integrating molecular phylogenetic evidence and paleobotanical evidence on the origin of the flower. *International Journal of Plant Sciences*, 167: 816–843.

Doyle, J. A. & Endress, P. K. (2000). Morphological phylogenetic analysis of basal angiosperms: comparison and combination with molecular data. *International Journal of Plant Sciences*, 161: S121–S153.

Doyle, J. A. & Endress, P. K. (2011). Tracing the evolutionary diversification of the flower in basal angiosperms. In *Flowers on the Tree of Life*, eds. L. Wanntorp & L. P. Ronse De Craene. Cambridge: Cambridge University Press, pp. 88–119.

Draghi, J. & Wagner, G. P. (2008). Evolution of evolvability in a developmental model. *Evolution*, 62: 301–315.

Dransfield, J. & Uhl, N.W. (1998). Palmae. In *The Families and Genera of Vascular Plants. Vol. 4*, ed. K. Kubitzki. Berlin: Springer, pp. 306–388.

Drea, S., Hileman, L. C., de Martino, G. & Irish, V. F. (2007). Functional analyses of genetic pathways controlling petal specification in poppy. *Development*, 134: 4157–4166.

Dreni, L., Jacchia, S., Fornara, F. *et al.* (2007). The D-lineage MADS-box gene *OsMADS13* controls ovule identity in rice. *The Plant Journal*, 52: 690–699.

Drinnan, A. N., Crane, P. R. & Hoot, S. B. (1994). Patterns of floral evolution in the early diversification of nonmagnoliid dicotyledons (eudicots). *Plant Systematics and Evolution, Supplement*, 8: 93–122.

Drost, H. G., Gabel, A., Grosse, I. & Quint, M. (2015). Evidence for active maintenance of phylotranscriptomic hourglass patterns in animal and plant embryogenesis. *Molecular Biology and Evolution*, 32: 1221–1231.

Du, X., Xiao, Q., Zhao, R. *et al.* (2008). *TrMADS3*, a new MADS-box gene, from a perennial species *Taihangia rupestris* (Rosaceae) is upregulated by cold and experiences seasonal fluctuation in expression level. *Development Genes and Evolution*, 218: 281–292.

Du, Z.-Y. & Wang, Y.-Z. (2008). Significance of RT-PCR expression patterns of *CYC*-like genes in *Oreocharis benthamii* (Gesneriaceae). *Journal of Systematics and Evolution*, 46: 23–31.

Duboule, D. (1994). Temporal colinearity and the phylotypic progression: a basis for the stability of a vertebrate Bauplan and the evolution of morphologies through heterochrony. *Development, Supplement*, 1994: 135–142.

Dubrovsky, J. G., Gambetta, G. A., Hernández-Barrera, A., Shishkova, S. & González, I. (2006). Lateral root initiation in *Arabidopsis*: developmental window, spatial patterning, density and predictability. *Annals of Botany*, 97: 903–915.

Dunwell, J. M. (2010). Haploids in flowering plants: origins and exploitation. *Plant Biotechnology Journal*, 8: 377–424.

Duthion, C., Ney, B. & Munier-Jolain, N. M. (1994). Development and growth of white lupin: implications for crop management. *Agronomy Journal*, 86: 1039–1045.

Eames, A. J. (1961). *Morphology of the Angiosperms*. New York, NY: McGraw-Hill.

Efroni, I., Blum, E., Goldshmidt, A. & Eshed, Y. (2008). A protracted and dynamic maturation schedule underlies *Arabidopsis* leaf development. *Plant Cell*, 20: 2293–2306.

Efroni, I., Eshed, Y. & Lifschitz, E. (2010). Morphogenesis of simple and compound leaves: a critical review. *Plant Cell*, 22: 1019–1032.

Egea-Cortines, M. & Weiss, J. (2013). Control of plant organ size. In *Encyclopedia of Life Sciences*. Chichester: Wiley.

Ehrenreich, I. M. & Pfennig, D. W. (2016). Genetic assimilation: a review of its potential proximate causes and evolutionary consequences. *Annals of Botany*, 117: 769–779.

El Ottra, J. H. L., Pirani, J. R. & Endress, P. K. (2013). Fusion within and between whorls of floral organs in Galipeinae (Rutaceae): structural features and evolutionary implications. *Annals of Botany*, 111: 821–837.

Elo, A., Lemmetyinen, J., Novak, A. *et al.* (2007). *BpMADS4* has a central role in inflorescence initiation in silver birch (*Betula pendula*). *Physiologia Plantarum*, 131: 149–158.

Elsner, J., Michalski, M. & Kwiatkowska, D. (2012). Spatiotemporal variation of leaf epidermal cell growth: a quantitative analysis of *Arabidopsis thaliana* wildtype and triple *cyclinD3* mutant plants. *Annals of Botany*, 109: 897–910.

Emery, J. F., Floyd, S. K., Alvarez, J. *et al.* (2003). Radial patterning of *Arabidopsis* shoots by class III HD-ZIP and KANADI genes. *Current Biology*, 13: 1768–1774.

Endress, M. E. (2001). Apocynaceae and Asclepiadaceae: united they stand. *Haseltonia*, 8: 2–9.

Endress, P. K. (1970). Die Infloreszenzen der apetalen Hamamelidaceen, ihre grundsätzliche morphologische und systematische Bedeutung. *Botanische Jahrbücher für Systematik, Pflanzengeschichte und Pflanzengeographie*, 90: 1–54.

Endress, P. K. (1976). Die Androeciumanlage bei polyandrischen Hamamelidaceen und ihre systematische Bedeutung. *Botanische Jahrbücher für Systematik, Pflanzengeschichte und Pflanzengeographie*, 97: 436–457.

Endress, P. K. (1978). Blütenontogenese, Blütenabgrenzung und systematische Stellung der perianthlosen Hamamelidoideae. *Botanische Jahrbücher für Systematik, Pflanzengeschichte und Pflanzengeographie*, 100: 249–317.

Endress, P. K. (1982). Syncarpy and alternative modes of escaping disadvantages of apocarpy in primitive angiosperms. *Taxon*, 31: 48–52.

Endress, P. K. (1984). The flowering process in the Eupomatiaceae (Magnoliales). *Botanische Jahrbücher für Systematik, Pflanzengeschichte und Pflanzengeographie*, 104: 297–319.

Endress, P. K. (1987a). The Chloranthaceae: reproductive structures and phylogenetic position. *Botanische Jahrbücher für Systematik, Pflanzengeschichte und Pflanzengeographie*, 109: 153–226.

Endress, P. K. (1987b). Floral phyllotaxis and floral evolution. *Botanische Jahrbücher für Systematik, Pflanzengeschichte und Pflanzengeographie*, 108: 417–438.

Endress, P. K. (1989). Chaotic floral phyllotaxis and reduced perianth in *Achlys* (Berberidaceae). *Botanica Acta*, 102: 159–163.

Endress, P. K. (1990a). Patterns of floral construction in ontogeny and phylogeny. *Biological Journal of the Linnean Society*, 39: 153–175.

Endress, P. K. (1990b). Evolution of reproductive structures and functions in primitive angiosperms (Magnoliidae). *Memoirs of the New York Botanical Garden*, 55: 5–34.

Endress, P. K. (1992). Evolution and floral diversity: the phylogenetic surroundings of *Arabidopsis* and *Antirrhinum*. *International Journal of Plant Sciences*, 153: S106–S122.

Endress, P. K. (1994a). *Diversity and Evolutionary Biology of Tropical Flowers*. Cambridge: Cambridge University Press.

Endress, P. K. (1994b). Floral structure and evolution of primitive angiosperms: recent advances. *Plant Systematics and Evolution*, 192: 79–97.

Endress, P. K. (1994c). Evolutionary aspects of the floral structure in *Ceratophyllum*. *Plant Systematics and Evolution, Supplement*, 8: 175–183.

Endress, P. K. (1995). Floral structure and evolution in Ranunculanae. *Plant Systematics and Evolution, Supplement*, 9: 47–61.

Endress, P. K. (1996). Homoplasy in angiosperm flowers. In *Homoplasy: The Recurrence of Similarity in Evolution*, eds. M. J. Sanderson & L. Hufford. San Diego, CA: Academic Press, pp. 303–325.

Endress, P. K. (1997). Evolutionary biology of flowers: prospects for the next century. In *Evolution and Diversification of Land Plants*, ed. K. Iwatsuki & P. H. Raven. Tokyo: Springer-Verlag, pp. 99–119.

Endress, P. K. (1998). *Antirrhinum* and Asteridae: evolutionary changes of floral symmetry. *Symposia of the Society for Experimental Biology*, 51: 133–140.

Endress, P. K. (1999). Symmetry in flowers: diversity and evolution. *International Journal of Plant Sciences*, 160: S3–S23.

Endress, P. K. (2001a). Evolution of floral symmetry. *Current Opinion in Plant Biology*, 4: 86–91.

Endress, P. K. (2001b). Origins of flower morphology. *Journal of Experimental Zoology (Molecular and Developmental Evolution)*, 291: 105–115.

Endress, P. K. (2001c). The flowers in extant basal angiosperms and inferences on ancestral flowers. *International Journal of Plant Sciences*, 162: 1111–1140.

Endress, P. K. (2004). Structure and relationships of basal relictual angiosperms. *Australian Systematic Botany*, 17: 343–366.

Endress, P. K. (2006). Angiosperm floral evolution: morphological and developmental framework. *Advances in Botanical Research*, 44: 1–61.

Endress, P. K. (2008a). Perianth biology in the basal grade of extant angiosperms. *International Journal of Plant Sciences*, 169: 844–862.

Endress, P. K. (2008b). The whole and the parts: relationships between floral architecture and floral organ shape, and their repercussions on the interpretation of fragmentary floral fossils. *Annals of the Missouri Botanical Garden*, 95: 101–120.

Endress, P. K. (2010a). Flower structure and trends of evolution in eudicots and their major subclades. *Annals of the Missouri Botanical Garden*, 97: 541–583.

Endress, P. K. (2010b). Disentangling confusions in inflorescence morphology: patterns and diversity of reproductive shoot ramification in angiosperms. *Journal of Systematics and Evolution*, 48: 225–239.

Endress, P. K. (2010c). The evolution of floral biology in basal angiosperms. *Philosophical Transactions of the Royal Society B*, 365: 411–421.

Endress, P. K. (2010d). Synorganisation without organ fusion in the flowers of *Geranium robertianum* (Geraniaceae) and its not so trivial obdiplostemony. *Annals of Botany*, 106: 687–695.

Endress, P. K. (2011). Evolutionary diversification of the flowers in angiosperms. *American Journal of Botany*, 98: 370–396.

Endress, P. K. (2012). The immense diversity of floral monosymmetry and asymmetry across angiosperms. *Botanical Review*, 78: 345–397.

Endress, P. K. (2014). Multicarpellate gynoecia in angiosperms - occurrence, development, organization and architectural constraints. *Botanical Journal of the Linnean Society*, 174: 1–43.

Endress, P. K. (2015). Patterns of angiospermy development before carpel sealing across living angiosperms: diversity, and morphological and systematic aspects. *Botanical Journal of the Linnean Society*, 178: 556–591.

Endress, P. K. & Doyle, J. A. (2007). Floral phyllotaxis in basal angiosperms: development and evolution. *Current Opinion in Plant Biology*, 10: 52–57.

Endress, P. K. & Doyle, J. A. (2015) Ancestral traits and specializations in the flowers of the basal grade of living angiosperms. *Taxon*, 64: 1093–1116.

Endress, P. K. & Igersheim, A. (2000a) Gynoecium structure and evolution in basal angiosperms. *International Journal of Plant Sciences*, 161: S211–S223.

Endress, P. K. & Igersheim, A. (2000b). The reproductive structures of the basal angiosperm *Amborella trichopoda* (Amborellaceae). *International Journal of Plant Sciences*, 161: S237–S248.

Endress, P. K., Igersheim, A., Sampson, F. B. & Schatz, G. E. (2000). Floral structure of *Takhtajania* and its systematic position in Winteraceae. *Annals of the Missouri Botanical Garden*, 87: 347–365.

Endress, P. K. & Lorence, D. H. (2004). Heterodichogamy of a novel type in *Hernandia* (Hernandiaceae) and its structural basis. *International Journal of Plant Sciences*, 165: 753–763.

Endress, P. K. & Matthews, M. L. (2006a). Elaborate petals and staminodes in eudicots: diversity, function, and evolution. *Organisms, Diversity and Evolution*, 6: 257–293.

Endress, P. K. & Matthews, M. L. (2006b). First steps towards a floral structural characterization of the major rosid subclades. *Plant Systematics and Evolution*, 260: 223–251.

Erbar, C. & Leins, P. (1996). Distribution of the character states 'early' and 'late sympetaly' within the 'Sympetalae Tetracyclicae' and presumably related groups. *Botanica Acta*, 109: 427–440.

Erbar, C. & Leins, P. (1997). Different patterns of floral development in whorled flowers, exemplified by Apiaceae and Brassicaceae. *International Journal of Plant Sciences*, 158: S49–S64.

Erwin, D. H., Laflamme, M., Tweedt, S. M. *et al.* (2011). The Cambrian conundrum: early divergence and later ecological success in the early history of animals. *Science*, 334: 1091–1097.

Eshed, Y., Baum, S. F. & Bowman, J. L. (1999). Distinct mechanisms promote polarity establishment in carpels of *Arabidopsis*. *Cell*, 99: 199–209.

Eshed, Y., Baum, S. F., Perea, J. V. & Bowman, J. L. (2001). Establishment of polarity in lateral organs of plants. *Current Biology*, 11: 1251–1260.

Eshed, Y., Izhaki, A., Baum, S. F., Floyd, S. K. & Bowman, J. L. (2004). Asymmetric leaf development and blade expansion in *Arabidopsis* are mediated by *KANADI* and *YABBY* activities. *Development*, 131: 2997–3006.

Espadaler, X. & Gómez, C. (2001). Female performance in *Euphorbia characias*: effect of flower position on seed quantity and quality. *Seed Science Research*, 11: 163–172.

Evered, D. & Marsh, J. (1989). *The Cellular Basis of Morphogenesis*. Chichester: Wiley.

Ewers, F. W. & Schmid, R. (1981). Longevity of needle fascicles of *Pinus longaeva* (bristlecone pine) and other North American pines. *Oecologia*, 51: 107–115.

Fahlgren, N., Jogdeo, S., Kasschau, K. D. *et al.* (2010). MicroRNA gene evolution in *Arabidopsis lyrata* and *Arabidopsis thaliana*. *Plant Cell*, 22: 1074–1089.

Fambrini, M., Salvini, M. & Pugliesi, C. (2011). A transposon-mediate inactivation of a *CYCLOIDEA*-like gene originates polysymmetric and androgynous ray flowers in *Helianthus annuus*. *Genetica*, 139: 1521–1529.

Fauron, C., Allen, J. O., Clifton, S. & Newton, K. J. (2004). Plant mitochondrial genomes. In *Molecular Biology and Biotechnology of Plant Organelles*, eds. H. Daniell & C. Chase. Dordrecht: Kluwer, pp. 155–171.

Feng, G., Qin, Z., Yan, J., Zhang, X. & Hu, Y. (2011). *Arabidopsis ORGAN SIZE RELATED1* regulates organ growth and final organ size in orchestration with *ARGOS* and *ARL*. *New Phytologist*, 191: 635–646.

Feng, M. & Lu, A.-M. (1998). Floral organogenesis and its systematic significance of the genus *Nandina* (Berberidaceae). *Acta Botanica Sinica*, 40: 102–108.

Feng, X., Zhao, Z., Tian, Z. *et al.* (2006). Control of petal shape and floral zygomorphy in *Lotus japonicus*. *Proceedings of the National Academy of Sciences of the United States of America*, 103: 4970–4975.

Ferjani, A., Horiguchi, G., Yano, S. & Tsukaya, H. (2007). Analysis of leaf development in *fugu* mutants of *Arabidopsis* reveals three compensation modes that modulate cell expansion in determinate organs. *Plant Physiology*, 144: 988–999.

Ferrándiz, C., Liljegren, S. J. & Yanofsky, M. F. (2000). Negative regulation of the *SHATTERPROOF* genes by *FRUITFULL* during *Arabidopsis* fruit development. *Science*, 289: 436–438.

Ferreira, B. G. & Isaias, R. M. S. (2014). Floral-like destiny induced by a galling Cecidomyiidae on the axillary buds of *Marcetia taxifolia* (Melastomataceae). *Flora*, 209: 391–400.

Ferrero, V., Arroyo, J., Vargas, P., Thompson, J. D. & Navarro, L. (2009). Evolutionary transitions of style polymorphisms in *Lithodora* (Boraginaceae). *Perspectives in Plant Ecology, Evolution and Systematics*, 11: 111–125.

Ferrero, V., Rojas, D., Vale, A. & Navarro, L. (2012). Delving into the loss of heterostyly in Rubiaceae: is there a similar trend in tropical and non-tropical zones? *Perspectives in Plant Ecology, Evolution and Systematics*, 14: 161–167.

Fink, W. L. (1982). The conceptual relationship between ontogeny and phylogeny. *Paleobiology*, 8: 254–264.

Fink, W. L. (1988). Phylogenetic analysis and the detection of ontogenetic patterns. In *Heterochrony in Evolution: A Multidisciplinary Approach*, ed. M. L. McKinney. New York, NY: Plenum Press, pp. 71–91.

Fischer, E. (2015). Magnoliopsida (Angiosperms) p.p.: Subclass Magnoliidae [Amborellanae to Magnolianae, Lilianae p.p. (Acorales to Asparagales)]. In *Syllabus of Plant Families*, 13th edn, Part 4, ed. W. Frey. Stuttgart: Borntraeger, pp. 111–495.

Fishbein, M. (2001). Evolutionary innovation and diversification in the flowers of Asclepiadaceae. *Annals of the Missouri Botanical Garden*, 88: 603–623.

Fisher, J. B. (2002). Indeterminate leaves of *Chisocheton* (Meliaceae): survey of structure and development. *Botanical Journal of the Linnean Society*, 139: 207–221.

Fisher, J. B. & Rutishauser, R. (1990). Leaves and epiphyllous shoots in *Chisocheton* (Meliaceae): a continuum of woody leaf and stem-axes. *Canadian Journal of Botany*, 68: 2316–2328.

Fisher, R. A. (1930). *The Genetical Theory of Natural Selection.* Oxford: Clarendon Press.

Fitting, H. (1921). Das Verblühen der Blüten. *Die Naturwissenschaften,* 9: 1–9.

Floyd, S. K. & Bowman, J. L. (2007). The ancestral developmental toolkit of land plants. *International Journal of Plant Sciences,* 168: 1–35.

Floyd, S. K. & Bowman, J. L. (2010). Gene expression patterns in seed plant shoot meristems and leaves: homoplasy or homology? *Journal of Plant Research,* 123: 43–55.

Forest, F., Chase, M. W., Persson, C., Crane, P. R. & Hawkins, J. A. (2007). The role of biotic and abiotic factors in the evolution of ant-dispersal in the milkwort family (Polygalaceae). *Evolution,* 61: 1675–1694.

Foucher, F., Morin, J., Courtiade, J. *et al.* (2003). *DETERMINATE* and *LATE FLOWERING* are two *TERMINAL FLOWER1/CENTRORADIALIS* homologs that control two distinct phases of flowering initiation and development in pea. *Plant Cell,* 15: 2742–2754.

Frajman, B. & Schönswetter, P. (2011). Giants and dwarfs: molecular phylogenies reveal multiple origins of annual spurges within *Euphorbia* subg. *Esula. Molecular Phylogenetics and Evolution,* 61: 413–424.

Francis, D. (2008). Apical meristems. In *Encyclopedia of Life Sciences.* Chichester: Wiley.

Franzke, A., Lysak, M. A., Al-Shehbaz, I. A., Koch, M. A. & Mummenhoff, K. (2011). Cabbage family affairs: the evolutionary history of Brassicaceae. *Trends in Plant Science,* 16: 108–116.

Friedman, W. E., Moore, R. C. & Purugganan, M. D. (2004). The evolution of plant development. *American Journal of Botany,* 91: 1726–1741.

Friis, E. M., Doyle, J. A., Endress, P. K. & Leng, Q. (2003). *Archaefructus:* angiosperm precursor or specialized early angiosperm? *Trends in Plant Science,* 8: 369–373.

Friis, E. M., Eklund, H., Pedersen, K. R. & Crane, P. R. (1994). *Virginianthus calycanthoides* gen. et sp. nov.: a calycanthaceous flower from the Potomac group (Early Cretaceous) of eastern North America. *International Journal of Plant Sciences,* 155: 772–785.

Friis, E. M., Pedersen, K. R. & Crane, P. R. (2000). Reproductive structure and organization of basal angiosperms from the early Cretaceous (Barremian or Aptian) of western Portugal. *International Journal of Plant Sciences,* 161: S169–S182.

Frohlich, M. W. & Parker, D. S. (2000). The mostly male theory of flower evolutionary origins: from genes to fossils. *Systematic Botany,* 25: 155–170.

Fruciano, C., Franchini, P. & Meyer, A. (2013). Resampling based approaches to study variation in morphological modularity. *PLoS ONE,* 8: e69376.

Fujikura, U., Horiguchi, G., Ponce, M. R., Micol, J. L. & Tsukaya, H. (2009). Coordination of cell proliferation and cell expansion mediated by ribosome-related processes in the leaves of *Arabidopsis thaliana*. *The Plant Journal*, 59: 499–508.

Fujinami, R., Ghogue, J. P. & Imaichi, R. (2013). Developmental morphology of the controversial ramulus organ of *Tristicha trifaria* (subfamily Tristichoideae, Podostemaceae): implications for evolution of a unique body plan in Podostemaceae. *International Journal of Plant Sciences*, 174: 609–618.

Fujinami, R. & Imaichi, R. (2015). Developmental morphology of flattened shoots in *Dalzellia ubonensis* and *Indodalzellia gracilis* with implications for the evolution of diversified shoot morphologies in the subfamily Tristichoideae (Podostemaceae). *American Journal of Botany*, 102: 848–859.

Fukuda, H. (2000). Programmed cell death of tracheary elements as a paradigm in plants. *Plant Molecular Biology*, 44: 245–253.

Fukuda, T., Yokoyama, J. & Maki, M. (2003). Molecular evolution of cycloidea-like genes in Fabaceae. *Journal of Molecular Evolution*, 57: 588–597.

Fukuda, Y. (1988). Phyllotaxis in two species of *Rubia*, *R. akane* and *R. sikkimensis*. *Botanical Magazine (Tokyo)*, 101: 25–38.

Fukuhara, T., Nagmasu, H. & Okada, H. (2003). Floral vasculature, sporogenesis and gametophyte development in *Pentastemona egregia* (Stemonaceae). *Systematics and Geography of Plants*, 73: 83–90.

Fukushima, K., Fujita, H., Yamaguchi, T. *et al.* (2015). Oriented cell division shapes carnivorous pitcher leaves of *Sarracenia purpurea*. *Nature Communications*, 6: 6450.

Fukushima, K. & Hasebe, M. (2014). Adaxial–abaxial polarity: the developmental basis of leaf shape diversity. *Genesis*, 52: 1–18.

Furumizu, C., Alvarez, J. P., Sakakibara, K. & Bowman, J. L. (2015). Antagonistic roles for KNOX1 and KNOX2 genes in patterning the land plant body plan following an ancient gene duplication. *PLoS Genetics*, 11: e1004980.

Furutani, I., Watanabe, Y., Prieto, R. *et al.* (2000). The SPIRAL genes are required for directional control of cell elongation in *Arabidopsis thaliana*. *Development*, 127: 4443–4453.

Fusco, G. & Minelli, A. (2010). Phenotypic plasticity in development and evolution. *Philosophical Transactions of the Royal Society B*, 365: 547–556.

Galego, L. & Almeida, J. (2002). Role of DIVARICATA in the control of dorsoventral asymmetry in *Antirrhinum* flowers. *Genes and Development*, 16: 880–891.

Galimba, K. D. & Di Stilio, V. S. (2015). Sub-functionalization to ovule development following duplication of a floral organ identity gene. *Developmental Biology*, 405: 158–172.

Galis, F. (1999). Why do almost all mammals have seven cervical vertebrae? Developmental constraints, Hox genes, and cancer. *Journal of Experimental Zoology (Molecular and Developmental Evolution)*, 285: 19–26.

Galis, F., Van Dooren, T. J., Feuth, J. D. *et al.* (2006). Extreme selection in humans against homeotic transformations of cervical vertebrae. *Evolution*, 60: 2643–2654.

Gallois, J. L., Woodward, C., Reddy, G. V. & Sablowski, R. (2002). Combined *SHOOT MERISTEMLESS* and *WUSCHEL* trigger ectopic organogenesis in *Arabidopsis. Development*, 129: 3207–3217.

Gao, J. Y., Ren, P. Y., Yang, Z. H. & Li, Q. J. (2006). The pollination ecology of *Paraboea rufescens* (Gesneriaceae): a buzz-pollinated tropical herb with mirror-image flowers. *Annals of Botany*, 97: 371–376.

Gao, Q., Tao, J.-H., Wang, Y.-Z. & Li, Z.-H. (2008). Expression differentiation of CYC-like floral symmetry genes correlated with their protein sequence divergence in *Chirita heterotricha* (Gesneriaceae). *Development, Genes and Evolution*, 218: 341–351.

Gao, X., Liang, W., Yin, C. *et al.* (2010). The *SEPALLATA*-like gene *OsMADS34* is required for rice inflorescence and spikelet development. *Plant Physiology*, 153: 728–740.

García, M. B. & Antor, R. J. (1995a). Age and size structure in populations of a long-lived dioecious geophyte: *Borderea pyrenaica* (Dioscoreaceae). *International Journal of Plant Sciences*, 156: 236–243.

García, M. B. & Antor, R. J. (1995b). Sex-ratio and sexual dimorphism in the dioecious *Borderea pyrenaica* (Dioscoreaceae). *Oecologia*, 101: 59–67.

García, M. B., Dahlgren, J. P. & Ehrlén, J. (2011). No evidence of senescence in a 300-year-old mountain herb. *Journal of Ecology*, 99: 1424–1430.

Garnock-Jones, P. J. & Johnson, P. N. (1987). *Iti lacustris* (Brassicaceae), a new genus and species from southern New Zealand. *New Zealand Journal of Botany*, 25: 603–610.

Gautheret, R. J. (1934). Culture du tissus cambial. *Comptes rendus hebdomadaires des séances de l'Académie des sciences*, 198: 2195–2196.

Geeta, R., Davalos, L. M., Levy, A. *et al.* (2012). Keeping it simple: flowering plants tend to retain, and revert to, simple leaves. *New Phytologist*, 193: 481–493.

Gehring, W. J., Affolter, M. & Bürglin, T. R. (1994). Homeodomain proteins. *Annual Review of Biochemistry*, 63: 487–526.

Gendron, J. M., Liu, J.-S., Fan, M. *et al.* (2012). Brassinosteroids regulate organ boundary formation in the shoot apical meristem of *Arabidopsis. Proceedings of the National Academy of Sciences of the United States of America*, 109: 21152–21157.

Gensel, P. G., Kotyk, M. E. & Basinger, J. E. (2001). Morphology of above- and below-ground structures in early Devonian (Pragian-Emsian) plants. In *Plants Invade the Land: Evolutionary and Environmental Perspectives*, eds. P. G. Gensel & D. Edwards. New York, NY: Columbia University Press, pp. 83–102.

Gerats, T. & Vandenbussche, M. (2005). A model system for comparative research: *Petunia*. *Trends in Plant Science*, 10: 251–256.

Gerrath, J. M. & Lacroix, C. R. (1997). Heteroblastic sequence and leaf development in *Leea guineensis*. *International Journal of Plant Sciences*, 158: 747–756.

Gerrath, J. M., Posluszny, U. & Dengler, N. G. (2001). Primary vascular patterns in the Vitaceae. *International Journal of Plant Sciences*, 162: 729–745.

Geuten, K., Becker, A., Kaufmann, K. *et al.* (2006). Petaloidy and petal identity MADS-box genes in the balsaminoid genera *Impatiens* and *Marcgravia*. *The Plant Journal*, 47: 501–518.

Geuten, K. & Irish, V. (2010). Hidden variability of floral homeotic B genes in Solanaceae provides a molecular basis for the evolution of novel functions. *Plant Cell*, 22: 2562–2578.

Gibson, G. & Dworkin, I. (2004). Uncovering cryptic genetic variation. *Nature Reviews Genetics*, 5: 1199–1212.

Gilbert, S. F. & Bolker, J. A. (2001). Homologies of process and modular elements of embryonic construction. *Journal of Experimental Zoology (Molecular and Developmental Evolution)*, 291: 1–12.

Gill, N., Buti, M., Kane, N. *et al.* (2014). Sequence-based analysis of structural organization and composition of the cultivated sunflower (*Helianthus annuus* L.) genome. *Biology*, 3: 295–319.

Gillies, A. C. M., Cubas, P., Coen, E. S. & Abbott R. J. (2002). Making rays in the Asteraceae: genetics and evolution of variation for radiate versus discoid flower heads. In *Developmental Genetics and Plant Evolution*, eds. Q. C. B. Cronk, R. M. Bateman & J. A. Hawkins. London: Taylor & Francis, pp. 237–246.

Giménez, E., Pineda, B., Capel, J. *et al.* (2010). Functional analysis of the Arlequin mutant corroborates the essential role of the Arlequin/TAGL1 gene during reproductive development of tomato. *PLoS ONE*, 5: e14427.

Gissi, C., Hastings, K. E. M., Gasparini, F. *et al.* (2017). An unprecedented taxonomic revision of a model organism: the paradigmatic case of *Ciona robusta* and *Ciona intestinalis*. *Zoologica Scripta*, 46: 521–522.

Givnish, T. J. (2010). Giant lobelias exemplify convergent evolution. *BMC Biology*, 8: 3.

Gleissberg, S. (1998a). Comparative analysis of leaf shape development in Papaveraceae-Papaveroideae. *Flora*, 193: 269–301.

Gleissberg, S. (1998b). Comparative analysis of leaf shape development in Papaver-aceae–Chelidonioideae. *Flora*, 193: 387–409.

Gleissberg, S., Groot, E. P., Schmalz, M. *et al.* (2005). Developmental events leading to peltate leaf structure in *Tropaeolum majus* (Tropaeolaceae) are associated with expression domain changes of a YABBY gene. *Development Genes and Evolution*, 215: 313–319.

Gleissberg, S. & Kadereit, J. W. (1999). Evolution of leaf morphogenesis: evidence from developmental and phylogenetic data in Papaveraceae. *International Journal of Plant Sciences*, 160: 787–794.

Godfrey-Smith, P. (2009). *Darwinian Populations and Natural Selection*. New York, NY: Oxford University Press.

Goebel, K. (1900–1905). *Organography of Plants*. English translation by I. Balfour. Parts 1 and 2. Oxford: Clarendon Press.

Goebel, K. (1920). *Die Entfaltungsbewegungen der Pflanzen und deren teleologische Deutung*. Jena: Fischer.

Goebel, K. (1924). *Die Entfaltungsbewegungen der Pflanzen und deren teleologische Deutung*, 2nd edn. Jena: Fischer.

Goebel, K. (1928). *Organographie der Pflanzen. 1. Allgemeine Organographie*, 3rd edn. Jena: Fischer.

Goebel, K. (1930). *Blütenbildung und Sprossgestaltung (Anthokladien und Inflor-eszenzen)*. Jena: G. Fischer.

Goebel, K. (1933). *Organographie der Pflanzen III. Samenpflanzen*. Jena: G. Fischer.

Goethe, J. W. von (1790). *Versuch die Metamorphose der Pflanzen zu erklären*. Gotha: Ettinger.

Goff, S. A., Ricke, D., Lan, T. H. *et al.* (2002). A draft sequence of the rice genome (*Oryza sativa* L. ssp. *japonica*). *Science*, 296: 92–100.

Goldschmidt, R. (1940). *The Material Basis of Evolution*. New Haven, CT: Yale University Press.

Golz, J. F., Roccaro, M., Kuzoff, R. & Hudson, A. (2004). *GRAMINIFOLIA* promotes growth and polarity of *Antirrhinum* leaves. *Development*, 131: 3661–3670.

González, F. & Bello, M. A. (2009). Intra-individual variation of flowers in *Gunnera* subgenus *Panke* (Gunneraceae) and proposed apomorphies for Gunnerales. *Botanical Journal of the Linnean Society*, 160: 262–283.

Gonzalez, N., Vanhaeren, H. & Inzé, D. (2012). Leaf size control: complex coordination of cell division and expansion. *Trends in Plant Science*, 17: 332–340.

Gooh, K., Ueda, M., Aruga, K. *et al.* (2015). Live-cell imaging and optical manipulation of *Arabidopsis* early embryogenesis. *Developmental Cell*, 34: 242–251.

Goto, K. & Meyerowitz, E. M. (1994). Function and regulation of the *Arabidopsis* floral homeotic gene *PISTILLATA*. *Genes and Development*, 8: 1548–1560.

Gould, K. S. (1993). Leaf heteroblasty in *Pseudopanax crassifolius*: functional significance of leaf morphology and anatomy. *Annals of Botany*, 71: 61–70.

Gould, S. J. (1977). *Ontogeny and Phylogeny*. Cambridge, MA: Harvard University Press.

Gould, S. J. (1988). The uses of heterochrony. In *Heterochrony in Evolution: A Multidisciplinary Approach*, ed. M. L. McKinney. New York, NY: Plenum Press, pp. 1–13.

Gourlay, C. W., Hofer, J. M. I. & Ellis, T. H. N. (2000). Pea compound leaf architecture is regulated by interactions among the genes *UNIFOLIATA, COCHLEATA, AFILA*, and *TENDRIL-LESS. Plant Cell*, 12: 1279–1294.

Govindan, B., Johnson, A. J., Nair, S. N. A. *et al.* (2016). Nutritional properties of the largest bamboo fruit *Melocanna baccifera* and its ecological significance. *Scientific Reports*, 6: 26135.

Graham, L. E., Cook, M. E. & Busse, J. S. (2000). The origin of plants: body plan changes contributing to a major evolutionary radiation. *Proceedings of the National Academy of Sciences of the United States of America*, 97: 4535–4540.

Granado-Yela, C., Balaguer, L., Cayuela, L. & Méndez, M. (2017). Unusual positional effects on flower sex in an andromonoecious tree: resource competition, architectural constraints, or inhibition by the apical flower? *American Journal of Botany*, 104: 608–615.

Greb, T., Clarenz, O., Schäfer, E. *et al.* (2003). Molecular analysis of the *LATERAL SUPPRESSOR* gene in *Arabidopsis* reveals a conserved control mechanism for axillary meristem formation. *Genes and Development*, 17: 1175–1187.

Greb, T. & Lohmann, J. U. (2016). Plant stem cells. *Current Biology*, 26: R816–R821.

Greenwood, M. S. (1995). Juvenility and maturation in conifers: current concepts. *Tree Physiology*, 15: 433–438.

Greilhuber, J., Borsch, T., Müller, K. *et al.* (2006). Smallest angiosperm genomes found in Lentibulariaceae with chromosomes of bacterial size. *Plant Biology*, 8: 770–777.

Grew, N. (1682). *The Anatomy of Plants*. London: Rawlings.

Griesemer, J. (2014). Reproduction and scaffolded developmental processes: an integrated evolutionary perspective. In *Towards a Theory of Development*, eds. A. Minelli & T. Pradeu. Oxford: Oxford University Press, pp. 183–202.

Griffith, M. E., da Silva Conceição, A. & Smyth, D. R. (1999). *PETAL LOSS* gene regulates initiation and orientation of second whorl organs in the *Arabidopsis* flower. *Development*, 126: 5635–5644.

Grimes, J. (1999). Inflorescence morphology, heterochrony, and phylogeny in the Mimosoid tribes Ingeae and Acacieae (Leguminosae: Mimosoideae). *Botanical Review*, 65: 317–347.

Grob, V., Moline, P., Pfeifer, E., Novelo, A. R. & Rutishauser, R. (2006). Developmental morphology of branching flowers in *Nymphaea prolifera*. *Journal of Plant Research*, 119: 561–570.

Groover, A., DeWitt, N., Heidel, A. & Jones, A. (1997). Programmed cell death of tracheary elements differentiating in vitro. *Protoplasma*, 196: 197–211.

Gu, Q., Ferrándiz, C., Yanofsky, M. F. & Martienssen, R. (1998). The *FRUITFULL* MADS-box gene mediates cell differentiation during *Arabidopsis* fruit development. *Development*, 125: 1509–1517.

Guédès, M. (1979). *Morphology of Seed-Plants*. Vaduz: Cramer.

Guerrant, E. O. (1982). Neotenic evolution *of Delphinium nudicaule* (Ranunculaceae): a hummingbird-pollinated larkspur. *Evolution*, 36: 699–712.

Guha, S. & Maheshwari, S. C. (1964). In vitro production of embryos from anthers of *Datura*. *Nature*, 204: 497.

Gunawardena, A. H. L. A. N. & Dengler, N. (2006). Alternative modes of leaf dissection in monocotyledons. *Botanical Journal of the Linnean Society*, 150: 25–44.

Gunawardena, A. H. L. A. N., Greenwood, J. S. & Dengler, N. G. (2004). Programmed cell death remodels lace plant leaf shape during leaf development. *Plant Cell*, 16: 60–73.

Guo, H. S., Xie, Q., Fei, J. F. & Chua, N. H. (2005). MicroRNA directs mRNA cleavage of the transcription factor NAC1 to downregulate auxin signals for *Arabidopsis* lateral root development. *Plant Cell*, 17: 1376–1386.

Haeckel, E. (1866). *Generelle Morphologie der Organismen. Allgemeine Grundzüge der organischen Formen-Wissenschaft, mechanisch begründet durch die von Charles Darwin reformirte Descendenz-Theorie, vol. 1: Allgemeine Anatomie der Organismen*. Berlin: Reimer.

Hagemann, W. & Gleissberg, W. (1996). Organogenetic capacity of leaves: the significance of marginal blastozones in angiosperms. *Plant Systematics and Evolution*, 199: 121–152.

Hake, S., Smith, H. M. S., Holtan, H. *et al.* (2004). The role of *KNOX* genes in plant development. *Annual Review of Cell and Developmental Biology*, 20: 125–151.

Hall, B. K. (1992). *Evolutionary Developmental Biology*. London: Chapman & Hall.

Hall, B. K. (1998). *Evolutionary Developmental Biology*, 2nd edn. London: Chapman and Hall.

Hall, B. K. (2005). Consideration of the neural crest and its skeletal derivatives in the context of novelty/innovation. *Journal of Experimental Zoology (Molecular and Developmental Evolution)*, 304B: 548–557.

Hall, J. C., Tisdale, T. E., Donohue, K. & Kramer, E. M. (2006). Developmental basis of an anatomical novelty: heteroarthrocarpy in *Cakile lanceolata* and *Erucaria erucarioides* (Brassicaceae). *International Journal of Plant Sciences*, 167: 771–789.

Hallgrímsson, B., Jamniczky, H., Young, N. M. *et al.* (2009). Deciphering the palimpsest: studying the relationship between morphological integration and phenotypic covariation. *Evolutionary Biology*, 36: 355–376.

Hallgrímsson, B., Jamniczky, H., Young, N. M. *et al.* (2012). The generation of variation and the developmental basis for evolutionary novelty. *Journal of Experimental Zoology (Molecular and Developmental Evolution)*, 318B: 501–517.

Hamada, S., Onouchi, H., Tanaka, H. *et al.* (2000). Mutations in the *WUSCHEL* gene of *Arabidopsis thaliana* result in the development of shoots without juvenile leaves. *The Plant Journal*, 24: 91–101.

Hamant, O., Heisler, M. G., Jönsson, H. *et al.* (2008). Developmental patterning by mechanical signals in *Arabidopsis*. *Science*, 322: 1650–1655.

Han, P., García-Ponce, B., Fonseca-Salazar, G., Alvarez-Buylla, E. R. & Yu, H. (2008). *AGAMOUS-LIKE 17*, a novel flowering promoter, acts in a *FT* independent photoperiod pathway. *The Plant Journal*, 55: 253–265.

Hannan, G. L. (1980). Heteromericarpy and dual seed germination modes in *Platystemon californicus* (Papaveraceae). *Madroño*, 27: 164–170.

Hansen, T. F. (2003). Is modularity necessary for evolvability? Remarks on the relationship between pleiotropy and evolvability. *BioSystems*, 69: 83–94.

Hansen, T. F. (2006). The evolution of genetic architecture. *Annual Review of Ecology and Systematics*, 37: 123–157.

Hareven, D., Gutfinger, T., Parnis, A. *et al.* (1996). The making of a compound leaf: genetic manipulation of leaf architecture in tomato. *Cell*, 84: 735–744.

Harling, G., Wilder, G. J. & Eriksson, R. (1998). Cyclanthaceae. In *The Families and Genera of Vascular Plants, Vol. 3*, ed. K. Kubitzki. Berlin: Springer, pp. 202–215.

Harper, J. L. & White, J. (1974). The demography of plants. *Annual Review of Ecology and Systematics*, 5: 419–463.

Harris, E. M. (1991). Floral initiation and early development in *Erigeron philadelphicus* (Asteraceae). *American Journal of Botany*, 78: 108–121.

Harris, E. M. (1995). Inflorescence and floral ontogeny in Asteraceae: a synthesis of historical and current concepts. *Botanical Review*, 61: 3–278.

Harris, M. A., Lock, A., Bühler, J., Oliver, S. G. & Wood, V. (2013). FYPO: the fission yeast phenotype ontology. *Bioinforma*, 29: 1671–1678.

Harrison, J., Möller, M., Langdale, J., Cronk, Q. C. B. & Hudson, A. (2005). The role of *KNOX* genes in the evolution of morphological novelty in *Streptocarpus*. *Plant Cell*, 17: 430–443.

Harrison, J. C. (2017). Development and genetics in the evolution of land plant body plans. *Philosophical Transactions of the Royal Society B*, 372: 20150490.

Hartmann, U., Höhmann, S., Nettesheim, K. *et al.* (2000). Molecular cloning of *SVP*: a negative regulator of the floral transition in *Arabidopsis*. *The Plant Journal*, 21: 351–360.

Hashimoto, T. (2002). Molecular genetic analysis of left–right handedness in plants. *Philosophical Transactions of the Royal Society B*, 357: 799–808.

Hasson, A., Blein, T. & Laufs, P. (2010). Leaving the meristem behind: the genetic and molecular control of leaf patterning and morphogenesis. *Comptes Rendus Biologies*, 333: 350–360.

Hawkins, J. A. (2002). Evolutionary developmental biology: impact on systematic theory and practice, and the contribution of systematics. In *Developmental Genetics and Plant Evolution*, eds. Q. C. B. Cronk, R. M. Bateman & J. A. Hawkins. London: Taylor & Francis, pp. 32–51.

Hay, A., Barkoulas, M. & Tsiantis, M. (2006). ASYMMETRIC LEAVES1 and auxin activities converge to repress *BREVIPEDICELLUS* expression and promote leaf development in *Arabidopsis*. *Development*, 133, 3955–3961.

Hay, A., Jackson, D., Ori, N. & Hake, S. (2003). Analysis of the competence to respond to *KNOTTED1* activity in *Arabidopsis* leaves using a steroid induction system. *Plant Physiology*, 131: 1671–1680.

Hay, A. & Tsiantis, M. (2006) The genetic basis for differences in leaf form between *Arabidopsis thaliana* and its wild relative *Cardamine hirsuta*. *Nature Genetics*, 38: 942–947.

Hay, A. & Tsiantis, M. (2010). KNOX genes: versatile regulators of plant development and diversity. *Development*, 137: 3153–3165.

Hay, A. S., Pieper, B., Cooke, E. *et al.* (2014). *Cardamine hirsuta*: a versatile genetic system for comparative studies. *The Plant Journal*, 78: 1–15.

He, C. Y. & Saedler, H. (2005). Heterotopic expression of *MPF2* is the key to the evolution of the Chinese lantern of *Physalis*, a morphological novelty in Solanaceae. *Proceedings of the National Academy of Sciences of the United States of America*, 102: 5779–5784.

He, Y., Doyle, M. R. & Amasino, R. M. (2004). PAF1-complex-mediated histone methylation of *FLOWERING LOCUS C* chromatin is required for the vernalization-responsive, winter-annual habit in *Arabidopsis*. *Genes and Development*, 18: 2774–2784.

Hecht, V., Foucher, F., Ferrándiz, C. *et al.* (2005). Conservation of *Arabidopsis* flowering genes in model legumes. *Plant Physiology*, 137: 1420–1434.

Heenan, P. B. (2002). *Cardamine lacustris*, a new combination for *Iti lacustris* (Brassicaceae). *New Zealand Journal of Botany*, 40: 563–569.

Heenan, P. B., Molloy, P. B. J. & Smissen, R. D. (2013). *Cardamine cubita* (Brassicaceae), a new species from New Zealand with a remarkable reduction in floral parts. *Phytotaxa*, 140: 43–50.

Heide-Jorgensen, H. S. (2013). Introduction: the parasitic syndrome in higher plants. In *Parasitic Orobanchaceae*, eds. D. M. Joel, J. Gressel & L. J. Musselman. Heidelberg: Springer, pp. 1–18.

Held, L. I. Jr. (2009). *Quirks of Human Anatomy: An Evo-Devo Look at the Human Body*. Cambridge: Cambridge University Press.

Hemerly, A., Engler, J. de A., Bergounioux, C. *et al.* (1995). Dominant negative mutants of the Cdc2 kinase uncouple cell division from iterative plant development. *EMBO Journal*, 14: 3925–3936.

Hempel, F. D., Weigel, D., Mandel, M. A. *et al.* (1997). Floral determination and expression of floral regulatory genes in *Arabidopsis*. *Development*, 124: 3845–3853.

Hendrikse, J. L., Parsons, T. E. & Hallgrímsson, B. (2007). Evolvability as the proper focus of evolutionary developmental biology. *Evolution and Development*, 9: 393–401.

Hennig, W. (1966). *Phylogenetic Systematics*. Urbana, IL: University of Illinois Press.

Henschel, K., Kofuji, R., Hasebe, M. *et al.* (2002). Two ancient classes of MIKC-type MADS-box genes are present in the moss *Physcomitrella patens*. *Molecular Biology and Evolution*, 19: 801–814.

Heyn, C. C., Dagan, O. & Nachman, B. (1974). The annual *Calendula* species, taxonomy and relationships. *Israel Journal of Botany*, 23: 169–201.

Hibara, K., Karim, M. R., Takada, S. *et al.* (2006). *Arabidopsis CUP-SHAPED COTYLEDON3* regulates postembryonic shoot meristem and organ boundary formation. *Plant Cell*, 18: 2946–2957.

Hileman, L. C. (2012). Flowers. In *Encyclopedia of Life Sciences*. Chichester: Wiley.

Hileman, L. C. (2014a). Trends in flower symmetry evolution revealed through phylogenetic and developmental genetic advances. *Philosophical Transactions of the Royal Society B*, 369: 20130348.

Hileman, L. C. (2014b). Bilateral flower symmetry: how, when and why? *Current Opinion in Plant Biology*, 17: 146–152.

Hileman, L. C. & Baum, D. A. (2003). Why do paralogs persist? Molecular evolution of *CYCLOIDEA* and related floral symmetry genes in Antirrhineae (Veronicaceae). *Molecular Biology and Evolution*, 20: 591–600.

Hileman, L. C. & Irish, V. F. (2009). More is better: the uses of developmental genetic data to reconstruct perianth evolution. *American Journal of Botany*, 96: 83–95.

Hileman, L. C., Kramer, E. M. & Baum, D. A. (2003). Differential regulation of symmetry genes and the evolution of floral morphologies. *Proceedings of the National Academy of Sciences of the United States of America*, 100: 12814–12819.

Hileman, L. C., Sundstrom, J. F., Litt, A. *et al.* (2006). Molecular and phylogenetic analyses of the MADS-box gene family in tomato. *Molecular Biology and Evolution*, 23: 2245–2258.

Hill, J. P. & Lord, E. M. (1989). Floral development in *Arabidopsis thaliana*: comparison of the wild type and the homeotic *pistillata* mutant. *Canadian Journal of Botany*, 67: 2922–2936.

Hillson, C. J. (1979). Leaf development in *Senecio rowleyanus* (Compositae). *American Journal of Botany*, 66: 59–63.

Hintz, M., Bartholmes, C., Nutt, P. *et al.* (2006). Catching a 'hopeful monster': shepherd's purse (*Capsella bursa-pastoris*) as a model system to study the evolution of flower development. *Journal of Experimental Botany*, 57: 3531–3542.

Hnatiuk, R. J. (1977). Population structure of *Livistona eastonii* Gardn., Mitchell Plateau, Western Australia. *Australian Journal of Ecology*, 2: 461–466.

Hódar, J. A. (2002). Leaf fluctuating asymmetry of holm oak in response to drought under contrasting climatic conditions. *Journal of Arid Environments*, 52: 233–243.

Hodges, S. A. (1997a). Rapid radiation due to a key innovation in columbines (Ranunculaceae: *Aquilegia*). In *Molecular Evolution and Adaptive Radiation*, eds. T. J. Givnish & K. J. Sytsma. Cambridge: Cambridge University Press, pp. 391–405.

Hodges, S. A. (1997b). Floral nectar spurs and diversification. *International Journal of Plant Sciences*, 158: S81–S88.

Hoehndorf, R., Ngonga Ngomo, A.-C. & Kelso, J. (2010). Applying the functional abnormality ontology pattern to anatomical functions. *Journal of Biomedical Semantics*, 1: 4.

Hofer, J., Gourlay, C., Michael, A. & Ellis, T. H. N. (2001a). Expression of a class-1 *knottedl-like* homeobox gene is downregulated in pea compound-leaf primordia. *Plant Molecular Biology*, 45: 387–398.

Hofer, J., Turner, L., Hellens, R. *et al.* (1997). *UNIFOLIATA* regulates leaf and flower morphogenesis in pea. *Current Biology*, 7: 581–587.

Hofer, J., Turner, L., Moreau, C. *et al.* (2009). *Tendril-less* regulates tendril formation in pea leaves. *Plant Cell*, 21: 420–428.

Hofer, J. M. I., Gourlay, C. W. & Ellis, T. H. N. (2001b). Genetic control of leaf morphology: a partial view. *Annals of Botany*, 88: 1129–1139.

Hofmeister, W. (1868). Allgemeine Morphologie der Gewächse. In *Handbuch der physiologischen Botanik*, vol. 1(2), ed. W. Hofmeister. Leipzig: Engelmann, pp. 405–664.

Hohmann, N., Wolf, E. M., Lysak, M. A. & Koch, M. A. (2015). A time-calibrated road map of Brassicaceae species radiation and evolutionary history. *Plant Cell*, 27: 2770–2784.

Hong, L., Dumond, M., Tsugawa, S. *et al.* (2016). Variable cell growth yields reproducible organ development through spatiotemporal averaging. *Developmental Cell*, 38: 15–32.

Honma, T. & Goto, K. (2001). Complexes of MADS-box proteins are sufficient to convert leaves into floral organs. *Nature*, 409: 525–529.

Horst, N. A., Katz, A., Pereman, I. *et al.* (2016). A single homeobox gene triggers phase transition, embryogenesis and asexual reproduction. *Nature Plants*, 2: 15209.

Horst, N. H. & Reski, R. (2016). Alternation of generations: unravelling the underlying molecular mechanism of a 165-year-old botanical observation. *Plant Biology*, 18: 549–551.

Howarth, D. G. & Donoghue, M. J. (2006). Phylogenetic analysis of the 'ECE' (CYC/ TB1) clade reveals duplications predating the core eudicots. *Proceedings of the National Academy of Sciences of the United States of America*, 103: 9101–9106.

Howarth, D. G., Martins, T., Chimney, E. & Donoghue, M. J. (2011). Diversification of *CYCLOIDEA* expression in the evolution of bilateral flower symmetry in Caprifoliaceae and *Lonicera* (Dipsacales). *Annals of Botany*, 107: 1521–1532.

Hu, T. T., Pattyn, P., Bakker, E. G. *et al.* (2011). The *Arabidopsis lyrata* genome sequence and the basis of rapid genome size change. *Nature Genetics*, 43: 476–481.

Hu, Y., Xie, Q. & Chua, N. H. (2003). The *Arabidopsis* auxin-inducible gene *ARGOS* controls lateral organ size. *Plant Cell*, 15: 1951–1961.

Huang, L.-J., Wang, X.-W. & Wang, X.-F. (2014). The structure and development of incompletely closed carpels in an apocarpous species, *Sagittaria trifolia* (Alismataceae). *American Journal of Botany*, 101: 1229–1234.

Huang, T. & Irish, V. F. (2016). Gene networks controlling petal organogenesis. *Journal of Experimental Botany*, 67: 61–68.

Hubbard, L., McSteen, P., Doebley, J. & Hake, S. (2002). Expression patterns and mutant phenotype of teosinte branched1 correlate with growth suppression in maize and teosinte. *Genetics*, 162: 1927–1935.

Huber, H. (1998). Dioscoreaceae. In *The Families and Genera of Vascular Plants, Vol. 3*, ed. K. Kubitzki. Berlin: Springer, pp. 216–235.

Hudson, A. & Jeffree, C. (2001). Leaf and internode. In *Encyclopedia of Life Sciences*. Chichester: Wiley.

Hudson, C. J., Freeman, J. S., Jones, R. C. *et al.* (2014). Genetic control of heterochrony in *Eucalyptus globulus*. *G3 (Bethesda)*, 4: 1235–1245.

Huether, C. A. (1968). Exposure of natural genetic variability underlying the pentamerous corolla constancy in *Linanthus androsaceus* ssp. *androsaceus*. *Genetics*, 60: 123–146.

Huether, C. A. (1969). Constancy of the pentamerous corolla phenotype in natural populations of *Linanthus*. *Evolution*, 23: 572–588.

Huijser, P., Klein, J., Lönnig, W. E. *et al.* (1992). Bracteomania, an inflorescence anomaly, is caused by the loss of function of the MADS-box gene *Squamosa* in *Antirrhinum majus*. *EMBO Journal*, 11: 1239–1249.

Huxley, J. S. (1942). *Evolution: The Modern Synthesis*. London: Allen and Unwin.

Iannelli, F., Pesole, G., Sordino, P. & Gissi, C. (2007). Mitogenomics reveals two cryptic species in *Ciona intestinalis*. *Trends in Genetics*, 23: 419–422.

Ichihashi, Y., Kawade, K., Usami, T. *et al.* (2011). Key proliferative activity in the junction between the leaf blade and leaf petiole of *Arabidopsis*. *Plant Physiology*, 157: 1151–1162.

Igersheim, A., Buzgo, M. & Endress, P. K. (2001). Gynoecium diversity and systematics in basal monocots. *Botanical Journal of the Linnean Society*, 136: 1–65.

Igersheim, A. & Endress, P. K. (1998). Gynoecium diversity and systematics of the paleoherbs. *Botanical Journal of the Linnean Society*, 127: 289–370.

Ikeuchi, M., Ogawa, Y., Iwase, A. & Sugimoto, K. (2016). Plant regeneration: cellular origins and molecular mechanisms. *Development*, 143: 1442–1451.

Ikezaki, M., Kojima, M., Sakakibara, H. *et al.* (2010). Genetic networks regulated by *ASYMMETRIC LEAVES1 (AS1)* and *AS2* in leaf development in *Arabidopsis thaliana*: KNOX genes control five morphological events. *The Plant Journal*, 61: 70–82.

Imaichi, R., Hiyama, Y. & Kato, M. (2004). Developmental morphology of foliose shoots and seedlings of *Dalzellia zeylanica* (Podostemaceae) with special reference to their meristems. *Botanical Journal of the Linnean Society*, 144: 289–302.

Imaichi, R., Hiyama, Y. & Kato, M. (2005). Leaf development in the absence of a shoot apical meristem in *Zeylanidium subulatum* (Podostemaceae). *Annals of Botany*, 96: 51–58.

Imaichi, R., Nagumo, S. & Kato, M. (2000). Ontogenic anatomy of *Streptocarpus grandis* (Gesneriaceae) with implications for evolution of monophyly. *Annals of Botany*, 86: 37–46.

Imbert, E. 2002. Ecological consequences and ontogeny of seed heteromorphism. *Perspectives in Plant Ecology, Evolution and Systematics*, 5: 13–36.

Immink, R. G. H., Ferrario, S., Busscher-Lange, J. *et al.* (2003). Analysis of the petunia MADS-box transcription factor family. *Molecular Genetics and Genomics*, 268: 598–606.

Immink, R. G. H., Hannapel, D. J., Ferrario, S. *et al.* (1999). A petunia MADS box gene involved in the transition from vegetative to reproductive development. *Development*, 126: 5117–5126.

Immink, R. G. H., Kaufmann, K. & Angenent, G. C. (2010). The 'ABC' of MADS domain protein behaviour and interactions. *Seminars in Cell and Developmental Biology*, 21: 87–93.

Immink, R. G. H., Tonaco, I. A. N., de Folter, S. *et al.* (2009). *SEPALLATA3*: the 'glue' for MADS box transcription factor complex formation. *Genome Biology*, 10: R24.

Ingram, G. C., Goodrich, J., Wilkinson, M. D. *et al.* (1995). Parallels between *UNUSUAL FLORAL ORGANS* and *FIMBRIATA*, genes controlling flower development in *Arabidopsis* and *Antirrhinum*. *Plant Cell*, 7: 1501–1510.

Inzé, D. (2003). Why should we study the plant cell cycle? *Journal of Experimental Botany*, 54: 1125–1126.

Irish, V. F. (2006). Duplication, diversification, and comparative genetics of angiosperm MADS-box genes. *Advances in Botanical Research*, 44: 129–161.

Irish, V. F. (2009). Evolution of petal identity. *Journal of Experimental Botany*, 60: 2517–2527.

Irish, V. F. & Sussex, I. M. (1990). Function of the *apetala-1* gene during *Arabidopsis* floral development. *Plant Cell*, 2: 741–753.

Ishida, T., Aida, M., Takada, S. & Tasaka, M. (2000). Involvement of *CUP-SHAPED COTYLEDON* genes in gynoecium and ovule development in *Arabidopsis thaliana*. *Plant and Cell Physiology*, 41: 60–67.

Ishii, H. S. & Harder, L. D. (2012). Phenological associations of within- and among-plant variation in gender with floral morphology and integration in protandrous *Delphinium glaucum*. *Journal of Ecology*, 100: 1029–1038.

Itkin, M., Seybold, H., Breitel, D. *et al.* (2009). *TOMATO AGAMOUS-LIKE 1* is a component of the fruit ripening regulatory network. *The Plant Journal*, 60: 1081–1095.

Itoh J.-I., Hasegawa A., Kitano H. & Nagato Y. (1998). A recessive heterochronic mutation, *plastochron1*, shortens the plastochron and elongates the vegetative phase in rice. *Plant Cell*, 10: 1511–1521.

Iwamoto, A., Shimuzu, A. & Ohba, H. (2003). Floral development and phyllotactic variation in *Ceratophyllum demersum* (Ceratophyllaceae). *American Journal of Botany*, 90: 1124–1130.

Jaakola, L., Poole, M., Jones, M. O. *et al.* (2010). A *SQUAMOSA* MADS box gene involved in the regulation of anthocyanin accumulation in bilberry fruits. *Plant Physiology*, 153: 1619–1629.

Jabbour, F., Cossard, G., Le Guilloux, M. *et al.* (2014). Specific duplication and dorsoventrally asymmetric expression patterns of *cycloidea*-like genes in zygomorphic species of Ranunculaceae. *PLoS ONE*, 9: e95727.

Jabbour, F., Ronse De Craene, L. P., Nadot, S. & Damerval, C. (2009). Establishment of zygomorphy on an ontogenetic spiral and evolution of perianth in the tribe Delphinieae (Ranunculaceae). *Annals of Botany*, 104: 809–822.

Jabbour, F., Udron, M., Le Guilloux, M. *et al.* (2015). Flower development schedule and *AGAMOUS*-like gene expression patterns in two morphs of *Nigella damascena* (Ranunculaceae) differing in floral architecture. *Botanical Journal of the Linnean Society*, 178: 608–619.

Jablonka, E. (2006). Genes as followers in evolution: a post-synthesis synthesis? *Biology and Philosophy*, 21: 143–154.

Jack, T. (2004). Molecular and genetic mechanisms of floral control. *Plant Cell*, 16: S1–S17.

Jack, T., Brockman, L. L. & Meyerowitz, E. M. (1992). The homeotic gene *APETALA3* of *Arabidopsis thaliana* encodes a MADS box and is expressed in petals and stamens. *Cell*, 68: 683–697.

Jack, T., Fox, G. L. & Meyerowitz, E. M. (1994). *Arabidopsis* homeotic gene *APETALA3* ectopic expression: transcriptional and posttranscriptional regulation determine floral organ identity. *Cell*, 76: 703–716.

Jäger-Zürn, I. (2007). The shoot apex of Podostemaceae: de novo structure or reduction of the conventional type? *Flora*, 202: 383–394.

Jaillon, O., Aury, J. M., Noel, B. *et al.* (2007). The grapevine genome sequence suggests ancestral hexaploidization in major angiosperm phyla. *Nature*, 449: 463–467.

James, P. J. (2009). 'Tree and leaf': a different angle. *The Linnean*, 25 (1): 13–19.

Janssen, B.-J., Lund, L. & Sinha, N. (1998). Overexpression of a homeobox gene, *LeT6*, reveals indeterminate features in the tomato compound leaf. *Plant Physiology*, 117: 771–786.

Janzen, D. H. (1976). Why bamboos wait so long to flower. *Annual Review of Ecology and Systematics*, 7: 347–391.

Jaramillo, M. A. & Kramer, E. M. (2004). *APETALA3* and *PISTILLATA* homologs exhibit novel expression patterns in the unique perianth of *Aristolochia* (Aristolochiaceae). *Evolution and Development*, 6: 449–458.

Jaramillo, M. A. & Kramer, E. M. (2007). The role of developmental genetics in understanding homology and morphological evolution in plants. *International Journal of Plant Sciences*, 168: 61–72.

Jasinski, S., Vialette-Guiraud, A. C. & Scutt, C. P. (2010). The evolutionary-developmental analysis of plant microRNAs. *Philosophical Transactions of the Royal Society B*, 365: 469–476.

Jaya, E., Kubien, D. S., Jameson, P. E. & Clemens, J. (2010). Vegetative phase change and photosynthesis in *Eucalyptus occidentalis*: architectural simplification prolongs juvenile traits. *Tree Physiology*, 30: 393–403.

Jesson, L. K. & Barrett, S. C. H. (2002a). Enantiostyly in *Wachendorfia* (Haemodoraceae); the influence of reproductive systems on the maintenance of the polymorphism. *American Journal of Botany*, 89: 253–263.

Jesson, L. K. & Barrett, S. C. H. (2002b). The genetics of mirror-image flowers. *Proceedings of the Royal Society B*, 269: 1835–1839.

Jesson, L. K. & Barrett, S. C. H. (2003). The comparative biology of mirror-image flowers. *International Journal of Plant Sciences*, 164: S237–S249.

Jeune, B. & Sattler, R. (1992). Multivariate analysis in process morphology of plants. *Journal of Theoretical Biology*, 156: 147–167.

Jeune, B. & Sattler, R. (1996). Quelques aspects d'une morphologie continuiste et dynamique. *Canadian Journal of Botany*, 74: 1023–1039.

Jiao, Y., Leebens-Mack, J., Ayyampalayam, S. *et al.* (2012). A genome triplication associated with early diversification of the core eudicots. *Genome Biology*, 13: R3.

Jiao, Y. N., Wickett, N. J., Ayyampalayam, S. *et al.* (2011). Ancestral polyploidy in seed plants and angiosperms. *Nature*, 473, 97–100.

Jibran, R., Hunter, D. A. & Dijkwel, P. P. (2013). Hormonal regulation of leaf senescence through integration of developmental and stress signals. *Plant Molecular Biology*, 82: 547–561.

Jiménez, S., Lawton-Rauh, A. L., Reighard, G. L., Abbott, A. G. & Bielenberg, D. G. (2009). Phylogenetic analysis and molecular evolution of the dormancy associated MADS-box genes from peach. *BMC Plant Biology*, 9: 81.

Jones, C. S. (1992). Comparative ontogeny of a wild cucurbit and its derived cultivar. *Evolution*, 46: 1827–1847.

Jong, K. (1973). *Streptocarpus* (Gesneriaceae) and the phyllomorph concept. *Acta Botanica Neerlandica*, 22: 243–255.

Jong, K. & Burtt, B. L. (1975). The evolution of morphological novelty exemplified in the growth patterns of some Gesneriaceae. *New Phytologist*, 75: 297–311.

Juarez, M. T., Twigg, R. W. & Timmermans, M. C. (2004). Specification of adaxial cell fate during maize leaf development. *Development*, 131: 4533–4544.

Juncosa, A. M. (1988). Floral development and character evolution in Rhizophoraceae. In *Aspects of Floral Development*, eds. P. Leins, S. C. Tucker & P. K. Endress. Berlin: Cramer, pp. 83–101.

Juntheikki-Palovaara, I., Tähtiharju, S., Lan, T. *et al.* (2014). Functional diversification of duplicated CYC2 clade genes in regulation of inflorescence development in *Gerbera hybrida* (Asteraceae). *The Plant Journal*, 79: 783–796.

Kagale, S., Robinson, S. J., Nixon, J. *et al.* (2014). Polyploid evolution of the Brassicaceae during the Cenozoic era. *Plant Cell*, 26: 2777–2791.

Kalinka, A. T., Varga, K. M., Gerrard, D. T. *et al.* (2010). Gene expression divergence recapitulates the developmental hourglass model. *Nature*, 468: 811–814.

Kampny, C. M., Dickinson, T. A. & Dengler, N. G. (1993). Quantitative comparison of floral development in *Veronica chamaedrys* and *Veronicastrum virginicum* (Scrophulariaceae). *American Journal of Botany*, 80: 449–460.

Kampny, C. M., Dickinson, T. A. & Dengler, N. G. (1994). Quantitative floral development in *Pseudolysimachion* (Scrophulariaceae): intraspecific variation and comparison with *Veronica* and *Veronicastrum*. *American Journal of Botany*, 81: 1343–1353.

Kang, H. G., Jeon, J. S., Lee, S. & An, G. H. (1998). Identification of class B and class C floral organ identity genes from rice plants. *Plant Molecular Biology*, 38: 1021–1029.

Kanno, A., Saeki, H., Kameya, T., Saedler, H. & Theissen, G. (2003). Heterotopic expression of class B floral homeotic genes supports a modified ABC model for tulip (*Tulipa gesneriana*). *Plant Molecular Biology*, 52: 831–841.

Kaplan, D. (1975). Comparative developmental evaluation of the morphology of unifacial leaves in the monocotyledons. *Botanische Jahrbücher für Systematik, Pflanzengeschichte und Pflanzengeographie*, 95: 1–105.

Kaplan, D. R. (1970). Comparative foliar histogenesis in *Acorus calamus* and its bearing on the phyllode theory of monocotyledonous leaves. *American Journal of Botany*, 57: 331–361.

Kaplan, D. R. (1973). The monocotyledons: their evolution and comparative biology. VII. The problem of leaf morphology and evolution in the monocotyledons. *Quarterly Review of Biology*, 48: 437–457.

Kaplan, D. R. (1984). Alternative modes of organogenesis in higher plants. In *Contemporary Problems in Plant Anatomy*, eds. R. A. White & W. C. Dickison. New York, NY: Academic Press, pp. 261–300.

Kaplan, D. R. (1992). The relationship of cells to organisms in plants: problem and implications of an organismal perspective. *International Journal of Plant Sciences*, 153: S28–S37.

Kaplan, D. R. (2001). Fundamental concepts of leaf morphology and morphogenesis: a contribution to the interpretation of molecular genetic mutants. *International Journal of Plant Sciences*, 162: 465–474.

Kaplan, D. R., Dengler, N. G. & Dengler, R. E. (1982). The mechanisms of plication inception in palm leaves: histogenetic observations of the palmate leaves of *Rhapis excelsa. Canadian Journal of Botany*, 60: 2999–3016.

Kaplan, D. R. & Hagemann, W. (1991). The relationship of cell and organism in vascular plants. *Bioscience*, 41: 693–703.

Karoly, K. & Conner, J. K. (2000). Heritable variation in a family-diagnostic trait. *Evolution*, 54: 1433–1438.

Kasha, K. J. & Kao, K. N. (1970). High frequency haploid production in barley (*Hordeum vulgare* L.). *Nature*, 225: 874–876.

Katayama, N., Kato, M., Nishiuchi, T. & Yamada, T. (2011). Comparative anatomy of embryogenesis in three species of Podostemaceae and evolution of the loss of embryonic shoot and root meristems. *Evolution and Development*, 13: 333–342.

Katayama, N., Kato, M. & Yamada, T. (2013). Origin and development of the cryptic shoot meristem in *Zeylanidium lichenoides*. *American Journal of Botany*, 100: 635–646.

Katayama, N., Koi, S. & Kato, M. (2010). Expression of *SHOOT MERISTEMLESS, WUSCHEL*, and *ASYMMETRIC LEAVES1* homologs in the shoots of Podostemaceae: implications for the evolution of novel shoot organogenesis. *Plant Cell*, 22: 2131–2140.

Kaufmann, K., Pajoro, A. & Angenent, G. C. (2010). Regulation of transcription in plants: mechanisms controlling developmental switches. *Nature Reviews Genetics*, 11: 830–842.

Kawade, K., Horiguchi, G. & Tsukaya, H. (2010). Non-cell-autonomously coordinated organ size regulation in leaf development. *Development*, 137: 4221–4227.

Kawamura, E., Horiguchi, G. & Tsukaya, H. (2010). Mechanisms of leaf tooth formation in *Arabidopsis*. *The Plant Journal*, 62: 429–441.

Kazama, Y., Fujiwara, M. T., Koizumi, A. *et al.* (2009). A *SUPERMAN*-like gene is exclusively expressed in female flowers of the dioecious plant *Silene latifolia*. *Plant Cell Physiology*, 50: 1127–1141.

Kellogg, E. A. (2000). A model of inflorescence development. In *Monocots: Systematics and Evolution*, eds. K. L. Wilson & D. A. Morrison. Melbourne: CSIRO, pp. 84–88.

Kellogg, E. A. (2001). Evolutionary history of the grasses. *Plant Physiology*, 125: 1198–1205.

Kellogg, E. A., Camara, P. E. A. S., Rudall, P. J. *et al.* (2013). Early inflorescence development in the grasses (Poaceae). *Frontiers in Plant Science*, 4: 250.

Kerr, A. D. (1972). Ephemeral means 'don't turn your head'. *American Orchid Society Bulletin*, 41: 208–211.

Kerstetter, R. A., Bollman, K., Taylor, R. A., Bomblies, K. & Poethig, R. S. (2001). *KANADI* regulates organ polarity in *Arabidopsis*. *Nature*, 411: 706–709.

Kerstetter, R. A. & Poethig, R. S. (1998). The specification of leaf identity during shoot development. *Annual Review of Cell and Developmental Biology*, 14: 373–398.

Kessler, S. & Sinha, N. (2004). Shaping up: the genetic control of leaf shape. *Current Opinion in Plant Biology*, 7: 65–72.

Khan, M. R., Hu, J. Y., Riss, S., He, C. & Saedler, H. (2009). *MPF2-like-A* MADS-box genes control the inflated calyx syndrome in *Withania* (Solanaceae): roles of Darwinian selection. *Molecular Biology and Evolution*, 26: 2463–2473.

Kidner, C. A. & Timmermans, M. C. P. (2010). Signaling sides: adaxial–abaxial patterning in leaves. *Current Topics in Developmental Biology*, 91: 141–168.

Kierzkowski, D., Nakayama, N., Routier-Kierzkowska, A.-L. *et al.* (2012). Elastic domains regulate growth and organogenesis in the plant shoot apical meristem. *Science*, 335: 1096–1099.

Kim, G., LeBlanc, M. L., Wafula, E. K., de Pamphilis, C. W. & Westwood, J. H. (2014). Genomic-scale exchange of mRNA between a parasitic plant and its hosts. *Science*, 345: 808–811.

Kim, K.-J., Choi, K.-S. & Jansen, R. K. (2005a). Two chloroplast DNA inversions originated simultaneously during the early evolution of the sunflower family (Asteraceae). *Molecular Biology and Evolution*, 22: 1783–1792.

Kim, M., Cui, M., Cubas, P. *et al.* (2008). Regulatory genes control a key morphological and ecological trait transferred between species. *Science*, 322: 1116–1119.

Kim, M., McCormick, S., Timmermans, M & Sinha, N. (2003). The expression domain of *PHANTASTICA* determines leaflet placement in compound leaves. *Nature*, 424: 438–443.

Kim, S., Koh, J., Ma, H. *et al.* (2005b). Sequence and expression studies of A-, B-, and E-class MADS-box homologues in *Eupomatia* (Eupomatiaceae): support for the bracteate origin of the calyptra. *International Journal of Plant Sciences*, 166: 185–198.

Kim, S., Koh, J., Yoo, M. J. *et al.* (2005c). Expression of floral MADS-box genes in basal angiosperms: implications for the evolution of floral regulators. *The Plant Journal*, 43: 724–744.

Kim, S., Yoo, M. J., Albert, V. A. *et al.* (2004). Phylogeny and diversification of B-function MADS-box genes in angiosperms: evolutionary and functional implications of a 260-million-year-old duplication. *American Journal of Botany*, 91: 2102–2118.

Kirchoff, B. K. (2000). Hofmeister's rule and primordium shape: influences on organ position in *Hedychium coronarium* (Zingiberaceae). In *Monocots: Systematics and Evolution*, eds. K. L. Wilson & D. A. Morrison. Melbourne: CSIRO, pp. 75–83.

Kirchoff, B. K. (2001). Character description in phylogenetic analysis: insights from Agnes Arber's concept of the plant. *Annals of Botany*, 88: 1203–1214.

Kirchoff, B. K. (2017). Inflorescence and flower development in *Musa velutina* H. Wendl. & Drude (Musaceae), with a consideration of developmental variability, restricted phyllotactic direction and hand initiation. *International Journal of Plant Sciences*, 178: 259–272.

Kirkpatrick, M. (2009). Patterns of quantitative genetic variation in multiple dimensions. *Genetica*, 136: 271–284.

Kirschner, M. & Gerhart, J. (1998). Evolvability. *Proceedings of the National Academy of Sciences of the United States of America*, 95: 8420–8427.

Kitomi, Y., Ogawa, A., Kitano, H. & Inukai, Y. (2008). CRL4 regulates crown root formation through auxin transport in rice. *Plant Root*, 2: 19–28.

Kliber, A. & Eckert, C. (2004). Sequential decline in allocation among flowers within inflorescences: proximate mechanisms and adaptive significance. *Ecology*, 85: 1675–1687.

Klingenberg, C. P. (1998). Heterochrony and allometry: the analysis of evolutionary change in ontogeny. *Biological Reviews*, 73: 79–123.

Klingenberg, C. P. (2005). Developmental constraints, modules and evolvability. In *Variation: A Central Concept in Biology*, eds. B. Hallgrímsson & B. K. Hall. Burlington, MA: Elsevier, pp. 219–247.

Klingenberg, C. P. (2008). Morphological integration and developmental modularity. *Annual Review of Ecology and Systematics*, 39: 115–132.

Knapp, S. (2010). On 'various contrivances': pollination, phylogeny and flower form in the Solanaceae. *Philosophical Transactions of the Royal Society B*, 365: 449–460.

Knight, C., Perroud, P. F. & Cove, D. (2009). *The Moss* Physcomitrella patens. London: Wiley-Blackwell.

Kobayashi, K., Maekawa, M., Miyao, A., Hirochika, H. & Kyozuka, J. (2010). *PANICLE PHYTOMER2 (PAP2)*, encoding a SEPALLATA subfamily MADS-box protein, positively controls spikelet meristem identity in rice. *Plant Cell Physiology*, 51: 47–57.

Koch, M. & Al-Shehbaz, I. A. (2002). Molecular data indicate complex intra- and intercontinental differentiation of American *Draba* (Brassicaceae). *Annals of the Missouri Botanical Garden*, 89: 88–109.

Koch, M., Al-Shehbaz, I. A. & Mummenhoff, K. (2003). Molecular systematics, evolution, and population biology in the mustard family (Brassicaceae). *Annals of the Missouri Botanical Garden*, 90: 151–171.

Koch, M., Haubold, B. & Mitchell-Olds, T. (2001). Molecular systematics of the Brassicaceae: evidence from coding plastidic matK and nuclear Chs sequences. *American Journal of Botany*, 88: 534–544.

Koenig, D., Bayer, E., Kang, J., Kuhlemeier, C. & Sinha, N. (2009). Auxin patterns *Solanum lycopersicum* leaf morphogenesis. *Development*, 136: 2997–3006.

Koenig, D. & Weigel, D. (2015). Beyond the thale: comparative genomics and genetics of *Arabidopsis* relatives. *Nature Reviews Genetics*, 16: 285–298.

Koes, R. (2008). Evolution and development of virtual inflorescences. *Trends in Plant Science*, 13: 1–3.

Kofuji, R., Sumikawa, N., Yamasaki, M. *et al.* (2003). Evolution and divergence of the MADS-box gene family based on genome-wide expression analyses. *Molecular Biology and Evolution*, 20: 1963–1977.

Kohn, J. R., Graham, S. W., Morton, B., Doyle, J. J. & Barrett, S. C. H. (1996). Reconstruction of the evolution of reproductive characters in Pontederiaceae using phylogenetic evidence from chloroplast DNA restriction-site variation. *Evolution*, 50: 1454–1469.

Koi, S., Imaichi, R. & Kato, M. (2005). Endogenous leaf initiation in the apical-meristemless shoot of *Cladopus queenslandicus* (Podostemaceae) and implications for evolution of shoot morphology. *International Journal of Plant Sciences*, 166: 199–206.

Kölsch, A. & Gleissberg, S. (2006). Diversification of *CYCLOIDEA*-like TCP genes in the basal eudicot families Fumariaceae end Papaveraceae s.str. *Plant Biology*, 8: 680–687.

Konishi, S., Izawa, T., Lin, S.Y. *et al.* (2006). An SNP caused loss of seed shattering during rice domestication. *Science*, 312: 1392–1396.

Kosuge, K. (1994). Petal evolution in Ranunculaceae. *Plant Systematics and Evolution, Supplement*, 8: 185–191.

Kozlov, M. V., Wilsey, B. J., Koricheva, J. & Haukioja, E. (1996). Fluctuating asymmetry of birch leaves increases under pollution impact. *Journal of Applied Ecology*, 33: 1489–1495.

Kozo Poljanski, B. (1936). On some 'third' conceptions in floral morphology. *New Phytologist*, 35: 479–492.

Kramer, E. M. (2009a). *Aquilegia*: a new model for plant development, ecology, and evolution. *Annual Review of Plant Biology*, 60: 261–277.

Kramer, E. M. (2009b). New model systems for the study of developmental evolution in plants. *Current Topics in Developmental Biology*, 86: 65–107.

Kramer, E. M., Di Stilio, V. S. & Schluter, P. M. (2003). Complex patterns of gene duplication in the *Apetala3* and *Pistillata* lineages of the Ranunculaceae. *International Journal of Plant Sciences*, 164: 1–11.

Kramer, E. M., Dorit, R. L. & Irish, V. F. (1998). Molecular evolution of genes controlling petal and stamen development: duplication and divergence within the *APETALA3* and *PISTILLATA* MADS-box gene lineages. *Genetics*, 149: 765–783.

Kramer, E. M. & Hodges, S. A. (2010). *Aquilegia* as a model system for the evolution and ecology of petals. *Philosophical Transactions of the Royal Society B*, 365: 477–490.

Kramer, E. M., Holappa, L., Gould, B. *et al.* (2007). Elaboration of B gene function to include the identity of novel floral organs in the lower eudicot *Aquilegia*. *Plant Cell*, 19: 750–766.

Kramer, E. M. & Irish, V. F. (1999). Evolution of genetic mechanisms controlling petal development. *Nature*, 399: 144–148.

Kramer, E. M. & Irish, V. F. (2000). Evolution of the petal and stamen developmental programs: evidence from comparative studies of the lower eudicots and basal angiosperms. *International Journal of Plant Sciences*, 161: S29–S40.

Kramer, E. M. & Jaramillo, M. A. (2005). Genetic basis for innovations in floral organ identity. *Journal of Experimental Zoology (Molecular and Developmental Evolution)*, 304B: 526–535.

Kramer, E. M., Jaramillo, M. A. & Di Stilio, V. S. (2004). Patterns of gene duplication and functional evolution during the diversification of the *AGAMOUS* subfamily of MADS box genes in angiosperms. *Genetics*, 166: 1011–1023.

Kramer, E. M., Su, H. J., Wu, C. C. & Hu, J. M. (2006). A simplified explanation for the frameshift mutation that created a novel C-terminal motif in the *APETALA3* gene lineage. *BMC Evolutionary Biology*, 6: 30.

Krizek, B. A. & Meyerowitz, E. M. (1996). The *Arabidopsis* homeotic genes *APETALA3* and *PISTILLATA* are sufficient to provide the B class organ identity function. *Development*, 122: 11–22.

Krolikowski, K. A., Victor, J. L., Wagler, T. N., Lolle, S. J. & Pruitt, R. J. (2003). Isolation and characterization of the *Arabidopsis* organ fusion gene *HOTHEAD*. *The Plant Journal*, 35: 501–511.

Kubitzki, K. (1998). Taccaceae. In *The Families and Genera of Vascular Plants, Vol. 3*, ed. K. Kubitzki. Berlin: Springer, pp. 425–428.

Kuhlemeier, C. & Reinhardt, D. (2001). Auxin and phyllotaxis. *Trends in Plant Science*, 6: 187–189.

Kuittinen, H., de Haan, A. A., Vogl, C. *et al.* (2004). Comparing the linkage maps of the close relatives *Arabidopsis lyrata* and *A. thaliana*. *Genetics*, 168: 1575–1584.

Kümpers, B. M. C., Richardson, J. E., Anderberg, A. A., Wilkie, P. & Ronse De Craene, L. (2016). The significance of meristic changes in the flowers of Sapotaceae. *Botanical Journal of the Linnean Society*, 180: 161–192.

Kutschera, U., Langguth, H., Kuo, D.-H., Weisblat, D. A. & Shankland, M. (2013). Description of a new leech species from North America, *Helobdella austinensis* n. sp. (Hirudinea: Glossiphoniidae), with observations on its feeding behaviour. *Zoosystematics and Evolution*, 89: 239–246.

Kuwabara, A., Tsukaya, H. & Nagata, T. (2001). Identification of factors that cause heterophylly in *Ludwigia arcuata* Walt. (Onagraceae). *Plant Biology*, 3: 98–105.

Kwiatkowska, D. (1995). Ontogenetic changes of phyllotaxis in *Anagallis arvensis* L. *Acta Societatis Botanicorum Poloniae*, 64: 319–325.

Kwiatkowska, D. (1999). Formation of pseudowhorls in *Peperomia verticillata* (L.) A. Dietr. shoots exhibiting various phyllotactic patterns. *Annals of Botany*, 83: 675–685.

Kyozuka, J., Kobayashi, T., Morita, M. & Shimamoto, K. (2000). Spatially and temporally regulated expression of rice MADS box genes with similarity to *Arabidopsis* class A, B and C genes. *Plant Cell Physiology*, 41: 710–718.

Labonne, J. D. J., Tamari, F. & Shore, J. S. (2010). Characterization of X-ray-generated floral mutants carrying deletions at the *S*-locus of distylous *Turnera subulata*. *Heredity*, 105: 235–243.

Lacroix, C., Jeune, B. & Purcell-Macdonald, S. (2003). Shoot and compound leaf comparisons in eudicots: dynamic morphology as an alternative approach. *Botanical Journal of the Linnean Society*, 143: 219–230.

Lacroix, C. & Sattler, R. (1988). Phyllotaxis theories and tepal-stamen superposition in *Basella rubra*. *American Journal of Botany*, 75: 906–917.

Lacroix, C. R. & Sattler, R. (1994). Expression of shoot features in early leaf development of *Murraya paniculata* (Rutaceae). *Canadian Journal of Botany*, 72: 678–687.

Lahti, D. C., Johnson, N. A., Ajie, B. C. *et al.* (2009). Relaxed selection in the wild. *Trends in Ecology and Evolution*, 24: 487–496.

Laland, K. N., Uller, T., Feldman, M. W. *et al.* (2015). The extended evolutionary synthesis: its structure, assumptions and predictions. *Proceedings of the Royal Society B*, 282: 20151019.

Lamont, B. B. (1980). Tissue longevity of the arborescent monocotyledon, *Kingia australis* (Xanthorrhoeaceae). *American Journal of Botany*, 67: 1262–1264.

Lamsdell, J. C. & Selden, P. A. (2013). Babes in the wood: a unique window into sea scorpion ontogeny. *BMC Evolutionary Biology*, 13, 98.

Landis, J. B., Barnett, L. L. & Hileman, L. C. (2012). Evolution of petaloid sepals independent of shifts in B-class MADS box gene expression. *Development Genes and Evolution*, 222: 19–28.

Landrein, B., Refahi, Y., Besnard, F. *et al.* (2015). Meristem size contributes to the robustness of phyllotaxis in *Arabidopsis*. *Journal of Experimental Botany*, 66: 1317–1324.

Lang, D., van Gessel, N., Ullrich, K. K. & Reski, R. (2016). The genome of the model moss *Physcomitrella patens*. *Advances in Botanical Research*, 78: 97–140.

REFERENCES

Langdale, J. & Harrison, J. C. (2008). Developmental transitions during the evolution of plant form. In *Evolving Pathways: Key Themes in Evolutionary Developmental Biology*, eds. A. Minelli & G. Fusco. Cambridge: Cambridge University Press, pp. 299–319.

Langer, R. H. & Wilson, D. (1965). Environmental control of cleistogamy in prairie grass (*Bromus unioloides* HBK). *New Phytologist*, 65: 80–85.

Langham, R. J., Walsh, J., Dunn, M. *et al.* (2004). Genomic duplication, fractionation and the origin of regulatory novelty. *Genetics*, 166: 935–945.

Lanner, R. M. & Connor, K. F. (2001). Does bristlecone pine senesce? *Experimental Gerontology*, 36: 675–685.

Lau, S., Slane, D., Herud, O., Kong, J. & Jürgens, G. (2012). Early embryogenesis in flowering plants: setting up the basic body pattern. *Annual Review of Plant Biology*, 63: 483–506.

Laufs, P., Grandjean, O., Jonak, C., Kieu, K. & Traas, J. (1998). Cellular parameters of the shoot apical meristem in *Arabidopsis*. *Plant Cell*, 10: 1375–1389.

Laux, T., Mayer, K. F. X., Berger, J. & Jürgens, G. (1996). The *WUSCHELL* gene is required for shoot and floral meristem integrity in *Arabidopsis*. *Development*, 122: 87–96.

Lavin, M., Herendeen, P. S. & Wojciechowski, M. F. (2005). Evolutionary rates analysis of Leguminosae implicates a rapid diversification of lineages during the Tertiary. *Systematic Biology*, 54: 575–594.

Lawrence, G. H. (1951). *Taxonomy of Flowering Plants*. New York, NY: Macmillan.

Layton, D. J. & Kellogg, E. A. (2014). Morphological, phylogenetic, and ecological diversity of the new model species *Setaria viridis* (Poaceae: Paniceae) and its close relatives. *American Journal of Botany*, 101: 539–557.

Le Rouzic, A. & Carlborg, O. (2008). Evolutionary potential of hidden genetic variation. *Trends in Ecology and Evolution*, 23: 33–37.

Lee, J., Park, J.-J., Kim, S. L., Yim, J. & An, G. (2007). Mutations in the rice *liguleless* gene result in a complete loss of the auricle, ligule and laminar joint. *Plant Molecular Biology*, 65: 487–499.

Lee, J.-H., Lin, H., Joo, S. & Goodenough, U. (2008). Early sexual origins of homeoprotein heterodimerization and evolution of the plant KNOX/BELL family. *Cell*, 133: 829–840.

Lee, J.-Y., Baum, S. F., Oh, S. H. *et al.* (2005). Recruitment of *CRABS CLAW* to promote nectary development within the eudicot clade. *Development*, 132: 5021–5032.

Lee, J.-Y., Mummenhoff, K. & Bowman, J. L. (2002). Allopolyploidization and evolution of species with reduced floral structures in *Lepidium* L. (Brassicaceae). *Proceedings of the National Academy of Sciences of The United States of America*, 99: 16835–16840.

Leinfellner, W. (1958). Über die peltaten Kronblätter der Sapindaceen. *Österreichische botanische Zeitschrift*, 105: 443–514.

Leins, P. & Erbar, C. (1997). Floral developmental studies: some old and new questions. *International Journal of Plant Sciences*, 158: S3–S12.

Leins, P. & Erbar, C. (2004). Floral organ sequences in Apiales (Apiaceae, Araliaceae, Pittosporaceae). *South African Journal of Botany*, 70: 468–474.

Leins, P. & Erbar, C. (2008). *Blüte und Frucht*, 2nd edn. Stuttgart: Schweizerbart.

Lewis, D. & Jones, D. A. (1992). The genetics of heterostyly. In *Evolution and Function of Heterostyly*, ed. S. C. H. Barrett. Berlin: Springer, pp. 129–150.

Li, D., Liu, C., Shen, L. *et al.* (2008). A repressor complex governs the integration of flowering signals in *Arabidopsis*. *Developmental Cell*, 15: 110–120.

Li, G. S., Meng, Z., Kong, H. Z. *et al.* (2005). Characterization of candidate class A, B and E floral homeotic genes from the perianthless basal angiosperm *Chloranthus spicatus* (Chloranthaceae). *Development Genes and Evolution*, 215: 437–449.

Li, H., Liang, W., Hu, Y. *et al.* (2011). Rice *MADS6* interacts with the floral homeotic genes *SUPERWOMAN1*, *MADS3*, *MADS58*, *MADS13*, and *DROOPING LEAF* in specifying floral organ identities and meristem fate. *Plant Cell*, 23: 2536–2552.

Li, J., Webster, M. A., Furuya, M. & Gilmartin, P. M. (2007). Identification and characterization of pin and thrum alleles of two genes that co-segregate with the *Primula* S locus. *The Plant Journal*, 51: 18–31.

Li, P. & Johnston, M. O. (2000). Heterochrony in plant evolutionary studies through the twentieth century. *Botanical Review*, 66: 57–88.

Li, P. & Johnston, M. O. (2010). Flower development and the evolution of self-fertilization in *Amsinckia*: the role of heterochrony. *Evolutionary Biology*, 37: 143–168.

Li, Z., Reighard, G. L., Abbott, A. G. & Bielenberg, D. G. (2009). Dormancy-associated MADS genes from the *EVG* locus of peach [*Prunus persica* (L.) Batsch] have distinct seasonal and photoperiodic expression patterns. *Journal of Experimental Botany*, 60: 3521–3530.

Liljegren, S. J., Ditta, G. S., Eshed, H. Y. *et al.* (2000). *SHATTERPROOF* MADS-box genes control seed dispersal in *Arabidopsis*. *Nature*, 404: 766–770.

Linnaeus, C. (1754). *Genera plantarum: eorumque characteres naturales secundum numerum, figuram, situm, et proportionem omnium fructificationis partium. Editio Quinta*. Holmiæ: Laurentius Salvius.

Litt, A. (2007). An evaluation of A-function: evidence from the *APETALA1* and *APETALA2* gene lineages. *International Journal of Plant Sciences*, 168: 73–91.

Litt, A. (2013). Comparative evolutionary genomics of land plants. *Annual Plant Reviews*, 45: 227–276.

Litt, A. & Irish, V. F. (2003). Duplication and diversification in the *APETALA1/FRUITFULL* floral homeotic gene lineage: implications for the evolution of floral development. *Genetics*, 165: 821–833.

Litt, A. & Kramer, E. M. (2010). The ABC model and the diversification of floral organ identity. *Seminars in Cell and Developmental Biology*, 21: 129–137.

Liu, C., Thong, Z. & Yu, H. (2009a). Coming into bloom: the specification of floral meristems. *Development*, 136: 3379–3391.

Liu, S., Wang, J., Wang, L. *et al.* (2009b) Adventitious root formation in rice requires *OsGNOM1* and is mediated by the OsPINs family. *Cell Research*, 19: 1110–1119.

Liu, Z. C., Franks, R. G. & Klink, V. P. (2000). Regulation of gynoecium marginal tissue formation by *LEUNIG* and *AINTEGUMENTA*. *Plant Cell*, 12: 1879–1891.

Liu, Z.-J. & Wang, X. (2016). A perfect flower from the Jurassic of China. *Historical Biology*, 28: 707–719.

Lloyd, D. G. & Webb, C. J. (1977). Secondary sex characters in plants. *Botanical Review*, 43: 177–216.

Lloyd, D. G. & Webb, C. J. (1986). The avoidance of interference between the presentation of pollen and stigmas in angiosperms. I. Dichogamy. *New Zealand Journal of Botany*, 24: 135–162.

Loiseau, J.-E. (1969). *La Phyllotaxie*. Paris: Masson.

Lolle, S. J., Cheung, A. Y. & Sussex, I. M. (1992). *Fiddlehead*: an *Arabidopsis* mutant constitutively expressing an organ fusion program that involves interactions between epidermal cells. *Developmental Biology*, 152: 383–392.

Long, J. & Barton, M. K. (2000). Initiation of axillary and floral meristems in *Arabidopsis*. *Developmental Biology*, 218: 341–353.

Long, J. A., Moan, E. I., Medford, J. I. *et al.* (1996). A member of the *KNOTTED* class of homeodomain proteins encoded by the *STM* gene of *Arabidopsis*. *Nature*, 379: 66–69.

Lord, E. M. (1979). The development of cleistogamous and chasmogamous flowers in *Lamium amplexicaule* (Labiatae): an example of heteroblastic inflorescence development. *Botanical Gazette*, 140: 39–50.

Lord, E. M. (1981). Cleistogamy: a tool for the study of floral morphogenesis, function and evolution. *Botanical Review*, 47: 421–449.

Lord, E. M. (1982). Floral morphogenesis in *L. amplexicaule* L. (Labiateae) with a model for the evolution of the cleistogamous flower. *Botanical Gazette*, 143: 63–72.

Lord, E. M. (2001). Heterochrony in plants. In *Encyclopedia of Life Sciences*. Chichester: Wiley.

Lord, E. M. & Hill, J. P. (1987). Evidence for heterochrony in the evolution of plant form. In *Development as an Evolutionary Process*, eds. R. A. Raff & E. C. Raff. New York, NY: Alan R. Liss, pp. 47–70.

Love, A. C. (2010). Idealization in evolutionary developmental investigation: a tension between phenotypic plasticity and normal stages. *Philosophical Transactions of the Royal Society B*, 365: 679–690.

Luo, D., Carpenter, R., Copsey, L. *et al.* (1999). Control of organ asymmetry in flowers of *Antirrhinum*. *Cell*, 99: 367–376.

Luo, D., Carpenter, R., Vincent, C., Copsey, L. & Coen, E. (1996). Origin of floral asymmetry in *Antirrhinum*. *Nature*, 383: 794–799.

Lynch, M. & Conery, J. S. (2000). The evolutionary fate and consequences of duplicate genes. *Science*, 290: 1151–1155.

Lyndon, R. F. (1998). *The Shoot Apical Meristem: Its Growth and Development*. Cambridge: Cambridge University Press.

Lysak, M. A., Berr, A., Pecinka, A. *et al.* (2006). Mechanisms of chromosome number reduction in *Arabidopsis thaliana* and related Brassicaceae species. *Proceedings of the National Academy of Sciences of the United States of America*, 103: 5224–5229.

Ma, H. (1998). To be, or not to be, a flower: control of floral meristem identity. *Trends in Genetics*, 14: 26–32.

Maizel, A. (2016). Plant organ growth: stopping under stress. *Current Biology*, 26: R417–R419.

Malcomber, S. T. & Kellogg, E. A. (2004). Heterogeneous expression patterns and separate roles of the *SEPALLATA* gene *LEAFY HULL STERILE1* in grasses. *Plant Cell*, 16: 1692–1706.

Malcomber, S. T. & Kellogg, E. A. (2005). *SEPALLATA* gene diversification: brave new whorls. *Trends in Plant Science*, 10: 427–435.

Malcomber, S. T. & Kellogg, E. A. (2006). Evolution of unisexual flowers in grasses (Poaceae) and the putative sex-determination gene, *TASSELSEED2* (*TS2*). *New Phytologist*, 170: 885–899.

Malcomber, S. T., Preston, J. C., Reinheimer, R., Kossuth, J. & Kellogg, E. A. (2006). Developmental gene evolution and the origin of grass inflorescence diversity. *Advances in Botanical Research*, 44: 425–481.

Manger, H. L. (1783). *Vollstaendige Anleitung zu einer Systematischen Pomologie wodurch die Genauigste Kentniß von der Natur, Beschaffenheit und den Unterschiedenen Merkmalen aller Obstarten Enthalten Werden Kann. Zweyter Theil von den Birnen*. Leipzig: J. F. Junius.

Manos, P. S. & Stone, D. E. (2001). Evolution, phylogeny, and systematics of the Juglandaceae. *Annals of the Missouri Botanical Garden*, 88: 231–262.

Mantegazza, R., Möller, M., Harrison, C. J. *et al.* (2007). Anisocotyly and meristem initiation in an unorthodox plant, *Streptocarpus rexii* (Gesneriaceae). *Planta*, 225: 653–663.

Mantegazza, R., Tononi, P., Möller, M. & Spada, A. (2009). *WUS* and *STM* homologues are linked to the expression of lateral dominance in the acaulescent *Streptocarpus rexii* (Gesneriaceae). *Planta*, 230: 529–542.

Marazzi, B. & Endress, P. K. (2008). Patterns and development of floral asymmetry in *Senna* (Leguminosae, Cassiinae). *American Journal of Botany*, 95: 22–40.

Marazzi, B., Endress, P. K., Paganucci de Queiroz, L. & Conti, E. (2006). Phylogenetic relationships within *Senna* (Leguminosae, Cassiinae) based on three chloroplast DNA regions: patterns in the evolution of floral symmetry and extrafloral nectaries. *American Journal of Botany*, 93: 288–303.

Martín-Trillo, M. & Cubas, P. (2010). TCP genes: a family snapshot ten years later. *Trends in Plant Science*, 15: 31–39.

Martinez, C. C., Chitwood, D. H., Smith, R. S. & Sinha, N. R. (2016). Left–right leaf asymmetry in decussate and distichous phyllotactic systems. *Philosophical Transactions of the Royal Society B*, 371: 20150412.

Martínez-Laborda, A. & Vera, A. (2009). *Arabidopsis* fruit development. *Annual Plant Reviews*, 38: 172–203.

Masel J. & Siegal, M. L. (2009). Robustness: mechanisms and consequences. *Trends in Genetics*, 25: 395–403.

Masel, J. & Trotter, M. V. (2010). Robustness and evolvability. *Trends in Genetics*, 26: 406–414.

Masiero, S., Li, M. A., Will, I. *et al.* (2004). *INCOMPOSITA*: a MADS-box gene controlling prophyll development and floral meristem identity in *Antirrhinum*. *Development*, 131: 5981–5990.

Matsuhashi, S., Sakai, S. & Kudoh, H. (2012). Temperature-dependent fluctuation of stamen number in *Cardamine hirsuta* (Brassicaceae). *International Journal of Plant Sciences*, 173: 391–398.

Matsui, K., Nishio, T. & Tetsuka, T. (2004). Genes outside the *S* supergene suppress *S* functions in buckwheat (*Fagopyrum esculentum*). *Annals of Botany*, 94: 805–809.

Matthews, M. L. & Endress, P. K. (2008). Comparative floral structure and systematics in Chrysobalanaceae s.l. (Chrysobalanaceae, Dichapetalaceae, Euphroniaceae, and Trigoniaceae; Malpighiales). *Botanical Journal of the Linnean Society*, 157: 249–309.

Mauseth, J. D. (1991). *Botany*. Orlando, FL: Holt Rhinehart & Winston.

Mayer, V., Möller, M., Perret, M. & Weber, A. (2003). Phylogenetic position and generic differentiation of Epithemateae (Gesneriaceae) inferred from plastid DNA sequence data. *American Journal of Botany*, 90: 321–329.

Mayers, A. M. & Lord, E. M. (1983). Comparative flower development in the cleistogamous species *Viola odorata*. I. A growth rate study. *American Journal of Botany*, 70: 1548–1555.

Mayers, A. M. & Lord, E. M. (1984). Comparative floral development in the cleistogamous species *Viola odorata*. III. A histological study. *Botanical Gazette*, 145: 83–91.

Mayo, S. J., Bogner, J. & Boyce, P. C. (1997). *The Genera of Araceae*. Kew: Royal Botanical Gardens.

Mayo, S. J., Bogner, J. & Boyce, P. C. (1998). Araceae. In *The Families and Genera of Vascular Plants, Vol. 4*, ed. K. Kubitzki. Berlin: Springer, pp. 26–73.

Mayr, E. (1982). *The Growth of Biological Thought: Diversity, Evolution, and Inheritance*. Cambridge, MA: Harvard University Press.

McConnell, J. R. & Barton, M. K. (1998). Leaf polarity and meristem formation in *Arabidopsis*. *Development*, 125: 2935–2942.

McConnell, J. R., Emery, J., Eshed, Y. *et al.* (2001). Role of *PHABULOSA* and *PHAVOLUTA* in determining radial patterning in shoots. *Nature*, 411: 709–713.

McCouch, S. R. (2008). Gene nomenclature system for rice. *Rice*, 1: 72–84.

McDill, J., Repplinger, M., Simpson, B. B. & Kadereit, J. W. (2009). The phylogeny of *Linum* and Linaceae subfamily Linoideae, with implications for their systematics, biogeography, and evolution of heterostyly. *Systematic Botany*, 34: 386–405.

McHale, N. A. & Koning, R. E. (2004). *PHANTASTICA* regulates development of the adaxial mesophyll in *Nicotiana* leaves. *Plant Cell*, 16: 1251–1262.

McIntosh, R. A. (1988). A catalogue of gene symbols for wheat. In *Proceedings of the 7th International Wheat Genetics Symposium*, eds. T. E. Miller & R. M. D. Koebner. Cambridge: IPSR, pp. 1225–1324.

McIntyre, G. I. & Best, K. F. (1975). Studies on the flowering of *Thlaspi arvense* L. II. A comparative study of early- and late-flowering strains. *Botanical Gazette*, 136: 151–158.

McIntyre, G. I. & Best, K. F. (1978). Studies on the flowering of *Thlaspi arvense* L. IV. Genetic and ecological differences between early- and late-flowering strains. *Botanical Gazette*, 139: 190–195.

McKone, M. J. & Tonkyn, D. W. (1986). Intrapopulation gender variation in common ragweed (Asteracae: *Ambrosia artemisiifolia* L.), a monecious, annual herb. *Oecologia*, 70: 63–67.

McNamara, K. J. (1986). A guide to the nomenclature of heterochrony. *Journal of Paleontology*, 60: 4–13.

McNeill, J., Barrie, F. R., Buck, W. R. *et al.* (eds.) (2012). *International Code of Nomenclature for Algae, Fungi, and Plants (Melbourne Code) Adopted by the Eighteenth International Botanical Congress Melbourne, Australia, July 2011.* Ruggell: Gantner.

M'Cosh, J. (1851). Some remarks on the plant morphologically considered. *Transactions of the Botanical Society*, 4: 127–132.

M'Cosh, J. & Dickie, G. (1856). *Typical Forms and Special Ends in Creation.* New York, NY: Carter.

Meeuse, A. D. J. (1972). Sixty-five years of theories of the multiaxial flower. *Acta Biotheoretica*, 21: 167–202.

Meijer, M. & Murray, J. A. (2001). Cell cycle controls and the development of plant form. *Current Opinion in Plant Biology*, 4: 44–49.

Meinke, D. W., Cherry, J. M., Dean, C., Rounsley, S. D. & Koornneef, M. (1998). *Arabidopsis thaliana*: a model plant for genome analysis. *Science*, 282: 662–682.

Melville, R. (1960). A new theory of the angiosperm flower. *Nature*, 118: 14–18.

Melzer, R. & Theißen, G. (2016). The significance of developmental robustness for species diversity. *Annals of Botany*, 117: 725–732.

Melzer, R., Verelst, W. & Theißen, G. (2009). The class E floral homeotic protein SEPALLATA3 is sufficient to loop DNA in 'floral quartet'-like complexes in vitro. *Nucleic Acids Research*, 37: 144–157.

Melzer, R., Wang, Y. Q. & Theißen, G. (2010). The naked and the dead: the ABCs of gymnosperm reproduction and the origin of the angiosperm flower. *Seminars in Cell and Developmental Biology*, 21: 118–128.

Mena, M., Ambrose, B. A., Meeley, R. B. *et al.* (1996). Diversification of C-function activity in maize flower development. *Science*, 274: 1537–1540.

Meng, A., Zhang, Z., Li, J., Ronse De Craene, L. & Wang, H. (2012). Floral development of *Stephania* (Menispermaceae): impact of organ reduction on symmetry. *International Journal of Plant Sciences*, 173: 861–874.

Merckx, V. (2013). *Mycoheterotrophy: The Biology of Plants Living on Fungi.* Berlin: Springer.

Merrill, E. K. (1979). Comparison of ontogeny of three types of leaf architecture in *Sorbus* L. (Rosaceae). *Botanical Gazette*, 140: 328–337.

Mestek Boukhibar, L. & Barkoulas, M. (2016). The developmental genetics of biological robustness. *Annals of Botany*, 117: 699–707.

Metzger, R. J. & Krasnow, M. A. (1999). Genetic control of branching morphogenesis. *Science*, 284: 1635–1639.

Meyen, S. V. (1973). Plant morphology in its nomothetical aspects. *Botanical Review*, 39: 205–260.

Meyerowitz, E. M. (1994). The genetics of flower development. *Scientific American*, 271: 40–47.

Meyerowitz, E. M. (1997). Control of cell division patterns in developing shoots and flowers of *Arabidopsis thaliana*. *Cold Spring Harbor Symposia on Quantitative Biology*, 62: 369–375.

Michaels, S. D. & Amasino, R. M. (1999). *FLOWERING LOCUS C* encodes a novel MADS domain protein that acts as a repressor of flowering. *Plant Cell*, 11: 949–956.

Miller, A. P. (1995). Leaf-mining insects and fluctuating asymmetry in elm *Ulmus glabra* leaves. *Journal of Animal Ecology*, 64: 697–707.

Minelli, A. (2000). Limbs and tail as evolutionarily diverging duplicates of the main body axis. *Evolution and Development*, 2: 157–165.

Minelli, A. (2003). *The Development of Animal Form: Ontogeny, Morphology, and Evolution*. Cambridge: Cambridge University Press.

Minelli, A. (2009a). *Forms of Becoming: The Evolutionary Biology of Development*. Princeton, NJ: Princeton University Press.

Minelli, A. (2009b). *Perspectives in Animal Phylogeny and Evolution*. Oxford: Oxford University Press.

Minelli, A. (2011). A principle of developmental inertia. In *Epigenetics: Linking Genotype and Phenotype in Development and Evolution*, eds. B. Hallgrímsson & B. K. Hall. San Francisco, CA: University of California Press, pp. 116–133.

Minelli, A. (2014). Developmental disparity. In *Towards a Theory of Development*, eds. A. Minelli & T. Pradeu. Oxford: Oxford University Press, pp. 227–245.

Minelli, A. (2015a). Biological systematics in the evo-devo era. *European Journal of Taxonomy*, 125: 1–23.

Minelli, A. (2015b). Evo devo and its significance for animal evolution and phylogeny. In *Evolutionary Developmental Biology of Invertebrates. 1. Introduction, Non-Bilateria, Acoelomorpha, Xenoturbellida, Chaetognatha*, ed. A. Wanninger. Wien: Springer, pp. 1–23.

Minelli, A. (2015c). Grand challenges in evolutionary developmental biology. *Frontiers in Ecology and Evolution*, 2: 85.

Minelli, A. (2015d). Constraints on animal (and plant) form in nature and art. *Art and Perception*, 3: 265–281.

Minelli, A. (2016a). Scaffolded biology. *Theory in Biosciences*, 135: 163–173.

Minelli, A. (2016b). Tracing homologies in an ever-changing world. *Rivista di estetica*, n.s., 56: 40–55.

Minelli, A. (2016c). Species diversity vs. morphological disparity in the light of evolutionary developmental biology. *Annals of Botany*, 117: 781–794.

Minelli, A. (2017). Evolvability and its evolvability. In *Challenging the Modern Synthesis: Adaptation, Development, and Inheritance*, eds. P. Huneman & D. Walsh. Oxford: Oxford University Press, pp. 211–238.

Minelli, A. & Fusco, G. (2005). Conserved vs. innovative features in animal body organization. *Journal of Experimental Zoology (Molecular and Developmental Evolution)*, 304B: 520–525.

Minelli, A. & Fusco, G. (2012). On the evolutionary developmental biology of speciation. *Evolutionary Biology*, 39: 242–254.

Minelli, A. & Fusco, G. (2013). Homology. In *The Philosophy of Biology: A Companion for Educators, History, Philosophy and Theory of the Life Sciences*, ed. K. Kampourakis. Dordrecht: Springer, pp. 289–322.

Minelli, A., Negrisolo, E. & Fusco, G. (2006). Reconstructing animal phylogeny in the light of evolutionary developmental biology. In *Reconstructing the Tree of Life: Taxonomy and Systematics of Species Rich Taxa*, eds. T. R. Hodkinson, J. A. N. Parnell & S. Waldren. Boca Raton, FL: Taylor & Francis/CRC Press, pp. 177–190.

Minelli, A. & Pradeu, T. (eds.) (2014). *Towards a Theory of Development*. Oxford: Oxford University Press.

Minter, T. C. & Lord, E. M. (1983). A comparison of cleistogamous and chasmogamous floral development in *Collomia grandiflora* Dougl. Ex Lindl. (Polemoniaceae). *American Journal of Botany*, 70: 1499–1508.

Mirabet, V., Das, P., Boudaoud, A. & Hamant, O. (2011). The role of mechanical forces in plant morphogenesis. *Annual Review of Plant Biology*, 62: 365–385.

Mitchell, C. H. & Diggle, P. K. (2005). The evolution of unisexual flowers: morphological and functional convergence results from diverse developmental transitions. *American Journal of Botany*, 92: 1068–1076.

Mitchell-Olds, T. (2001). *Arabidopsis thaliana* and its wild relatives: a model system for ecology and evolution. *Trends in Ecology and Evolution*, 16: 693–700.

Mitchell-Olds, T., Al-Shehbaz, I. A., Koch, M. & Sharbel, T. F. (2005). Crucifer evolution in the post-genomic era. In *Plant Diversity and Evolution: Genotypic and Phenotypic Variation in Higher Plants*, ed. R. J. Henry. Cambridge, MA: CAB International, pp. 119–137.

Mitsuda, N. & Ohme-Takagi, M. (2009). Functional analysis of transcription factors in *Arabidopsis*. *Plant Cell Physiology*, 50: 1232–1248.

Miwa, H., Kinoshita, A., Fukuda, H. & Sawa, S. (2009). Plant meristems: *CLAVATA3/ESR*-related signaling in the shoot apical meristem and the root apical meristem. *Journal of Plant Research*, 122: 31–39.

Mizukami, Y. & Fischer, R. L. (2000). Plant organ size control: *AINTEGUMENTA* regulates growth and cell numbers during organogenesis. *Proceedings of the National Academy of Sciences of the United States of America*, 97: 942–947.

Moczek, A. P. (2008). On the origins of novelty in development and evolution. *BioEssays*, 30: 432–447.

Moczek, A. P. (2010). Phenotypic plasticity and diversity in insects. *Philosophical Transactions of the Royal Society B*, 365: 593–603.

Molinero-Rosales, N., Jamilena, M., Zurita, S. *et al.* (1999). *FALSIFLORA*, the tomato orthologue of *FLORICAULA* and *LEAFY*, controls flowering time and floral meristem identity. *The Plant Journal*, 20: 685–693.

Möller, M., Pfosser, M., Jang, C. G. *et al.* (2009). A preliminary phylogeny of the 'didymocarpoid Gesneriaceae' based on three molecular data sets: incongruence with available tribal classifications. *American Journal of Botany*, 96: 989–1010.

Mondragón-Palomino, M., Hiese, L., Harter, A., Koch, M. A. & Theißen, G. (2009). Positive selection and ancient duplications in the evolution of class B floral homeotic genes of orchids and grasses. *BMC Evolutionary Biology*, 9: 81.

Mondragón-Palomino, M. & Theißen, G. (2008). MADS about the evolution of orchid flowers. *Trends in Plant Science*, 13: 51–59.

Mondragón-Palomino, M. & Theißen, G. (2009). Why are orchid flowers so diverse? Reduction of evolutionary constraints by paralogues of class B floral homeotic genes. *Annals of Botany*, 104: 583–594.

Mondragón-Palomino, M. & Theißen, G. (2011). Conserved differential expression of paralogous *DEFICIENS*- and *GLOBOSA*-like MADS-box genes in the flowers of Orchidaceae: refining the 'orchid code'. *The Plant Journal*, 66: 1008–1019.

Mondragón-Palomino, M. & Trontin, C. (2011). High time for a roll call: gene duplication and phylogenetic relationships of *TCP*-like genes in monocots. *Annals of Botany*, 107: 1533–1544.

Monniaux, M., Pieper, B. & Hay, A. (2016). Stochastic variation in *Cardamine hirsuta* petal number. *Annals of Botany*, 117: 881–887.

Moody, A., Diggle, P. K. & Steingraeber, D. A. (1999). Developmental analysis of the evolutionary origin of vegetative propagules in *Mimulus gemmiparus* (Scrophulariaceae). *American Journal of Botany*, 86: 1512–1522.

Mordhorst, A. P., Toonen, M. A. J., de Vries, S. C. & Meinke, D. (1997). Plant embryogenesis. *Critical Reviews in Plant Sciences*, 16: 535–576.

Moreno-Risueno, M. A., Busch, W. & Benfey, P. N. (2010). Omics meet networks: using systems approaches to-infer regulatory networks in plants. *Current Opinion in Plant Biology*, 13: 126–131.

Moreno-Risueno, M. A., Van Norman, J. M. & Benfey, P. N. (2012). Transcriptional switches direct plant organ formation and patterning. *Current Topics in Developmental Biology*, 98: 229–257.

Mouradov, A., Glassick, T., Hamdorf, B. *et al.* (1998). *NEEDLY*, a *Pinus radiata* ortholog of *FLORICAULA/LEAFY* genes, expressed in both reproductive and vegetative meristems. *Proceedings of the National Academy of Sciences of the United States of America*, 95: 6537–6542.

Mower, J. P., Stefanovic, S., Young, G. J. & Palmer, J. D. (2004). Gene transfer from parasitic to host plants. *Nature*, 432: 165–166.

Moyroud, E., Kusters, E., Monniaux, M., Koes, R. & Parcy, F. (2010). *LEAFY* blossoms. *Trends in Plant Science*, 15: 346–352.

Mueller, A.L., Solow, T. H., Taylor, N. *et al.* (2005). The SOL Genomics Network (SGN): a comparative resource for solanaceous biology and beyond. *Plant Physiology*, 138: 1310–1317.

Mukherjee, K. & Brocchieri, L. (2010). Evolution of plant homeobox genes. In *Encyclopedia of Life Sciences*. Chichester: Wiley.

Mukherjee, K., Brocchieri, L. & Burglin, T. R. (2009). A comprehensive classification and evolutionary analysis of plant homeobox genes. *Molecular Biology and Evolution*, 26: 2775–2794.

Müller, M. & Cronk, Q. C. B. (2001). Evolution of morphological novelty: a phylogenetic analysis of growth patterns in *Streptocarpus* (Gesneriaceae). *Evolution*, 55: 918–929.

Mummenhoff, K., Al-Shehbaz, I. A., Bakker, F. T., Linder, H. P. & Mühlhausen, A. (2005). Phylogeny, morphological evolution, and speciation of endemic Brassicaceae genera in the Cape flora of southern Africa. *Annals of the Missouri Botanical Garden*, 92: 400–424.

Mummenhoff, K., Brüggemann, H. & Bowman, J. L. (2001). Chloroplast DNA phylogeny and biogeography of *Lepidium* (Brassicaceae). *American Journal of Botany*, 88: 2051–2063.

Mummenhoff, K., Polster, A., Mühlhausen, A. & Theißen, G. (2009). *Lepidium* as a model system for studying the evolution of fruit development in Brassicaceae. *Journal of Experimental Botany*, 60: 1503–1513.

Mungall, C., Gkoutos, G. V., Smith, C., Haendel, M. & Ashburner, M. (2010). Integrating phenotype ontologies across multiple species. *Genome Biology*, 11: R2.

Munné-Bosch, S. (2008). Do perennials really senesce? *Trends in Plant Science*, 13: 216–220.

Münster, T., Pahnke, J., Di Rosa, A *et al.* (1997). Floral homeotic genes were recruited from homologous MADS-box genes preexisting in the common ancestor of ferns and seed plants. *Proceedings of the National Academy of Sciences of the United States of America*, 94: 2415–2420.

Münster, T., Wingen, L. U., Faigl, W. *et al.* (2001). Characterization of three *GLOBOSA*-like MADS-box genes from maize: evidence for ancient paralogy in one class of floral homeotic B-function genes of grasses. *Gene*, 262: 1–13.

Müntzing, A. (1936). The evolutionary significance of autopolyploidy. *Hereditas*, 21: 263–378.

Murai, K., Miyamae, M., Kato, H., Takumi, S. & Ogihara, Y. (2003). *WAP1*, a wheat *APETALA1* homolog, plays a central role in the phase transition from vegetative to reproductive growth. *Plant and Cell Physiology*, 44: 1255–1265.

Murray, N. A. & Johnson, D. M. (1987). Synchronous dichogamy in a Mexican anonillo *Rollinia jimenezi* var. *nelsonii*. *Contributions from the University of Michigan Herbarium*, 16: 173–178.

Nagasawa, N., Miyoshi, M., Sano, Y. *et al.* (2003). *SUPERWOMAN1* and *DROOPING LEAF* genes control floral organ identity in rice. *Development*, 130: 705–718.

Nah, G. & Chen, J. (2010). Tandem duplication of the FLC locus and the origin of a new gene in *Arabidopsis* related species and their functional implications in allopolyploids. *New Phytologist*, 186: 228–238.

Naiki, A. (2012). Heterostyly and the possibility of its breakdown by polyploidization. *Plant Species Biology*, 27: 3–29.

Nakada, M., Komatsu, M., Ochiai, T. *et al.* (2006). Isolation of *MaDEF* from *Muscari armeniacum* and analysis of its expression using laser microdissection. *Plant Science*, 170: 143–150.

Nakamura, T., Fukuda, T., Nakano, M. *et al.* (2005). The modified ABC model explains the development of the petaloid perianth of *Agapanthus praecox* ssp. *orientalis* (Agapanthaceae) flowers. *Plant Molecular Biology*, 58: 435–445.

Nakano, T., Kimbara, J., Fujisawa, M., Kitagawa, M. *et al.* (2012). *MACROCALYX* and *JOINTLESS* interact in the transcriptional regulation of tomato fruit abscission zone development. *Plant Physiology*, 158: 439–450.

Nakata, M., Matsumoto, N., Tsugeki, R. *et al.* (2012). Roles of the middle domain-specific *WUSCHEL-RELATED* HOMEOBOX genes in early development of leaves in *Arabidopsis*. *Plant Cell*, 24: 519–535.

Nakayama, H., Nakayama, N., Seiki, S. *et al.* (2014). Regulation of the KNOX-GA gene module induces heterophyllic alteration in North American lake cress. *Plant Cell*, 26: 4733–4748.

Nakayama, H., Yamaguchi, T. & Tsukaya, H. (2012). Acquisition and diversification of cladodes: leaf-like organs in the genus *Asparagus*. *Plant Cell*, 24: 929–940.

Nardmann, J. & Werr, W. (2006). The shoot stem cell niche in angiosperms: expression patterns of *WUS* orthologues in rice and maize imply major modifications in the course of mono- and dicot evolution. *Molecular Biology and Evolution*, 23: 2492–2504.

Nath, U., Crawford, B. C., Carpenter, R. & Coen, E. (2003). Genetic control of surface curvature. *Science*, 299: 1404–1407.

Nickrent, D. L., Blarer, A., Qiu, Y. L., Vidal-Russell, R. & Anderson, F. E. (2004). Phylogenetic inference in Rafflesiales: the influence of rate heterogeneity and horizontal gene transfer. *BMC Evolutionary Biology*, 4: 40.

Nicolas, M. & Cubas, P. (2016). The role of TCP transcription factors in shaping flower structure, leaf morphology, and plant architecture. In *Plant Transcription Factors: Evolutionary, Structural and Functional Aspects*, ed. D. H. Gonzalez. Amsterdam: Elsevier, pp. 249–267.

Niklas, K. J. (2016). *Plant Evolution: An Introduction to the History of Life*. Chicago, IL: University of Chicago Press.

Nikolov, L. A., Staedler, Y. M., Manickam, S. *et al.* (2014). Floral structure and development in Rafflesiaceae with emphasis on their exceptional gynoecia. *American Journal of Botany*, 101: 225–243.

Nikovics, K., Blein, T., Peaucelle, A. *et al.* (2006). The balance between the *MIR164A* and *CUC2* genes controls leaf margin serration in *Arabidopsis*. *Plant Cell*, 18: 2929–2945.

Nishiyama, T., Fujita, T., Shin-I, T. *et al.* (2003). Comparative genomics of *Physcomitrella patens* gametophytic transcriptome and *Arabidopsis thaliana*: implication for land plant evolution. *Proceedings of the National Academy of Sciences of the United States of America*, 100: 8007–8012.

Nodine, M. D. & Bartel, D. P. (2012). Maternal and paternal genomes contribute equally to the transcriptome of early plant embryos. *Nature*, 482: 94–97.

Nowak, M. D., Russo, G., Schlapbach, R. *et al.* (2015). The draft genome of *Primula veris* yields insights into the molecular basis of heterostyly. *Genome Biology*, 16: 12.

Nuraliev, M. S., Degtajareva, G. V., Sokoloff, D. D. *et al.* (2014). Flower morphology and relationships of *Schefflera subintegra* (Araliaceae, Apiales): an evolutionary step towards extreme floral polymery. *Botanical Journal of the Linnean Society*, 175: 553–597.

Nürnberger, T. & Brunner, F. (2002). Innate immunity in plants and animals: emerging parallels between the recognition of general elicitors and pathogen-associated molecular patterns. *Current Opinion in Plant Biology*, 5: 318–324.

Nutt, P., Ziermann, J., Hintz, M., Neuffer, B. & Theißen, G. (2006). *Capsella* as a model system to study the evolutionary relevance of floral homeotic mutants. *Plant Systematics and Evolution*, 259: 217–235.

Ocarez, N. & Mejía, N. (2016). Suppression of the D-class MADS-box *AGL11* gene triggers seedlessness in fleshy fruits. *Plant Cell Reports*, 35: 239–254.

Ochando, I., Jover-Gil, S., Ripoll, J. J. *et al.* (2006). Mutations in the microRNA complementarity site of the *INCURVATA4* gene perturb meristem function and adaxialize lateral organs in *Arabidopsis*. *Plant Physiology*, 141: 607–619.

Ochiai, T., Nakamura, T., Mashiko, Y. *et al.* (2004). The differentiation of sepal and petal morphologies in Commelinaceae. *Gene*, 343: 253–262.

Ogura, T. & Busch, W. (2016). Genotypes, networks, phenotypes: moving toward plant systems genetics. *Annual Review of Cell and Developmental Biology*, 32: 103–126.

Ohmori, S., Kimizu, M., Sugita, M. *et al.* (2009). *MOSAIC FLORAL ORGANS1*, an *AGL6*-like MADS box gene, regulates floral organ identity and meristem fate in rice. *Plant Cell*, 21: 3008–3025.

Okada, K., Komaki, M. K. & Shimura, Y. (1989). Mutational analysis of pistil structure and development of *Arabidopsis thaliana*. *Cell Differentiation and Development*, 28: 27–38.

Olmstead, R. G., Bohs, L., Migid, H. A. *et al.* (2008). A molecular phylogeny of the Solanaceae. *Taxon*, 57: 1159–1181.

Olmstead, R. G., Michaels, H. J., Scott, K. M. & Palmer, J. D. (1992). Monophyly of the Asteridae sensu lato and identification of their major lineages inferred from DNA sequences of *rbcL*. *Annals of the Missouri Botanical Garden*, 79: 249–265.

Olmstead, R. G. & Palmer, J. D. (1992). A chloroplast DNA phylogeny of the Solanaceae: subfamilial relationships and character evolution. *Annals of the Missouri Botanical Garden*, 79: 346–360.

Olmstead, R. G. & Reeves, P. A. (1995). Evidence for the polyphyly of the Scrophulariaceae based on chloroplast rbcL and ndhF sequences. *Annals of the Missouri Botanical Garden*, 82: 176–193.

Olson, M. E. (2003). Ontogenetic origins of floral bilateral symmetry in Moringaceae (Brassicales). *American Journal of Botany*, 90: 49–71.

Orkwiszewski, J. A. & Poethig, R. S. (2000). Phase identity of the maize leaf is determined after leaf initiation. *Proceedings of the National Academy of Sciences of the United States of America*, 97: 10631–10636.

Otsuga, D., DeGuzman, B., Prigge, M. J., Drews, G. N. & Clark, S. E. (2001). *REVOLUTA* regulates meristem initiation at lateral positions. *The Plant Journal*, 25: 223–236.

Owen, R. (1843). *Lectures on the Comparative Anatomy and Physiology of the Invertebrate Animals, Delivered at the Royal College of Surgeons*. London: Longman, Brown, Green and Longmans.

Oyama, S. (2000). *The Ontogeny of Information*, 2nd edn. Durham, NC: Duke University Press.

Ozerova, L. V. & Timonin, A. C. (2009). On the evidence of subunifacial and unifacial leaves: developmental studies in leaf-succulent *Senecio* L. species (Asteraceae). *Wulfenia*, 16: 61–77.

Pabón-Mora, N. & González, F. (2012). Leaf development, metamorphic hetero-blasty and heterophylly in *Berberis* s.l. (Berberidaceae). *Botanical Review*, 74: 463–489.

Palauqui, J. C. & Laufs, P. (2011). Phyllotaxis: in search of the golden angle. *Current Biology*,: 21: R502-R504.

Pallakies, H. & Simon, R. (2010). Positional information in plant development. In *Encyclopedia of Life Sciences*. Chichester: Wiley.

Panero, J. L. & Funk, V. A. (2008). The value of sampling anomalous taxa in phylogenetic studies: major clade of the Asteraceae revealed. *Molecular Phylogenetics and Evolution*, 47: 757–782.

Pang, H.-B., Sun, Q.-W., He, S.-Z. & Wang, Y.-Z. (2010). Expression pattern of *CYC*-like genes relating to a dorsalized actinomorphic flower in *Tengia* (Gesneriaceae). *Journal of Systematics and Evolution*, 48: 309–317.

Parcy, F. (2005). Flowering: a time for integration. *International Journal of Developmental Biology*, 49: 585–593.

Parcy, F., Nilsson, O., Busch, M. A., Lee, I. & Weigel, D. (1998). A genetic frame-work for floral patterning. *Nature*, 395: 561–566.

Park, J.-H., Ishikawa, Y., Ochiai, T., Kanno, A. & Kameya, T. (2004). Two *GLOBOSA*-like genes are expressed in second and third whorls of homochlamydeous flowers in *Asparagus officinalis* L. *Plant Cell Physiology*, 45: 325–332.

Park, J.-H., Ishikawa, Y., Yoshida, R., Kanno, A. & Kameya, T. (2003). Expression of *AODEF*, a B-functional MADS-box gene, in stamens and inner tepals of the dioecious species *Asparagus officinalis* L. *Plant Molecular Biology*, 51: 867–875.

Paterson, A. H., Bowers, J. E. & Chapman, B. A. (2004). Ancient polyploidization predating divergence of the cereals, and its consequences for comparative genomics. *Proceedings of the National Academy of Sciences of the United States of America*, 101: 9903–9908.

Pauw, A. (2005). Inversostyly: a new stylar polymorphism in an oil-secreting plant, *Hemimeris racemosa* (Scrophulariaceae). *American Journal of Botany*, 92: 1878–1886.

Pavlicev, M. & Hansen, T. H. (2011). Genotype-phenotype maps maximizing evolvability: modularity revisited. *Evolutionary Biology*, 38: 371–389.

Peaucelle, A., Louvet, R., Johansen, J. N. *et al.* (2008). *Arabidopsis* phyllotaxis is controlled by the methylesterification status of cell-wall pectins. *Current Biology*, 18: 1943–1948.

Pelaz, S., Ditta, G. S., Baumann, E., Wisman, E. & Yanofsky, M. F. (2000). B and C floral organ identity functions require *SEPALLATA* MADS-box genes. *Nature*, 405: 200–203.

Pelaz, S., Gustafson-Brown, C., Kohalmi, S. E., Crosby, W. L. & Yanofsky, M. F. (2001a). *APETALA1* and *SEPALLATA3* interact to promote flower development. *The Plant Journal*, 26: 385–394.

Pelaz, S., Tapia-Lopez, R., Alvarez-Buylla, E. R. & Yanofsky, M. F. (2001b). Conversion of leaves into petals in *Arabidopsis*. *Current Biology*, 11: 182–184.

Pellicer, J., Fay, M. F. & Leitch, I. J. (2010). The largest eukaryotic genome of them all? *Botanical Journal of the Linnean Society*, 164: 10–15.

Pennington, R. T., Klitgaard, B. B., Ireland, H. & Lavin, M. (2000). New insights into floral evolution of basal Papilionoideae from molecular phylogenies. In *Advances in Legume Systematics, 9*, eds. P. S. Herendeen & A. Bruneau. Kew: Royal Botanic Gardens, pp. 233–248.

Perez-Rodriguez, P., Riano-Pachon, D. M., Correa, L. G. G. *et al.* (2010). PlnTFDB: updated content and new features of the plant transcription factor database. *Nucleic Acids Research*, 38: D822–D827.

Peris, C. I., Rademacher, E. H. & Weijers, D. (2010). Green beginnings: pattern formation in the early plant embryo. *Current Topics in Developmental Biology*, 91: 1–27.

Peterson, R. L. (1992). Adaptations of root structure in relation to biotic and abiotic factors. *Canadian Journal of Botany*, 70: 661–675.

Petricka, J. J., Winter, C. M. & Benfey, P. N. (2012). Control of *Arabidopsis* root development. *Annual Review of Plant Biology*, 63: 563–590.

Pham, T. & Sinha, N. (2003). Role of KNOX genes in shoot development of *Welwitschia mirabilis*. *International Journal of Plant Sciences*, 164: 333–343.

Piazza, P., Bailey, C. D., Cartolano, M. *et al.* (2010). *Arabidopsis thaliana* leaf form evolved via loss of KNOX expression in leaves in association with a selective sweep. *Current Biology*, 20: 2223–2228.

Pigliucci, M. (2001). *Phenotypic Plasticity: Beyond Nature and Nurture*. Baltimore, MD: Johns Hopkins University Press.

Pigliucci, M. (2008). Is evolvability evolvable? *Nature Reviews Genetics*, 9: 75–82.

Pigliucci, M. & Müller, G. (eds.) (2010). *Evolution: The Extended Synthesis*. Cambridge, MA: MIT Press.

Pigliucci, M. & Murren, C. (2003). Genetic assimilation and a possible evolutionary paradox: can macroevolution sometimes be so fast as to pass us by? *Evolution*, 57: 1455–1464.

Pigliucci, M., Murren, C. J. & Schlichting, C. D. (2006). Phenotypic plasticity and evolution by genetic assimilation. *Journal of Experimental Biology*, 209: 2362–2367.

Piñeyro-Nelson, A., Almeida, A. M. R., Sass, C., Iles, W. J. D. & Specht, C. D. (2017). Change of fate and staminodial laminarity as potential agents of floral diversification in the Zingiberales. *Journal of Experimental Zoology (Molecular and Developmental Evolution)*, 328: 41–54.

Pires, N. D. & Dolan, L. (2012). Morphological evolution in land plants: new designs with old genes. *Philosophical Transactions of the Royal Society B*, 367: 508–518.

Poethig, R. S. (1988). Heterochronic mutations affecting shoot development in maize. *Genetics*, 119: 959–973.

Poethig, R. S. (2009). Small RNAs and developmental timing in plants. *Current Opinion in Genetics and Development*, 19: 374–378.

Poethig, R. S. & Sussex, I. M. (1985). The developmental morphology and growth dynamics of the tobacco leaf. *Planta*, 165: 158–169.

Porras, R. & Muñoz, J. M. (2000). Cleistogamous capitulum in *Centaurea melitensis* (Asteraceae): heterochronic origin. *American Journal of Botany*, 87: 925–933.

Posé, D., Yant, L. & Schmid, M. (2012). The end of innocence: flowering networks explode in complexity. *Current Opinion in Plant Biology*, 15: 45–50.

Povilus, R. A., Losada, J. M. & Friedman, W. E. (2015). Floral biology and ovule and seed ontogeny of *Nymphaea thermarum*, a water lily at the brink of extinction with potential as a model system for basal angiosperms. *Annals of Botany*, 115: 211–226.

Powell, A. E. & Lenhard, M. (2012). Control of organ size in plants. *Current Biology*, 22: R360–R367.

Powell, R. (2007). Is convergence more than an analogy? Homoplasy and its implications for macroevolutionary predictability. *Biology and Philosophy*, 22: 565–578.

Pradeu, T. (2012). *The Limits of the Self: Immunology and Biological Identity*. Oxford: Oxford University Press.

Pradeu, T. (2016). Organisms or biological individuals? Combining physiological and evolutionary individuality. *Biology and Philosophy*, 31: 797–817.

Pradeu, T., Laplane, L., Prévot, K. *et al.* (2016). Defining 'development'. *Current Topics in Developmental Biology*, 117: 171–183.

Prenner, G. (2004). Floral development in *Polygala myrtifolia* (Polygalaceae) and its similarities with Leguminosae. *Plant Systematics and Evolution*, 249: 67–76.

Prenner, G. (2014). Floral ontogeny in *Passiflora lobata* (Malpighiales, Passifloraceae) reveals a rare pattern in petal formation and provides new evidence for interpretation of the tendril and corona. *Plant Systematics and Evolution*, 300: 1285–1297.

Prenner, G. & Klitgaard, B. B. (2008). Towards unlocking the deep nodes of Leguminosae: floral development and morphology of the enigmatic *Duparquetia orchidacea* (Leguminosae, Caesalpinioideae). *American Journal of Botany*, 95: 1349–1365.

Prenner, G. & Rudall. P. J. (2007). Comparative ontogeny of the cyathium in *Euphorbia* and its allies: exploring the organ-flower-inflorescence boundaries. *American Journal of Botany*, 94: 1612–1629.

Preston, J. C. & Hileman, L. C. (2009). Developmental genetics of floral symmetry evolution. *Trends in Plant Science*, 14: 147–154.

Preston, J. C. & Hileman, L. C. (2010). SQUAMOSA-PROMOTER BINDING PROTEIN 1 initiates flowering in *Antirrhinum majus* through the activation of meristem identity genes. *The Plant Journal*, 62: 704–712.

Preston, J. C. & Hileman, L. C. (2012). Parallel evolution of TCP and B-class genes in Commelinaceae flower bilateral symmetry. *Evo Devo*, 3: 6.

Preston, J. C., Hileman, L. C. & Cubas, P. (2011a). Reduce, reuse, and recycle: developmental evolution of trait diversification. *American Journal of Botany*, 98: 397–403.

Preston, J. C. & Kellogg, E. A. (2006). Reconstructing the evolutionary history of paralogous *APETALA1/FRUITFULL*-like genes in grasses (Poaceae). *Genetics*, 174: 421–437.

Preston, J. C. & Kellogg, E. A. (2007). Conservation and divergence of *APETALA1/FRUITFULL*-like gene function in grasses: evidence from gene expression analyses. *The Plant Journal*, 52: 69–81.

Preston, J. C. & Kellogg, E. A. (2008). Discrete developmental roles for temperate cereal grass *VERNALIZATION1/FRUITFULL*-like genes in flowering competency and the transition to flowering. *Plant Physiology*, 146: 265–276.

Preston, J. C., Kost, M. A. & Hileman, L. C. (2009). Conservation and diversification of the symmetry developmental program among close relatives of snapdragon with divergent floral morphologies. *New Phytologist*, 182: 751–762.

Preston, J. C., Martinez, C. C. & Hileman, L. C. (2011b). Gradual disintegration of the floral symmetry gene network is implicated in the evolution of a wind-pollination syndrome. *Proceedings of the National Academy of Sciences of the United States of America*, 108: 2343–2348.

Prigge, M. J., Otsuga, D., Alonso, J. M. *et al.* (2005). Class III homeodomain-leucine zipper gene family members have overlapping, antagonistic, and distinct roles in *Arabidopsis* development. *Plant Cell*, 17: 61–76.

Primack, R. B. (1985). Longevity of individual flowers. *Annual Reviews of Ecology and Systematics*, 16: 15–37.

Pruitt, R. E., Vielle-Calzada, J. P., Ploense, S. E., Grossniklaus, U. & Lolle, S. J. (2000). *FIDDLEHEAD*, a gene required to suppress epidermal cell interactions in *Arabidopsis*, encodes a putative lipid biosynthetic enzyme. *Proceedings of the National Academy of Sciences of the United States of America*, 97: 1311–1316.

Prusinkiewicz, P. (2004). Modeling plant growth and development. *Current Opinion in Plant Biology*, 7: 79–84.

Prusinkiewicz, P., Erasmus, Y., Lane, B., Harder, J. D. & Coen, E. (2007). Evolution and development of inflorescence architectures. *Science*, 316: 1452–1456.

Prusinkiewicz, P. & Lindenmayer, A. (1990). *The Algorithmic Beauty of Plants*. Berlin: Springer.

Quint, M., Drost, H.-G., Gabel, A. *et al.* (2012). A transcriptomic hourglass in plant embryogenesis. *Nature*, 490: 98–101.

Quodt, V., Faigl, W., Saedler, H. & Münster, T. (2007). The MADS-domain protein PPM2 preferentially occurs in gametangia and sporophytes of the moss *Physcomitrella patens*. *Gene*, 400: 25–34.

Raff, R. A. & Kaufman, T. C. (1983). *Embryos, Genes, and Evolution*. New York, NY: Macmillan.

Rao, N. N., Prasad, K., Kumar, P. R. & Vijayraghavan, U. (2008). Distinct regulatory role for *RFL*, the rice *LFY* homolog, in determining flowering time and plant architecture. *Proceedings of the National Academy of Sciences of the United States of America*, 105: 3646–3651.

Rasmussen, D. A., Kramer, E. M. & Zimmer, E. A. (2009). One size fits all? Molecular evidence for a commonly inherited petal identity program in Ranunculales. *American Journal of Botany*, 96: 96–109.

Rast, M. I. & Simon, R. (2008). The meristem-to-organ boundary: more than an extremity of anything. *Current Opinion in Genetics and Development*, 18: 287–294.

Raven, J. A. & Edwards, D. (2001). Roots: evolutionary origins and biogeochemical significance. *Journal of Experimental Botany*, 52: 381–401.

Ravi, M. & Chan, S. W. (2010). Haploid plants produced by centromere-mediated genome elimination. *Nature*, 464: 615–618.

Reardon, W., Fitzpatrick, D. A., Fares, M. A. & Nugent, J. M. (2009). Evolution of flower shape in *Plantago lanceolata*. *Plant Molecular Biology*, 71: 241–250.

Rebocho, A. B., Bliek, M., Kusters, E. *et al.* (2008). Role of *EVERGREEN* in the development of the cymose petunia inflorescence. *Developmental Cell*, 15: 437–447.

Ree, R. H. & Donoghue, M. J. (1999). Inferring rates of change in flower symmetry in asterid angiosperms. *Systematic Botany*, 48: 633–641.

Reeves, P. A., He, Y., Schmitz, R. J. *et al.* (2007). Evolutionary conservation of the *FLOWERING LOCUS C* mediated vernalization response: evidence from the sugar beet (*Beta vulgaris*). *Genetics*, 176: 295–307.

Reeves, P. A. & Olmstead, R. G. (1998). Evolution of novel morphological and reproductive traits in a clade containing *Antirrhinum majus* (Scrophulariaceae). *American Journal of Botany*, 85: 1047–1056.

Reilly, S. M. (1997). An integrative approach to heterochrony: the distinction between interspecific and intraspecific phenomena. *Biological Journal of the Linnean Society*, 60: 119–143.

Reinhardt, D., Mandel, T. & Kuhlemeier, C. (2000). Auxin regulates the initiation and radial position of plant lateral organs. *Plant Cell*, 12: 507–518.

Reinhardt, D., Pesce, E. R., Stieger, P. *et al.* (2003). Regulation of phyllotaxis by polar auxin transport. *Nature*, 426: 255–260.

Reinhart, B. J., Weinstein, E. G., Rhoades, M. W., Bartel, B. & Bartel, D. P. (2002). MicroRNAs in plants. *Genes and Development*, 16: 1616–1626.

Reiser, L., Sanchez-Baracaldo, P. & Kake, S. (2000). Knots in the family tree: evolutionary relationships and functions of *knox* homeobox genes. *Plant Molecular Biology*, 42: 151–166.

Remizowa, M. V., Sokoloff, D. D. & Rudall, P. J. (2010). Evolutionary history of the monocot flower. *Annals of the Missouri Botanical Garden*, 97: 617–645.

Ren, J.-B. & Guo, Y.-P. (2015). Behind the diversity: ontogenies of radiate, disciform, and discoid capitula of *Chrysanthemum* and its allies. *Journal of Systematics and Evolution*, 53: 520–528.

Ren, Y., Chang, H.-L. & Endress, P. K. (2010). Floral development in Anemoneae (Ranunculaceae). *Botanical Journal of the Linnean Society*, 162: 77–100.

Reski, R. (2003). *Physcomitrella patens* as a novel tool for plant functional genomics. In *Plant Biotechnology 2002 and Beyond*, ed. I. K. Vasil. Dodrecht: Kluwer, pp. 205–209.

Reski, R. & Cove, D. J. (2004). Quick guide: *Physcomitrella patens*. *Current Biology*, 14: R261–R262.

Reut, M. S. & Fineran, B. A. (2000). Ecology and vegetative morphology of the carnivorous plant *Utricularia dichotoma* (Lentibulariaceae) in New Zealand. *New Zealand Journal of Botany*, 38: 433–450.

Rhoades, M. W., Reinhart, B. J., Lim, L. P. *et al.* (2002). Prediction of plant microRNA targets. *Cell*, 110: 513–520.

Rice, D. W., Alverson, A. J., Richardson, A. O. *et al.* (2013). Horizontal transfer of entire genomes via mitochondrial fusion in the angiosperm *Amborella*. *Science*, 342: 1468–1473.

Richards, A. J. (1997). *Plant Breeding Systems*, 2nd edn. London: Chapman & Hall.

Richardson, A. O. & Palmer, J. D. (2007). Horizontal gene transfer in plants. *Journal of Experimental Botany*, 58: 1–9.

Rigato, E. & Minelli, A. (2013). The great chain of being is still here. *Evolution: Education and Outreach*, 6: 18.

Rijpkema, A. S., Royaert, S., Zethof, J. *et al.* (2006). Analysis of the *Petunia TM6* MADS box gene reveals functional divergence within the *DEF/AP3* lineage. *Plant Cell*, 18: 1819–1832.

Rijpkema, A. S., Vandenbussche, M., Koes, R., Heijmans, K. & Gerats, T. (2010). Variations on a theme: changes in the floral ABCs in angiosperms. *Seminars in Cell and Developmental Biology*, 21: 100–107.

Rijpkema, A. S., Zethof, J., Gerats, T. & Vandenbussche, M. (2009). The petunia *AGL6* gene has a *SEPALLATA*-like function in floral patterning. *The Plant Journal*, 60: 1–9.

Roberts, J. A., Elliott, K. A. & González-Carranza, Z. H. (2002). Abscission, dehiscence, and other cell separation processes. *Annual Review of Plant Biology*, 53: 131–158.

Roberts, J. A. & González-Carranza, Z. H. (2013). Abscission. In *Encyclopedia of Life Sciences*. Chichester: Wiley.

Robinson, B. W. & Dukas, R. (1999). The influence of phenotypic modifications on evolution: the Baldwin effect and modern perspectives. *Oikos*, 85: 528–589.

Roeder, A. H. K. (2010). Sepals. In *Encyclopedia of Life Sciences*. Chichester: Wiley.

Roeder, A. H. K., Ferrándiz, C. & Yanofsky, M. F. (2003). The role of the REPLUMLESS homeodomain protein in patterning the *Arabidopsis* fruit. *Current Biology*, 13: 1630–1635.

Roeder, A. H. K. & Yanofsky, M. F. (2006). Fruit development in *Arabidopsis*. In *The* Arabidopsis *Book*. Rockville, MD: American Society of Plant Biologists, 4: e0075.

Rohde, A. & Bhalerao, R. P. (2007). Plant dormancy in the perennial context. *Trends in Plant Science*, 12: 217–223.

Rolland-Lagan, A.-G., Bangham, J. A. & Coen, E. (2003). Growth dynamics underlying petal shape and asymmetry. *Nature*, 422: 161–163.

Ronse De Craene, L. P. (2003). The evolutionary significance of homeosis in flowers: a morphological perspective. *International Journal of Plant Sciences*, 164: S225–S235.

Ronse De Craene, L. P. (2004). Floral development of *Berberidopsis corallina*: a crucial link in the evolution of flowers in the core eudicots. *Annals of Botany*, 94: 1–11.

Ronse De Craene, L. P. (2007). Are petals sterile stamens or bracts? The origin and evolution of petals in the core eudicots. *Annals of Botany*, 100: 621–630.

Ronse De Craene, L. P. (2008). Homology and evolution of petals in the core eudicots. *Systematic Botany*, 33: 301–325.

Ronse De Craene, L. P. (2010). *Floral Diagrams: An Aid to Understanding Flower Morphology and Evolution*. Cambridge: Cambridge University Press.

Ronse De Craene, L. P. (2011) Floral development of *Napoleonaea* (Lecythidaceae), a deceptively complex flower. In *Flowers on the Tree of Life*, ed. L. Wanntorp & L. P. Ronse De Craene. Cambridge: Cambridge University Press, pp. 279–295.

Ronse De Craene, L. P. (2013). Reevaluation of the perianth and androecium in Caryophyllales: implications for flower evolution. *Plant Systematics and Evolution*, 299: 1599–1636.

Ronse De Craene, L. P. (2016). Meristic changes in flowering plants: how flowers play with numbers. *Flora*, 221: 22–37.

Ronse De Craene, L. P. (2017). Floral development of *Berberidopsis beckleri* – can an additional species of the Berberidopsidaceae add evidence to floral evolution in the core eudicots? *Annals of Botany*, 119: 599–610.

Ronse De Craene, L. P. & Brockington, S. F. (2013). Origin and evolution of petals in angiosperms. *Plant Ecology and Evolution*, 146: 5–25.

Ronse De Craene, L. P., Linder, H. P. & Smets, E. F. (2002). Ontogeny and evolution of the flower of South African Restionaceae with special emphasis on the gynoecium. *Plant Systematics and Evolution*, 231: 225–258.

Ronse De Craene, L. P. & Smets, E. F. (1990). The floral development of *Popowia whitei* (Annonaceae). *Nordic Journal of Botany*, 10: 411–420. (Correction: *Nordic Journal of Botany*, 11: 420 (1991)).

Ronse De Craene, L. P. & Smets, E. F. (1994). Merosity in flowers: definition, origin, and taxonomic significance. *Plant Systematics and Evolution*, 191: 83–104.

Ronse De Craene, L. P. & Smets, E. F. (1995). Evolution of the androecium in the Ranunculiflorae. *Plant Systematics and Evolution, Supplement*, 9: 63–70.

Ronse De Craene, L. P. & Smets, E. F. (1998). Meristic changes in gynoecium morphology, exemplified by floral ontogeny and anatomy. In *Reproductive Biology in Systematics, Conservation and Economic Botany*, eds. S. J. Owens & P. J. Rudall. Kew: Royal Botanic Gardens, pp. 85–112.

Ronse De Craene, L. P. & Smets, E. F. (2000). Floral development of *Galopina tomentosa* with a discussion of sympetaly and placentation in the Rubiaceae. *Systematics and Geography of Plants*, 70: 155–170.

Ronse De Craene, L. P. & Smets, E. F. (2001). Staminodes: their morphological and evolutionary significance. *Botanical Review*, 67: 351–402.

Ronse De Craene, L. P., Soltis, P. S. & Soltis, D. E. (2003). Evolution of floral structure in basal angiosperms. *International Journal of Plant Sciences*, 164: S329–S363.

Ronse De Craene, L. P. & Wanntorp, L. (2008). Morphology and anatomy of the flower of *Meliosma* (Sabiaceae): implications for pollination biology. *Plant Systematics and Evolution*, 271: 79–91.

Roquet, C., Coissac, E., Cruaud, C. *et al.* (2016). Understanding the evolution of holoparasitic plants: the complete plastid genome of the holoparasite *Cytinus hypocistis* (Cytinaceae). *Annals of Botany*, 118: 885–896.

Rosin, F. M. & Kramer, E. M. (2009). Old dogs, new tricks: regulatory evolution in conserved genetic modules leads to novel morphologies in plants. *Developmental Biology*, 332: 25–35.

Rudall, P. J. (2008). Fascicles and filamentous structures: comparative ontogeny of morphological novelties in the mycoheterotrophic family Triuridaceae. *International Journal of Plant Sciences*, 169: 1023–1037.

Rudall, P. J. (2010). All in a spin: centrifugal organ formation and floral patterning. *Current Opinion in Plant Biology*, 13: 108–114.

Rudall, P. J. (2013). Identifying key features in the origin and early diversification of angiosperms. *Annual Plant Reviews*, 45: 163–188.

Rudall, P. J., Alves, M. & das Graças Sajo, M. (2016). Inside-out flowers of *Lacandonia brasiliana* (Triuridaceae) provide new insights into fundamental aspects of floral patterning. *PeerJ*, 4: e1653.

Rudall, P. J. & Bateman, R. M. (2002). Roles of synorganisation, zygomorphy and heterotopy in floral evolution: the gynostemium and labellum of orchids and other lilioid monocots. *Biology Reviews*, 77: 403–441.

Rudall, P. J. & Bateman, R. M. (2003). Evolutionary change in flowers and inflorescences: evidence from naturally occurring terata. *Trends in Plant Science*, 8: 76–82.

Rudall, P. J. & Bateman, R. M. (2004). Evolution of zygomorphy in monocot flowers: iterative patterns and developmental constraints. *New Phytologist*, 162: 25–44.

Rudall, P. J. & Bateman, R. M. (2006). Morphological phylogenetic analysis of Pandanales: testing contrasting hypotheses of floral evolution. *Systematic Botany*, 31: 223–238.

Rudall, P. J. & Bateman, R. M. (2010). Defining the limits of flowers: the challenge of distinguishing between the evolutionary products of simple versus compound strobili. *Philosophical Transactions of the Royal Society B*, 365: 397–409.

Rudall, P. J. & Buzgo, M. (2002). Evolutionary history of the monocot leaf. In *Developmental Genetics and Plant Evolution*, eds. Q. C. B. Cronk, R. M. Bateman & J. A. Hawkins. London: Taylor & Francis, pp. 431–458.

Rudall, P. J., Cunniff, J., Wilkin, P. & Caddick, L. R. (2005a). Evolution of dimery, pentamery and the monocarpellary condition in the monocot family Stemonaceae (Pandanales). *Taxon*, 54: 701–711.

Rudall, P. J., Remizowa, M. V., Prenner, G. *et al.* (2009). Non-flowers near the base of extant angiosperms? Spatiotemporal arrangement of organs in reproductive units of Hydatellaceae, and its bearing on the origin of the flower. *American Journal of Botany*, 96: 67–82.

Rudall, P. J., Sokoloff, D. D., Remizowa, M. V. *et al.* (2007). Morphology of Hydatellaceae, an anomalous aquatic family recently recognized as an early-divergent angiosperm lineage. *American Journal of Botany*, 94: 1073–1092.

Rudall, P. J., Stuppy, W., Cunniff, J., Kellogg, E. A. & Briggs, B. G. (2005b). Evolution of reproductive structures in grasses (Poaceae) inferred by sister-group comparison with their putative closest living relatives, Ecdeiocoleaceae. *American Journal of Botany*, 92: 1432–1443.

Ruiz de Clavijo, E. (1994). Heterocarpy and seed polymorphism in *Ceratocapnos heterocarpa* (Fumariaceae). *International Journal of Plant Sciences*, 155: 196–202.

Ruiz-Sanchez, E. & Sosa, V. (2015). Origin and evolution of fleshy fruit in woody bamboos. *Molecular Phylogenetics and Evolution*, 91: 123–134.

Ruskin, J. (1900). *Modern Painters*. New York, NY: The Kelmscott Society.

Rutishauser, R. (1981). Blattstellung und Sprossentwicklung bei Blütenpflanzen unter besonderer Berücksichtigung der Nelkengewächse (Caryophyllaceen s.l.). *Dissertationes Botanicae*, 62: 1–165.

Rutishauser, R. (1984). Blattquirle, Stipeln und Kolleteren bei den Rubieae (Rubiaceae) im Vergleich mit anderen Angiospermen. *Beiträge zur Biologie der Pflanzen*, 59: 375–424.

Rutishauser, R. (1995). Developmental patterns of leaves in Podostemonaceae as compared to more typical flowering plants: saltational evolution and fuzzy morphology. *Canadian Journal of Botany*, 73: 1305–1317.

Rutishauser, R. (1997). Structural and developmental diversity in Podostemaceae (river-weeds). *Aquatic Botany*, 57: 29–70.

Rutishauser, R. (1998). Plastochrone ratio and leaf arc as parameters of a quantitative phyllotaxis analysis in vascular plants. In *Symmetry in Plants*, eds. R.V. Jean & D. Barabé. Singapore: World Scientific, pp. 171–212.

Rutishauser, R. (1999). Polymerous leaf whorls in vascular plants: developmental morphology and fuzziness of organ identity. *International Journal of Plant Sciences*, 160: S81–S103.

Rutishauser, R. (2016a). Evolution of unusual morphologies in Lentibulariaceae (bladderworts and allies) and Podostemaceae (river-weeds): a pictorial report at the interface of developmental biology and morphological diversification. *Annals of Botany*, 117: 811–832.

Rutishauser, R. (2016b). *Acacia* (wattle) and *Cananga* (ylang-ylang): from spiral to whorled and irregular (chaotic) phyllotactic patterns – a pictorial report. *Acta Societatis Botanicorum Poloniae*, 85 (4): 3531.

Rutishauser, R., Grob, V. & Pfeifer, E. (2008). Plants are used to having identity crises. In *Evolving Pathways. Key Themes in Evolutionary Developmental*

Biology, eds. A. Minelli & G. Fusco. Cambridge: Cambridge University Press, pp. 194–213.

Rutishauser, R. & Grubert, M. (1999). The architecture of *Mourera fluviatilis* (Podostemaceae). Developmental morphology of inflorescences, flowers, and seedlings. *American Journal of Botany*, 86: 907–922.

Rutishauser, R. & Huber, K. A. (1991). The developmental morphology of *Indotristicha ramosissima* (Podostemaceae, Tristichoideae). *Plant Systematics and Evolution*, 178: 195–223.

Rutishauser, R. & Isler, B. (2001). Developmental genetics and morphological evolution of flowering plants, especially bladderworts (*Utricularia*): fuzzy Arberian morphology complements classical morphology. *Annals of Botany*, 88: 1173–1202.

Rutishauser, R. & Moline, P. (2005). Evo-devo and the search for homology ('sameness') in biological systems. *Theory in Biosciences*, 124: 213–241.

Rutishauser, R., Pfeifer, E., Moline, P. & Philbrick, C. T. (2003). Developmental morphology of roots and shoots of *Podostemum ceratophyllum* (Podostemaceae-Podostemoideae). *Rhodora*, 105: 337–353.

Rutishauser, R., Ronse De Craene, L. P., Smets, E. & Mendoza-Heuer, I. (1998). *Theligonum cynocrambe*: developmental morphology of a peculiar rubaceous herb. *Plant Systematics and Evolution*, 210: 1–24.

Rutishauser, R. & Sattler, R. (1986). Architecture and development of the phyllode–stipule whorls in *Acacia longipedunculata*: controversial interpretations and continuum approach. *Canadian Journal of Botany*, 64: 1987–2019.

Rutishauser, R. & Sattler, R. (1997). Expression of shoot processes in leaf development of *Polemonium caeruleum* as compared to other dicotyledons. *Botanische Jahrbücher für Systematik, Pflanzengeschichte und Pflanzengeographie*, 119: 563–582.

Saddic, L. A., Huvermann, B., Bezhani, S. *et al.* (2006). The LEAFY target LMI1 is a meristem identity regulator and acts together with LEAFY to regulate expression of *CAULIFLOWER*. *Development*, 133: 1673–1682.

Sajo, M. G., Mello-Silva, R. & Rudall, P. J. (2010). Homologies of floral structures in Velloziaceae, with particular reference to the corona. *International Journal of Plant Sciences*, 171: 595–606.

Sakai, H., Medrano, L. J. & Meyerowitz, E. M. (1995). Role of *SUPERMAN* in maintaining *Arabidopsis* floral whorl boundaries. *Nature*, 378: 199–203.

Sakakibara, K., Ando, S., Yip, H. K. *et al.* (2013). KNOX2 genes regulate the haploid-to-diploid morphological transition in land plants. *Science*, 339: 1067–1070.

Salomé, P. A., Bomblies, K., Laitinen, R. A. *et al.* (2011). Genetic architecture of flowering-time variation in *Arabidopsis thaliana*. *Genetics*, 188: 421–433.

Sander, K. (1983). The evolution of patterning mechanisms: gleanings from insect embryogenesis and spermatogenesis. In *Development and Evolution: the Sixth Symposium of the British Society for Developmental Biology*, eds. B. C. Goodwin, N. Holder & C. C. Wylie. Cambridge: Cambridge University Press, pp. 137–160.

Sarojam, R., Sappl, P. G., Goldshmidt, A. *et al.* (2010). Differentiating *Arabidopsis* shoots from leaves by combined YABBY activities. *Plant Cell*, 22: 2113–2130.

Sato, S., Nakamura, Y., Kaneko, T. *et al.* (2008). Genome structure of the legume, *Lotus japonicus*. *DNA Research*, 15: 227–239.

Sattler, R. (1972). Centrifugal primordial inception in floral development. *Advances in Plant Morphology*, 1972: 170–178.

Sattler, R. (1992). Process morphology: structural dynamics in development and evolution. *Canadian Journal of Botany*, 70: 708–714.

Sattler, R. (1994). Homology, homeosis, and process morphology in plants. In *Homology: The Hierarchical Basis of Comparative Biology*, ed. B. K. Hall. London: Academic Press, pp. 423–475.

Sattler, R. (1996). Classical morphology and continuum morphology: opposition and continuum. *Annals of Botany*, 78: 577–581.

Sattler, R. & Jeune, B. (1992). Multivariate analysis confirms the continuum view of plant form. *Annals of Botany*, 69: 249–262.

Sattler, R. & Rutishauser, R. (1990). Structural and dynamic descriptions of the development of *Utricularia foliosa* and *U. australis*. *Canadian Journal of Botany*, 68: 1989–2003.

Sattler, R. & Rutishauser, R. (1992). Partial homology of pinnate leaves and shoots: orientation of leaflet inception. *Botanische Jahrbücher für Systematik, Pflanzengeschichte und Pflanzengeographie*, 114: 61–79.

Sattler, R. & Rutishauser, R. (1997). The fundamental relevance of morphology and morphogenesis to plant research. *Annals of Botany*, 80: 571–582.

Saunders, R. M. K. (2010). Floral evolution in the Annonaceae: hypotheses of homeotic mutations and functional convergence. *Biological Reviews*, 85: 571–591.

Savage, A. J. P. & Ashton, P. S. (1983). The population structure of the double coconut and some other Seychelles palms. *Biotropica*, 15: 15–25.

Sawa, S., Ito, T., Shimura, Y. & Okada, K. (1999b). *FILAMENTOUS FLOWER* controls the formation and development of *Arabidopsis* inflorescences and floral meristems. *Plant Cell*, 11: 69–86.

Sawa, S., Watanabe, K., Goto, K. *et al.* (1999a). *FILAMENTOUS FLOWER*, a meristem and organ identity gene of *Arabidopsis*, encodes a protein with a zinc finger and HMG-related domains. *Genes and Development*, 13: 1079–1088.

Scanlon, M. J. (2000). Developmental complexities of simple leaves. *Current Opinion in Plant Biology*, 3: 31–36.

Scarpella, E., Barkoulas, M. & Tsiantis, M. (2010). Control of leaf and vein development by auxin. *Cold Spring Harbor Perspectives in Biology*, 2: a001511.

Scarpella, E., Marcos, D., Friml, J. & Berleth, T. (2006). Control of leaf vascular patterning by polar auxin transport. *Genes and Development*, 20: 1015–1027.

Schäferhoff, B., Fleischmann, A., Fischer, E. *et al.* (2010). Towards resolving Lamiales relationships: insights from rapidly evolving chloroplast sequences. *BMC Evolutionary Biology*, 10: 352.

Scheres, B., Wolkenfelt, H., Willemsen, V. *et al.* (1994). Embryonic origin of the *Arabidopsis* primary root and root meristem initials. *Development*, 120: 2475–2487.

Schlichting, C. D. & Murren, C. J. (2004). Evolvability and the raw materials for adaptation. In *Plant Adaptation: Molecular Genetics and Ecology*, eds. Q. C. B. Cronk, J. Whitton, R. H. Ree & I. E. P. Taylor. Ottawa: NRC Research Press, pp. 18–29.

Schlichting, C. D. & Pigliucci, M. (1998). *Phenotypic Evolution: A Reaction Norm Perspective*. Sunderland, MA: Sinauer Associates.

Schlichting, C. D. & Wund, M. A. (2014). Phenotypic plasticity and epigenetic marking: an assessment of evidence for genetic accommodation. *Evolution*, 68: 656–672.

Schmid, M., Davison, T. S., Henz, S. R. *et al.* (2005). A gene expression map of *Arabidopsis thaliana* development. *Nature Genetics*, 37: 501–506.

Schmuths, H., Meister, A., Horres, R. & Bachmann, K. (2004). Genome size variation among accessions of *Arabidopsis thaliana*. *Annals of Botany*, 93: 317–321.

Schneider, H., Pryer, K. M., Cranfill, R., Smith, A. R. & Wolf, P. G. (2002). Evolution of vascular plant body plans: a phylogenetic perspective. In *Developmental Genetics and Plant Evolution*, eds. Q. C. B. Cronk, R. M. Bateman & J. A. Hawkins. London: Taylor & Francis, pp. 1–14.

Schneitz, K. & Balasubramanian, S. (2009). Floral meristems. In *Encyclopedia of Life Sciences*. Chichester: Wiley.

Schoen, D. J., Johnston, M. O., L'Heureux, A. M. & Marsolais, J. V. (1997). Evolutionary history of the mating system in *Amsinckia* (Boraginaceae). *Evolution*, 51: 1090–1099.

Schönenberger, J., Anderberg, A. A. & Sytsma, K. J. (2005). Molecular phylogenetics and patterns of floral evolution in the Ericales. *International Journal of Plant Sciences*, 166: 265–288.

Schoof, H., Lenhard, M., Haeker, A. *et al.* (2000). The stem cell population of *Arabidopsis* shoot meristems is maintained by a regulatory loop between the *CLAVATA* and *WUSCHEL* genes. *Cell*, 100: 635–644.

Schranz, M. E., Quijada, P., Sung, S. B. *et al.* (2002). Characterization and effects of the replicated flowering time gene *FLC* in *Brassica rapa*. *Genetics*, 162: 1457–1468.

Schultz, E. A. & Haughn, G. W. (1993). Genetic analysis of the floral initiation process (FLIP) in *Arabidopsis*. *Development*, 119: 745–765.

Schumacher, K., Schmitt, T., Rossberg, M., Schmitz, C. & Theres, K. (1999). The *lateral suppressor* (*ls*) gene of tomato encodes a new member of the vhiid protein family. *Proceedings of the National Academy of Sciences of the United States of America*, 96: 290–295.

Schwander, T. & Leimar, O. (2011). Genes as leaders and followers in evolution. *Trends in Ecology and Evolution*, 26: 143–151.

Schwarz, S., Grande, A. V., Bujdoso, N., Saedler, H. & Huijser, P. (2008). The microRNA regulated SBP-box genes *SPL9* and *SPL15* control shoot maturation in *Arabidopsis*. *Plant Molecular Biology*, 67: 183–195.

Schwarz-Sommer, Z., Davies, B. & Hudson, A., (2003). An everlasting pioneer: the story of *Antirrhinum* research. *Nature Reviews Genetics*, 4: 657–666.

Schwarz-Sommer, Z., Huijser, P., Nacken, W., Saedler, H. & Sommer, H. (1990). Genetic control of flower development by homeotic genes in *Antirrhinum majus*. *Science*, 250: 931–936.

Scofield, S. & Murray, J. A. (2006). *KNOX* gene function in plant stem cell niches. *Plant Molecular Biology*, 60: 929–946.

Scribailo, R. W. & Tomlinson, P. B. (1992). Shoot and floral development in *Calla palustris* (Araceae-Calloideae). *International Journal of Plant Sciences*, 153: 1–13.

Searle, I., He, Y., Turck, F. *et al.* (2006). The transcription factor FLC confers a flowering response to vernalization by repressing meristem competence and systemic signaling in *Arabidopsis*. *Genes and Development*, 20: 898–912.

Seifriz, W. (1950). Gregarious flowering of *Chusquea*. *Nature*, 165: 635–636.

Semiarti, E., Ueno, Y., Tsukaya, H. *et al.* (2001). The *ASYMMETRIC LEAVES2* gene of *Arabidopsis thaliana* regulates formation of a symmetric lamina, establishment of venation and repression of meristem-related homeobox genes in leaves. *Development*, 128: 1771–1783.

Sentoku, N., Sato, Y., Kurata, N. *et al.* (1999). Regional expression of the rice *KN1*-type homeobox gene family during embryo, shoot, and flower development. *Plant Cell*, 11: 1651–1663.

Seymour, G. B., Østergaard, L., Chapman, N. H., Knapp, S. & Martin, C. (2013). Fruit development and ripening. *Annual Review of Plant Biology*, 64: 219–241.

Seymour, G. B., Ryder, C. D., Cevik, V. *et al.* (2011). A *SEPALLATA* gene is involved in the development and ripening of strawberry (*Fragaria × ananassa*

Duch.) fruit, a non-climacteric tissue. *Journal of Experimental Botany*, 62: 1179–1188.

Shah, J. J. & Dave, Y. S. (1970). Tendrils of *Passiflora foetida*: histogenesis and morphology. *American Journal of Botany*, 57: 786–793.

Shan, H., Su, K., Lu, W. *et al.* (2006). Conservation and divergence of candidate class B genes in *Akebia trifoliata* (Lardizabalaceae). *Development Genes and Evolution*, 216: 785–795.

Shannon, S. & Meeks-Wagner, D. R. (1991). A mutation in the *Arabidopsis Tfl1* gene affects inflorescence meristem development. *Plant Cell*, 3: 877–892.

Sharma, P. P., Clouse, R. M. & Wheeler, W. C. (2017). Hennig's semaphoront concept and the use of ontogenetic stages in phylogenetic reconstruction. *Cladistics*, 33: 93–108.

Sheffield, E. & Bell, P. R. (1987). Current studies of the pteridophyte life cycle. *Botanical Reviews*, 53: 442–490.

Sheldon, C. C., Burn, J. E., Perez, P. P. *et al.* (1999). The *FLF* MADS box gene: a repressor of flowering in *Arabidopsis* regulated by vernalization and methylation. *Plant Cell*, 11: 445–458.

Shepard, K. A. & Purugganan, M. D. (2002). The genetics of plant morphological evolution. *Current Opinion in Plant Biology*, 5: 49–55.

Shitsukawa, N., Ikari, C., Shimada, S. *et al.* (2007). The einkorn wheat (*Triticum monococcum*) mutant, maintained vegetative phase, is caused by a deletion in the *VRN1* gene. *Genes and Genetic Systems*, 82: 167–170.

Siddall, M. E. & Borda, E. (2003). Phylogeny and revision of the leech genus *Helobdella* (Glossiphoniidae) based on mitochondrial gene sequences and morphological data and a special consideration of the *triserialis* complex. *Zoologica Scripta*, 32: 23–33.

Sieber, P., Wellmer, F., Gheyselinck, J., Riechmann, J. L. & Meyerowitz, E. M. (2007). Redundancy and specialization among plant microRNAs: role of the *MIR164* family in developmental robustness. *Development*, 134: 1051–1060.

Siegfried, K. R., Eshed, Y., Baum, S. F. *et al.* (1999). Members of the *YABBY* gene family specify abaxial cell fate in *Arabidopsis*. *Development*, 126: 4117–4128.

Simon, R., Carpenter, R., Doyle, S. & Coen, E. (1994). *Fimbriata* controls flower development by mediating between meristem and organ identity genes. *Cell*, 78: 99–107.

Singer, S. D., Krogan, N. T. & Ashton, N. W. (2007). Clues about the ancestral roles of plant MADS-box genes from a functional analysis of moss homologues. *Plant Cell Reports*, 26: 1155–1169.

Sinha, N. (1999). Leaf development in angiosperms. *Annual Review of Plant Physiology and Plant Molecular Biology*, 50: 419–446.

Skippington, E., Barkman, T. J., Rice, D. W. & Palmer, J. D. (2015). Miniaturized mitogenome of the parasitic plant *Viscum scurruloideum* is extremely divergent and dynamic and has lost all *nad* genes. *Proceedings of the National Academy of Sciences of the United States of America*, 112: E3515–E3524.

Slotte, T., Hazzouri, K. M., Agren, J. A. *et al.* (2013). The *Capsella rubella* genome and the genomic consequences of rapid mating system evolution. *Nature Genetics*, 45: 831–835.

Smaczniak, C., Immink, R. G. H., Angenent, G. C. & Kaufmann, K. (2012). Developmental and evolutionary diversity of plant MADS-domain factors: insights from recent studies. *Development*, 139: 3081–3098.

Smith, J. F., Brown, K. D., Carroll, C. L. & Denton, D. S. (1997). Familial placement of *Cyrtandromoea*, *Titanotrichum* and *Sanango*, three problematic genera of the Lamiales. *Taxon*, 46: 65–74.

Smith, J. F., Hileman, L. C., Powell, M. P. & Baum, D. A. (2004). Evolution of GCYC, a Gesneriaceae homolog of *CYCLOIDEA*, within Gesnerioideae (Gesneriaceae). *Molecular Phylogenetics and Evolution*, 31: 765–779.

Smith, K. K. (2001). Heterochrony revisited: the evolution of developmental sequences. *Biological Journal of the Linnean Society*, 73: 169–186.

Smith, R.S., Guyomarc'h, S., Mandel, T. *et al.* (2006). A plausible model of phyllotaxis. *Proceedings of the National Academy of Sciences of the United States of America*, 103: 1301–1306.

Smith, Z. R. & Long, J. A. (2010). Control of *Arabidopsis* apical–basal embryo polarity by antagonistic transcription factors. *Nature*, 464: 423–426.

Smýkal, P., Aubert, G., Burstin, J. *et al.* (2012). Pea (*Pisum sativum* L.) in the genomic era. *Agronomy*, 2: 74–115.

Smyth, D. R., Bowman, J. L. & Meyerowitz, E. M. (1990). Early flower development in *Arabidopsis*. *Plant Cell*, 2: 755–768.

Sokoloff, D. D., Rudall, P. J. & Remizowa, M. (2006). Flower-like terminal structures in racemose inflorescences: a tool in morphogenetic and evolutionary research. *Journal of Experimental Botany*, 57: 3517–3530.

Sokoloff, D. D., Sokolski, A. A., Remizowa, M. V. & Nuraliev, M. S. (2007). Flower structure and development in *Tupidanthus calyptratus* (Araliaceae): an extreme case of polymery among asterids. *Plant Systematics and Evolution*, 268: 209–234.

Soltis, D. E. (2007). Saxifragaceae. In *The Families and Genera of Vascular Plants, Vol. 9*, ed. K. Kubitzki. Berlin: Springer, pp. 418–435.

Soltis, D. E., Albert, V. A., Leebens-Mack, J. *et al.* (2009). Polyploidy and angiosperm diversification. *American Journal of Botany*, 96: 336–348.

Soltis, D. E., Ma, H., Frohlich, M. W. *et al.* (2007). The floral genome: an evolutionary history of gene duplication and shifting patterns of gene expression. *Trends in Plant Science*, 12: 358–367.

Soltis, D. E., Senters, A. E., Zanis, M. J. *et al.* (2003). Gunnerales are sister to other core eudicots: implications for the evolution of pentamery. *American Journal of Botany*, 90: 461–470.

Soltis, D. E., Soltis, P. S., Endress, P. K. & Chase, M. W. (2005). *Phylogeny and Evolution of Angiosperms*. Sunderland, MA: Sinauer.

Soltis, P. S., Soltis, D. E., Kim, S., Chanderbali, A. & Buzgo, M. (2006). Expression of floral regulators in basal angiosperms and the origin and evolution of ABC function. In *Developmental Genetics of the Flower*, eds. D. E. Soltis, J. H. Leebens-Mack & P. S. Soltis. San Diego, CA: Elsevier, pp. 483–506.

Somerville, C. R. & Meyerowitz, E. M. (eds.) (2002–) *The* Arabidopsis *Book*. Rockville, MD: American Society of Plant Biologists.

Song, Y. H., Ito, S. & Imaizumi, T. (2013). Flowering time regulation: photoperiod- and temperature-sensing in leaves. *Trends in Plant Science*, 18: 575–583.

Souer, E., Rebocho, A. B., Bliek, M. *et al.* (2008). Patterning of inflorescences and flowers by the F-box protein *DOUBLE TOP* and the *LEAFY* homolog *ABERRANT LEAF AND FLOWER* in *Petunia*. *Plant Cell*, 20: 2033–2048.

Souer, E., van der Krol, A., Kloos, D. *et al.* (1998). Genetic control of branching pattern and floral identity during *Petunia* inflorescence development. *Development*, 125: 733–742.

Sousa-Baena, M. S., Lohmann, L. G., Rossi, M. & Sinha, N. R. (2014a). Acquisition and diversification of tendrilled leaves in Bignonieae (Bignoniaceae) involved changes in expression patterns of SHOOTMERISTEMLESS (STM), LEAFY/FLORICAULA (LFY/FLO), and PHANTASTICA (PHAN). *New Phytologist*, 201: 993–1008.

Sousa-Baena, M. S., Sinha, N. R. & Lohmann, L. G. (2014b). Evolution and development of tendrils in Bignonieae (Lamiales, Bignoniaceae). *Annals of the Missouri Botanical Garden*, 99: 323–347.

Specht, C. D. & Bartlett, M. E. (2009). Flower evolution: the origin and subsequent diversification of the angiosperm flower. *Annual Reviews in Ecology, Evolution and Systematics*, 40: 217–243.

Specht, C. D., Yockteng, R., Almeida, A. M., Kirchoff, B. K. & Kress, W. J. (2012). Homoplasy, pollination, and emerging complexity during the evolution of floral development in the tropical gingers (Zingiberales). *Botanical Review*, 78: 440–462.

Stahle, M. I., Kuehlich, J., Staron, L., von Arnim, A. G. & Golz, J. F. (2009). YABBYs and the transcriptional corepressors LEUNIG and LEUNIG_HOMOLOG maintain leaf polarity and meristem activity in *Arabidopsis*. *Plant Cell*, 21: 3105–3118.

Stebbins, G. L. (1974). *Flowering Plants. Evolution Above the Species Level.* Cambridge, MA: Belknap Press.

Steeves, T. A. & Sussex, I. M. (1989). *Patterns in Plant Development,* 2nd edn. Cambridge: Cambridge University Press.

Steingraeber, D. A. & Fisher, J. B. (1986). Indeterminate growth of leaves in *Guarea* (Meliaceae): a twig analogue. *American Journal of Botany,* 73: 852–862.

Stevens, P. F. (1975). Review of *Chisocheton* (Meliaceae) in Papuasia. *Contributions from Herbarium Australiense,* 11: 1–55.

Strable, J. & Scanlon, M. J. (2009). Maize (*Zea mays*): a model organism for basic and applied research in plant biology. *Cold Spring Harbor Protocols,* 2009 (10): pdb. emo132.

Stuessy, T. F. & Urtubey, E. (2006). Phylogenetic implications of corolla morphology in subfamily Barnadesioideae (Asteraceae). *Flora,* 201: 340–352.

Suárez-Baron, H., Pérez-Mesa, P., Ambrose, B. A., González, F. & Pabón-Mora, N. (2017). Deep into the aristolochia flower: expression of C, D, and E-class genes in *Aristolochia fimbriata* (Aristolochiaceae). *Journal of Experimental Zoology (Molecular and Developmental Evolution),* 328B: 55–71.

Sulman, J. D., Drew, B. T., Drummond, C., Hayasaka, E. & Sytsma, K. J. (2013). Systematics, biogeography, and character evolution of *Sparganium* (Typhaceae): diversification of a widespread, aquatic lineage. *American Journal of Botany,* 100: 2023–2039.

Sun, G., Dilcher, D. L., Zheng, S. & Zhou, Z. (1998). In search of the first flower: a Jurassic angiosperm, *Archaefructus,* from northeast China. *Science,* 282: 1692–1695.

Sun, G., Ji, Q., Dilcher, D. L. *et al.* (2002). Archaefructaceae, a new basal angiosperm family. *Science,* 296: 899–904.

Szymkowiak, E. J. & Sussex, I. M. (1996). What chimeras can tell us about plant development. *Annual Review of Plant Physiology and Plant Molecular Biology,* 47: 351–376.

Tadege, M., Sheldon, C. C., Helliwell, C. A. *et al.* (2001). Control of flowering time by *FLC* orthologues in *Brassica* napus. *The Plant Journal,* 28: 545–553.

Takada, S., Hibara, K., Ishida, T. & Tasaka, M. (2001). The *CUP-SHAPED COTYLEDON1* gene of *Arabidopsis* regulates shoot apical meristem formation. *Development,* 128: 1127–1135.

Takhtajan, A. (1969). *Flowering Plants: Origin and Dispersal.* Edinburgh: Oliver & Boyd.

Takhtajan, A. (1991). *Evolutionary Trends in Flowering Plants.* New York, NY: Columbia University Press.

Tamura, M. (1995). Ranunculaceae. In *Die natürlichen Pflanzenfamilien, 17a, Part IV*, eds. A. Engler & K. Prantl. Berlin: Ducker & Humblot.

Tanabe, Y., Hasebe, M., Sekimoto, H. *et al.* (2005). Characterization of MADS-box genes in charophycean green algae and its implication for the evolution of MADS-box genes. *Proceedings of the National Academy of Sciences of the United States of America*, 102: 2436–2441.

Tang, G., Reinhart, B. J., Bartel, D. P. & Zamore, P. D. (2003). A biochemical framework for RNA silencing in plants. *Genes and Development*, 17: 49–63.

Tank, D. C. & Olmstead, R. G. (2008). From annuals to perennials: phylogeny of subtribe Castillejinae (Orobanchaceae). *American Journal of Botany*, 95: 608–625.

Tattersall, A. D., Turner, L., Knox, M. R. *et al.* (2005). The mutant *crispa* reveals multiple roles for *PHANTASTICA* in pea compound leaf development. *Plant Cell*, 17: 1046–1060.

Taylor, D. W. & Hickey, L. J. (1996). Evidence for and implications of an herbaceous origin of angiosperms. In *Flowering Plant Origin, Evolution and Phylogeny*, eds. D. W. Taylor & L. J. Hickey. New York, NY: Chapman & Hall, pp. 232–266.

Taylor, P. (1989). *The Genus* Utricularia: *A Taxonomic Monograph*. London: HMSO.

Taylor, S., Hofer, J. & Murfet, I. (2001). *Stamina pistilloida*, the pea ortholog of *Fim* and *UFO*, is required for normal development of flowers, inflorescences, and leaves. *Plant Cell*, 13: 31–46.

Taylor, S. A., Hofer, J. M., Murfet, I. C. *et al.* (2002). *PROLIFERATING INFLORESCENCE MERISTEM*, a MADS-box gene that regulates floral meristem identity in pea. *Plant Physiology*, 129: 1150–1159.

Teeri, T. H., Elomaa, P., Kotilainen, M. & Albert, V. A. (2006a). Mining plant diversity: *Gerbera* as a model system for plant developmental and biosynthetic research. *BioEssays*, 28: 756–767.

Teeri, T. H., Uimari, A., Kotilainen, M. *et al.* (2006b). Reproductive meristem fates in Gerbera. *Journal of Experimental Botany*, 57: 3445–3455.

Telfer, A. & Poethig, R. S. (1998) *HASTY*, a gene that regulates the timing of shoot maturation in *Arabidopsis thaliana*. *Development*, 125: 1889–1898.

Terpstra, I. & Heidstra, R. (2009). Stem cells: the root of all cells. *Seminars in Cell and Developmental Biology*, 20: 1089–1096.

Theißen, G. (2000). Evolutionary developmental genetics of floral symmetry: the revealing power of Linnaeus' monstrous flower. *Bioessays*, 22: 209–213.

Theißen, G. (2006). The proper place of hopeful monsters in evolutionary biology. *Theory in Biosciences*, 124: 349–369.

Theißen, G. (2009). Saltational evolution: hopeful monsters are here to stay. *Theory in Biosciences*, 128: 43–51.

Theißen, G., Becker, A., Winter, K.-U. *et al.* (2002). How the land plants learned their floral ABCs: the role of MADS box genes in the evolutionary origin of flowers. In *Developmental Genetics and Plant Evolution*, eds. Q. C. B. Cronk, R. M. Bateman & J. A. Hawkins. London: Taylor & Francis, pp. 173–206.

Theißen, G., Kim, J. T. & Saedler, H. (1996). Classification and phylogeny of the MADS-box multigene family suggest defined roles of MADS-box gene subfamilies in the morphological evolution of eukaryotes. *Journal of Molecular Evolution*, 43: 484–516.

Theissen, G. & Melzer, R. (2007). Molecular mechanisms underlying origin and diversification of the angiosperm flower. *Annals of Botany*, 100: 603–619.

Theißen, G. & Melzer, R. (2016). Robust views on plasticity and biodiversity. *Annals of Botany*, 117: 693–697.

Theißen, G. & Saedler, H. (2001). Floral quartets. *Nature*, 409: 469–471.

Thien, L. B. (1980). Patterns of pollination in the primitive angiosperms. *Biotropica*, 12: 1–13.

Thomas, H. (2003). Do green plants age, and if so, how? *Topics in Current Genetics*, 3: 145–171.

Thomas, M. M., Rudall, P. J., Ellis, A. G., Savolainen, V. & Glover, B. J. (2009). Development of a complex floral trait: the pollinator-attracting petal spots of the beetle daisy, *Gorteria diffusa* (Asteraceae). *American Journal of Botany*, 96: 2184–2196.

Thompson, B. E., Bartling, L., Whipple, C. *et al.* (2009). *bearded-ear* encodes a MADS box transcription factor critical for maize floral development. *Plant Cell*, 21: 2578–2590.

Thorpe, T. A. (2007). History of plant tissue culture. *Molecular Biotechnology*, 37: 169–180.

Timmermans, M. C. P., Hudson, A., Becraft, P. W. & Nelson, T. (1999). Rough sheath2: a Myb protein that represses *knox* homeobox genes in maize lateral organ primordia. *Science*, 284: 151–153.

Tomescu, A. M. F. (2009). Megaphylls, microphylls and the evolution of leaf development. *Trends in Plant Science*, 14: 5–12.

Tomlinson, P. B. (1990). *The Structural Biology of Palms*. Oxford: Clarendon Press.

Tomlinson, P. B. & Huggett, B. A. (2012). Cell longevity and sustained primary growth in palm stems. *American Journal of Botany*, 99: 1891–1902.

Tooke, F., Ordidge, M., Chiurugwi, T. & Battey, N. (2005). Mechanisms and function of flower and inflorescence reversion. *Journal of Experimental Botany*, 56: 2587–2599.

Toriba, T., Harada, K., Takamura, A. *et al.* (2007). Molecular characterization the *YABBY* gene family in *Oryza sativa* and expression analysis of *OsYABBY1*. *Molecular Genetics and Genomics*, 277: 457–468.

Townsley, B. T. & Sinha, N. R. (2012). A new development: evolving concepts in leaf ontogeny. *Annual Review of Plant Biology*, 63: 535–562.

Trevaskis, B., Bagnall, D. J., Ellis, M. H., Peacock, W. J. & Dennis, E. S. (2003). MADS box genes control vernalization-induced flowering in cereals. *Proceedings of the National Academy of Sciences of the United States of America*, 100: 13099–13104.

Tröbner, W., Ramirez, L., Motte, P. *et al.* (1992). *GLOBOSA*: a homeotic gene which interacts with *DEFICIENS* in the control of *Antirrhinum* floral organogenesis. *EMBO Journal*, 11: 4693–4704.

Troll, W. (1937, 1939, 1943). *Vergleichende Morphologie der höheren Pflanzen*. Berlin: Borntraeger.

Troll, W. (1964, 1969). *Die Infloreszenzen*. Jena: Fischer.

True, J. R. & Haag, E. S. (2001). Developmental system drift and flexibility in evolutionary trajectories. *Evolution and Development*, 3: 109–119.

Tsai, W. C., Chuang, M. H., Kuoh, C. S., Chen, W. H. & Chen, H. H. (2004). Four *DEF*-like MADS box genes displayed distinct floral morphogenetic roles in *Phalaenopsis* orchid. *Plant and Cell Physiology*, 45: 831–844.

Tsai, W. C., Pan, Z. J., Hsiao, Y. Y., Chen, L. J. & Liu, Z. J. (2014). Evolution and function of MADS-box genes involved in orchid floral development. *Journal of Systematics and Evolution*, 52: 397–410.

Tsiantis, M., Brown, M. I. N., Skibinski, G. & Langdale, J. A. (1999). Disruption of auxin transport is associated with aberrant leaf development in maize. *Plant Physiology*, 121: 1163–1168.

Tsuda, K. & Katagiri, F. (2010). Comparing signaling mechanisms engaged in pattern-triggered and effector-triggered immunity. *Current Opinion in Plant Biology*, 13: 459–465.

Tsuda, K., Sato, M., Stoddard, T., Glazebrook, J. & Katagiri, F. (2009). Network properties of robust immunity in plants. *PLoS Genetics*, 5: e1000772.

Tsukaya, H. (1995). Developmental genetics of leaf morphogenesis in dicotyledonous plants. *Journal of Plant Research*, 108: 407–416.

Tsukaya, H. (1997). Determination of the unequal fate of cotyledons of a one-leaf plant, *Monophyllaea*. *Development*, 124: 1275–1280.

Tsukaya, H. (2000). The role of meristematic activities in the formation of leaf blades. *Journal of Plant Research*, 113: 119–126.

Tsukaya, H. (2002). Interpretation of mutants in leaf morphology: genetic evidence for a compensatory system in leaf morphogenesis that provides a new link between cell and organismal theory. *International Review of Cytology*, 217: 1–39

Tsukaya, H. (2003). Organ shape and size: a lesson from studies of leaf morphogenesis. *Current Opinion in Plant Biology*, 6: 57–62.

Tsukaya, H. (2006). Mechanism of leaf-shape determination. *Annual Review of Plant Biology*, 57: 477–496.

Tsukaya, H. (2008). Controlling size in multicellular organs: focus on the leaf. *PLoS Biology*, 6: 1373–1376.

Tsukaya, H. (2013). Leaf development. In *The Arabidopsis Book*. Rockville, MD: American Society of Plant Biologists, 11: e0163.

Tsukaya, H. (2014). Comparative leaf development in angiosperms. *Current Opinion in Plant Biology*, 17: 103–109.

Tsukaya, H., Inaba-Higano, K. & Komeda, Y. (1995). Phenotypic and molecular mapping of an *acaulis2* mutant of *Arabidopsis thaliana* with flower stalks of much reduced length. *Plant Cell Physiology*, 36: 239–246.

Tucker, S. C. (1984a). Origin of symmetry in flowers. In *Contemporary Problems in Plant Anatomy*, eds. R. A. White & W. C. Dickison. New York, NY: Academic Press, pp. 351–395.

Tucker, S. C. (1984b). Unidirectional organ initiation in leguminous flowers. *American Journal of Botany*, 71: 1139–1148.

Tucker, S. C. (1987). Floral initiation and development in legumes. In *Advances in Legume Systematics, 3*, ed. C. H. Stirton. Kew: Royal Botanic Gardens, pp. 183–239.

Tucker, S. C. (1991). Helical floral organogenesis in *Gleditsia*, a primitive caesalpinioid legume. *American Journal of Botany*, 78: 1130–1149.

Tucker, S. C. (1996). Trends in evolution of floral ontogeny in *Cassia* sensu stricto, *Senna*. and *Chamaecrista* (Leguminosae: Caesalpinoideae: Cassieae: Cassiinae): a study in convergence. *American Journal of Botany*, 83: 687–711.

Tucker, S. C. (1999). Evolutionary lability of symmetry in early floral development. *International Journal of Plant Sciences*, 160: S25–S39.

Tucker, S. C. (2000). Floral development in tribe Detarieae (Leguminosae: Caesalpinioideae): *Amherstia, Brownea,* and *Tamarindus. American Journal of Botany*, 87: 1385–1407.

Tucker, S. C. (2001). Floral development in *Schotia* and *Cynometra* (Leguminosae: Caesalpinioideae: Detarieae). *American Journal of Botany*, 88: 1164–1180.

Tucker, S. C. (2002). Floral ontogeny in Sophoreae (Leguminosae: Papilionoideae). III. *Cadia purpurea* with radial symmetry and random petal aestivation. *American Journal of Botany*, 89: 748–757.

Tucker, S. C. (2003). Floral development in legumes. *Plant Physiology*, 131: 911–926.

Tucker, S. C. & Douglas, A. W. (1996). Floral structure, development, and relationships of paleoherbs: *Saruma*, *Cabomba*, *Lactoris*, and selected Piperales. In *Flowering Plant Origin, Evolution, and Phylogeny*, eds. D. W. Taylor & L. J. Hickey. New York, NY: Chapman & Hall, pp. 141–175.

Tuskan, G. A., Difazio, S., Jansson, S. *et al.* (2006). The genome of black cottonwood, *Populus trichocarpa* (Torr. & Gray). *Science*, 313: 1596–1604.

Tzeng, T.-Y., Chen, H.-Y. & Yang, C.-H. (2002). Ectopic expression of carpel-specific MADS box genes from lily and *Lisianthus* causes similar homeotic conversion of sepal and petal in *Arabidopsis*. *Plant Physiology*, 130: 1827–1836.

Tzeng, T.-Y. & Yang, C.-H. (2001). A MADS box gene from lily (*Lilium longiflorum*) is sufficient to generate dominant negative mutation by interacting with *PISTILLATA* (PI) in *Arabidopsis thaliana*. *Plant Cell Physiology*, 42: 1156–1168.

Überlacker, B., Klinge, B. & Werr, W. (1996). Ectopic expression of the maize homeobox genes *ZmHox1a* or *ZmHox1b* causes pleiotropic alterations in the vegetative and floral development of transgenic tobacco. *Plant Cell*, 8: 349–362.

Uhl, N. W. & Dransfield, J. (1987). *Genera Palmarum*. Lawrence, KS: Allen Press.

Uimari, A., Kotilainen, M., Elomaa, P. *et al.* (2004). Integration of reproductive meristem fates by a *SEPALLATA*-like MADS box gene. *Proceedings of the National Academy of Sciences of the United States of America*, 101: 15817–15822.

Uittien, H. (1928). Uber den Zusammenhang zwischen Blattnervatur und Sprossverzweigung. *Recueil des Travaux Botaniques Neerlandais*, 25: 390–412.

Uller, T. & Helanterä, H. (2011). When are genes 'leaders' or 'followers' in evolution? *Trends in Ecology and Evolution*, 26: 435–436.

Usami, T., Horiguchi, G., Yano, S. & Tsukaya, H. (2009). The more and smaller cells mutants of *Arabidopsis thaliana* identify novel roles for *SQUAMOSA PROMOTER BINDING PROTEIN-LIKE* genes in the control of heteroblasty. *Development*, 136: 955–964.

Uyttewaal, M., Burian, A., Alim, K. *et al.* (2012). Mechanical stress acts via katanin to amplify differences in growth rate between adjacent cells in *Arabidopsis*. *Cell*, 149: 439–451.

Vallejo-Marín, M., Manson, J. S., Thomson, J. D. & Barrett, S. C. H. (2009). Division of labour within flowers: heteranthery, a floral strategy to reconcile contrasting pollen fates. *Journal of Evolutionary Biology*, 22: 828–839.

Vallius, E. (2000). Position-dependent reproductive success of flowers in *Dactylorhiza maculata* (Orchidaceae). *Functional Ecology*, 14: 573–579.

Van de Peer, Y., Fawcett, J. A., Proost, S., Sterck, L. & Vandepoele, K. (2009). The flowering world: a tale of duplications. *Trends in Plant Science*, 14: 680–688.

van der Graaff, E., Dulk-Ras, A. D., Hooykaas, P. J. J. & Keller, B. (2000). Activation tagging of the *LEAFY PETIOLE* gene affects leaf petiole development in *Arabidopsis thaliana*. *Development*, 127: 4971–4980.

van der Maesen, L. J. G. (1970). Primitiae Africanae VIII. A revision of the genus *Cadia* Forskål (Caes.) and some remarks regarding *Dicraeopetalum* Harms (Pap.) and *Platycelyphium* (Harms) (Pap.). *Acta Botanica Neerlandica*, 19: 227–248.

van Doorn, W. G. & Stead, A. D. (1997). Abscission of flowers and floral parts. *Journal of Experimental Botany*, 48: 821–837.

Van Dyken, J. & Wade, M. J. (2010). The genetic signature of conditional expression. *Genetics*, 84: 557–570.

Vandenbussche, M., Horstman, A., Zethof, J. *et al.* (2009). Differential recruitment of WOX transcription factors for lateral development and organ fusion in *Petunia* and *Arabidopsis*. *Plant Cell*, 21: 2269–2283.

Vandenbussche, M., Zethof, J., Royaert, S., Weterings, K. & Gerats, T. (2004). The duplicated B-class heterodimer model: whorl-specific effects and complex genetic interactions in *Petunia hybrida* flower development. *Plant Cell*, 16: 741–754.

Vandenbussche, M., Zethof, J., Souer, E. *et al.* (2003). Toward the analysis of the petunia MADS box gene family by reverse and forward transposon insertion mutagenesis approaches: B, C, and D floral organ identity functions require *SEPALLATA*-like MADS box genes in petunia. *Plant Cell*, 15: 2680–2693.

Vazquez, F. (2009). Small RNAs in plants. In *Encyclopedia of Life Sciences*. Chichester: Wiley.

Vázquez-Lobo, A., Carlsbecker, A., Vergara-Silva, F. *et al.* (2007). Characterization of the expression patterns of LEAFY/FLORICAULA and NEEDLY orthologs in female and male cones of the conifer genera *Picea*, *Podocarpus*, and *Taxus*: implications for current evo-devo hypotheses for gymnosperms. *Evolution and Development*, 9: 446–459.

Vekemans, D., Proost S., Vanneste K. *et al.* (2012a). Gamma paleohexaploidy in the stem lineage of core eudicots: significance for MADS-box gene and species diversification. *Molecular Biology and Evolution*, 29: 3793–3806.

Vekemans, D., Viaene, T., Caris, P. & Geuten, K. (2012b). Transference of function shapes organ identity in the dove tree inflorescence. *New Phytologist*, 193: 216–228.

Velhagen, W. A. (1997). Analyzing developmental sequences using sequence units. *Systematic Biology*, 46: 204–210.

Viaene, T., Vekemans, D., Irish, V. F. et al. (2009). *Pistillata*: duplications as a mode for floral diversification in (basal) asterids. *Molecular Biology and Evolution*, 26: 2627–2645.

Vialette-Guiraud, A. C. M., Adam, H., Finet, C. et al. (2011). Insights from ANA-grade angiosperms into the early evolution of *CUP-SHAPED COTYLEDON* genes. *Annals of Botany*, 107: 1511–1519.

Vialette-Guiraud, A. C. M. & Scutt, C. P. (2009). Carpel evolution. *Annual Plant Reviews*, 38: 1–34.

Vijayraghavan, U., Prasad, K. & Meyerowitz, E. (2005). Specification and maintenance of the floral meristem: interactions between positively acting promoters of flowering and negative regulators. *Current Science*, 89: 1835–1843.

Vlad, D., Kierzkowski, D., Rast, M. I. et al. (2014). Leaf shape evolution through duplication, regulatory diversification, and loss of a homeobox gene. *Science*, 343: 780–783.

Vollbrecht, E., Veit, B., Sinha, N. & Hake, S. (1991). The developmental gene *Knotted-1* is a member of a maize homeobox gene family. *Nature*, 350: 241–243.

von Baer, K. E. (1828). *Über Entwicklungsgeschichte der Thiere: Beobachtung und Reflexion, Vol. 1*. Königsberg: Bornträger.

von Balthazar, M. & Endress, P. K. (2002). Development of inflorescences and flowers in Buxaceae and the problem of perianth interpretation. *International Journal of Plant Sciences*, 163: 847–876.

von Hagen, K. B. & Kadereit, J. W. (2002). Phylogeny and flower evolution of the Swertiinae (Gentianaceae-Gentianeae): homoplasy and the principle of variable proportions. *Systematic Botany*, 27: 548–572.

von Wangenheim, D., Fangerau, J., Schmitz, A. et al. (2016). Rules and self-organizing properties of post-embryonic plant organ cell division patterns. *Current Biology*, 26: 439–449.

Vrebalov, J., Pan, I. L., Arroyo, A. J. et al. (2009). Fleshy fruit expansion and ripening are regulated by the tomato *SHATTERPROOF* gene *TAGL1*. *Plant Cell*, 21: 3041–3062.

Vrebalov, J., Ruezinsky, D., Padmanabhan, V. et al. (2002). A MADS-box gene necessary for fruit ripening at the tomato *ripeninginhibitor* (*rin*) locus. *Science*, 296: 343–346.

Vroemen, C. W., Mordhorst, A. P., Albrecht, C., Kwaaitaal, M. A. & de Vries, S. C. (2003). The *CUP-SHAPED COTYLEDON3* gene is required for boundary and shoot meristem formation in *Arabidopsis*. *Plant Cell*, 15: 1563–1577.

Waddington, C. H. (1953). Genetic assimilation of an acquired character. *Evolution*, 7: 118–126.

Wagner, A. (2005). *Robustness and Evolvability in Living Systems*. Princeton, NJ: Princeton University Press.

Wagner, A. (2011). *The Origins of Evolutionary Innovations: A Theory of Transformative Change in Living Systems*. Oxford: Oxford University Press.

Wagner, A. (2014). *Arrival of the Fittest: Solving Evolution's Greatest Puzzle*. New York, NY: Penguin.

Wagner, G. P. (1996). Homologues, natural kinds and the evolution of modularity. *American Zoologist*, 36: 36–43.

Wagner, G. P. & Altenberg, L. (1996). Complex adaptations and evolution of evolvability. *Evolution*, 50: 967–976.

Wagner, G. P., Pavlicev, M. & Cheverud, J. M. (2007). The road to modularity. *Nature Reviews Genetics*, 8: 921–931.

Waites, R. & Hudson, A. (1995). *phantastica*: a gene required for dorsoventrality of leaves in *Antirrhinum majus*. *Development*, 121: 2143–2154.

Waites, R. & Hudson, A. (2001). The *Handlebars* gene is required with *Phantastica* for dorsoventral asymmetry of organs and for stem cell activity in *Antirrhinum*. *Development*, 128: 1923–1931.

Waites, R., Selvadurai, H. R. N., Oliver, I. R. & Hudson, A. (1998). The *Phantastica* gene encodes a MYB transcription factor involved in growth and dorsoventrality of lateral organs in *Antirrhinum*. *Cell*, 93: 779–789.

Wake, D. (2003). Homology and homoplasy. In *Keywords and Concepts in Evolutionary Developmental Biology*, eds. K. Hall & W. M. Olson. Cambridge, MA: Harvard University Press, pp. 191–201.

Walbot, V. (1996). Sources and consequences of phenotypic and genotypic plasticity in flowering plants. *Trends in Plant Science*, 1: 27–32.

Walker, J. W. & Walker, A. G. (1984). Ultrastructure of lower Cretaceous angiosperm pollen and the origin and early evolution of flowering plants. *Annals of the Missouri Botanical Garden*, 71: 464–521.

Walker-Larsen, J. & Harder, L. D. (2000). The evolution of staminodes in angiosperms: patterns of stamen reduction, loss, and functional reinvention. *American Journal of Botany*, 87: 1367–1384.

Walsh, B. & Blows, M. W. (2009). Abundant genetic variation + strong selection = multivariate genetic constraints: a geometric view of adaptation. *Annual Review of Ecology, Evolution and Systematics*, 40: 41–59.

Wang, J. W., Czech, B. & Weigel, D. (2009a). miR156-regulated SPL transcription factors define an endogenous flowering pathway in *Arabidopsis thaliana*. *Cell*, 138: 738–749.

Wang, J. W., Park, M. Y., Wang, L. J. et al. (2011). MiRNA control of vegetative phase change in trees. *PLoS Genetics*, 7: e1002012.

Wang, R., Farrona, S., Vincent, C. *et al.* (2009b). *PEP1* regulates perennial flowering in *Arabis alpina. Nature,* 459: 423–428.

Wang, R. L., Stec, A., Hey, J., Lukens, L. & Doebley, J. (1999). The limits of selection during maize domestication. *Nature,* 398: 236–239.

Wang, Z., Luo, Y., Li, X. *et al.* (2008). Genetic control of floral zygomorphy in pea (*Pisum sativum* L.). *Proceedings of the National Academy of Sciences of the United States of America,* 105: 10414–10419.

Wanntorp, L. & Ronse De Craene, L. P. (2005). The *Gunnera* flower: key to eudicot diversification or response to pollination mode? *International Journal of Plant Sciences,* 166: 945–953.

Wardlaw, C. W. (1955). *Embryogenesis in Plants.* London: Methuen.

Warner, K. A., Rudall, P. J. & Frohlich, M. W. (2008). Differentiation of perianth organs in Nymphaeales. *Taxon,* 57: 1096–1109.

Warner, K. A., Rudall, P. J. & Frohlich, M. W. (2009). Environmental control of sepalness and petalness in perianth organs of waterlilies: a new Mosaic Theory on the evolutionary origin of a differentiated perianth. *Journal of Experimental Botany,* 60: 3559–3574.

Washburn, J. D., Bird, K. A., Conant, G. C. & Pires, J. C. (2016). Convergent evolution and the origin of complex phenotypes in the age of systems biology. *International Journal of Plant Sciences,* 177: 305–318.

Watanabe, K. & Okada, K. (2003). Two discrete cis elements control the abaxial side-specific expression of the *FILAMENTOUS FLOWER* gene in *Arabidopsis. Plant Cell,* 15: 2592–2602.

Weber, A. (2003). What is morphology and why is it time for its renaissance in plant systematics? In *Deep Morphology: Towards a Renaissance of Morphology in Plant Systematics,* eds. T. F. Stuessy, V. Mayer & E. Hörandl. Ruggell: Gantner, pp. 3–32.

Weber, A., Clark, J. L. & Möller, M. (2013). A new formal classification of Gesneriaceae. *Selbyana,* 31: 68–94.

Weberling, F. (1989). *Morphology of Flowers and Inflorescences.* Cambridge: Cambridge University Press.

Webster, M. A. & Gilmartin, P. M. (2006). Analysis of late stage flower development in *Primula vulgaris* reveals novel differences in cell morphology and temporal aspects of floral heteromorphy. *New Phytologist,* 171: 591–603.

Wei, L., Wang, Y.-Z. & Li, Z.-Y. (2011). Floral ontogeny of Ruteae (Rutaceae) and its systematic implications. *Plant Biology,* 14: 190–197.

Weigel, D. (2012). Natural variation in *Arabidopsis*: from molecular genetics to ecological genomics. *Plant Physiology,* 158: 2–22.

Weigel, D., Alvarez, J., Smyth, D. R., Yanofsky, M. F. & Meyerowitz, E. M. (1992). *LEAFY* controls floral meristem identity in *Arabidopsis. Cell,* 69: 843–859.

Weiss, M. R. (1995). Floral color change: a widespread functional convergence. *American Journal of Botany*, 82: 167–185.

Wendel, J. F. (2015). The wondrous cycles of polyploidy in plants. *American Journal of Botany*, 102: 1753–1756.

West-Eberhard, M. J. (2003). *Developmental Plasticity and Evolution*. New York, NY: Oxford University Press.

Westerkamp, C. & Weber, A. (1999). Keel flowers of the Polygalaceae and Fabaceae: a functional comparison. *Botanical Journal of the Linnean Society*, 129: 207–221.

Weston, P. H. (2000). Process morphology from a cladistic perspective. In *Homology and Systematics*, eds. R. Scotland & R. T. Pennington. London: Taylor & Francis, pp. 124–144.

Westwood, J. H., Yoder, J. I., Timko, M. P. & dePamphilis, C. W. (2010). The evolution of parasitism in plants. *Trends in Plant Science*, 15: 227–235.

Whipple, C. J., Ciceri, P., Padilla, C. M. *et al.* (2004). Conservation of B-class floral homeotic gene function between maize and *Arabidopsis*. *Development*, 131: 6083–6091.

Whipple, C. J., Zanis, M. J., Kellogg, E. A. & Schmidt, R. J. (2007). Conservation of B-class gene expression in the second whorl of a basal grass and outgroups links the origin of lodicules and petals. *Proceedings of the National Academy of Sciences of the United States of America*, 104: 1081–1086.

White, D. W. (2006). *PEAPOD* regulates lamina size and curvature in *Arabidopsis*. *Proceedings of the National Academy of Sciences of the United States of America*, 103: 13238–13243.

Wigge, P. A., Kim, M. C., Jaeger, K. E. *et al.* (2005). Integration of spatial and temporal information during floral induction in *Arabidopsis*. *Science*, 309: 1056–1059.

Wiley, E. O. (1981). *Phylogenetics: The Theory and Practice of Phylogenetic Systematics*. New York, NY: Wiley.

Wilkinson, M., de Andrade Silva, E., Zachgo, S., Saedler, H. & Schwarz-Sommer, Z. (2000). *CHORIPETALA* and *DESPENTEADO*: general regulators during plant development and potential floral targets of FIMBRIATA-mediated degradation. *Development*, 127: 3725–3734.

Williams, D. M. & Ebach, M. C. (2007). Heterology: the shadows of a shade. *Cladistics*, 23: 64–83.

Williston, S.W. (1914). *Water Reptiles of the Past and Present*. Chicago, IL: University of Chicago Press.

Willmann, M. R. & Poethig, R. S. (2007). Conservation and evolution of miRNA regulatory programs in plant development. *Current Opinion in Plant Biology*, 10: 503–511.

Wiltshire, R. J. E., Murfet, I. C. & Reid, J. B. (1994). The genetic control of heterochrony: evidence from developmental mutants of *Pisum sativum* L. *Journal of Evolutionary Biology*, 7: 447–465.

Wiltshire, R. J. E., Potts, B. M. & Reid, J. B. (1998). Genetic control of reproductive and vegetative phase change in the *Eucalyptus risdonii–E. tenuiramis* complex. *Australian Journal of Botany*, 46: 45–63.

Winter, K.-U., Becker, A., Münster, T. *et al.* (1999). MADS-box genes reveal that gnetophytes are much more closely related to conifers than to flowering plants. *Proceedings of the National Academy of Sciences of the United States of America*, 96: 7342–7347.

Winther, R. G. (2015). Evo-devo as a trading zone. In *Conceptual Change in Biology: Scientific and Philosophical Perspectives on Evolution and Development*, ed. A. C. Love. Dordrecht: Springer, pp. 459–482.

Wojciechowski, M. F., Lavin, M. & Sanderson, M. J. (2004). A phylogeny of legumes (Leguminosae) based on analysis of the plastid *matK* gene resolves many well-supported subclades within the family. *American Journal of Botany*, 91: 1846–1862.

Wolfe, J. M. & Hegna, T. A. (2014). Testing the phylogenetic position of Cambrian pancrustacean larval fossils by coding ontogenetic stages. *Cladistics*, 30: 366–390.

Woloszynska, M., Bocer, T., Mackiewicz, P. & Janska, H. (2004). A fragment of chloroplast DNA was transferred horizontally, probably from non-eudicots, to mitochondrial genome of *Phaseolus*. *Plant Molecular Biology*, 56: 811–820.

Woodrick, R., Martin, P. R., Birman, I. & Pickett, F. B. (2000). The *Arabidopsis* embryonic shoot fate map. *Development*, 127: 813–820.

Worley, A., Baker, A., Thompson, J. & Barrett, S. C. H. (2000). Floral display in *Narcissus*: variation in flower size and number at the species, population, and individual levels. *International Journal of Plant Sciences*, 161: 69–79.

Wörz, A. (1996) Rubiaceae. Rötegewächse. In *Die Farn und Blütenpflanzen Baden-Württembergs*. Band 5, ed. O. Sebald, S. Seybold, G. Philippi & A. Wörz. Stuttgart: Ulmer, pp. 449–484.

Wróblewska, M., Dołzbłasz, A. & Zagórska-Marek, B. (2016). The role of ABC genes in shaping perianth phenotype in the basal angiosperm *Magnolia*. *Plant Biology*, 18: 230–238.

Wu, C. A., Lowry, D. B., Cooley, A. M. *et al.* (2008). *Mimulus* is an emerging model system for the integration of ecological and genomic studies. *Heredity*, 100: 220–230.

Wu, G., Park, M. Y., Conway, S. R. *et al.* (2009). The sequential action of miR156 and miR172 regulates developmental timing in *Arabidopsis*. *Cell*, 138: 750–759.

Wu, G. & Poethig, R. S. (2006). Temporal regulation of shoot development in *Arabidopsis thaliana* by *miR156* and its target *SPL3*. *Development*, 133: 3539–3547.

Wund, M. A. (2012). Assessing the impacts of phenotypic plasticity on evolution. *Integrative and Comparative Biology*, 52: 5–15.

Wunderlin, R. P. (1983). Revision of the arborescent *Bauhinia*s (Fabaceae: Caesalpinioideae: Cercideae) native to middle America. *Annals of the Missouri Botanical Garden*, 70: 95–127.

Xi, Z., Bradley, R. K., Wurdack, K. J. *et al.* (2012). Horizontal transfer of expressed genes in a parasitic flowering plant. *BMC Genomics*, 13: 227.

Xi, Z., Wang, Y., Bradley, R. K. *et al.* (2013). Massive mitochondrial gene transfer in a parasitic flowering plant clade. *PLoS Genetics*, 9: e1003265.

Xiao, H., Wang, Y., Liu, D. *et al.* (2003). Functional analysis of the rice AP3 homologue *OsMADS16* by RNA interference. *Plant Molecular Biology*, 52: 957–966.

Xu, L., Xu, Y., Dong, A., Sun, Y. *et al.* (2003). Novel *as1* and *as2* defects in leaf adaxial–abaxial polarity reveal the requirement for *ASYMMETRIC LEAVES1* and 2 and *ERECTA* functions in specifying leaf adaxial identity. *Development*, 130: 4097–4107.

Xu, Y., Sun, Y., Liang, W. Q. & Huang, H. (2002). The *Arabidopsis AS2* gene encoding a predicted leucine-zipper protein is required for the leaf polarity formation. *Acta Botanica Sinica*, 44: 1194–1202.

Xu, Y., Teo, L. L., Zhou, J., Kumar, P. P. & Yu, H. (2006). Floral organ identity genes in the orchid *Dendrobium crumenatum*. *The Plant Journal*, 46: 54–68.

Yamada, T., Yokota, S., Hirayama, Y. *et al.* (2011). Ancestral expression patterns and evolutionary diversification of *YABBY* genes in angiosperms. *The Plant Journal*, 67: 26–36.

Yamaguchi, A., Wu, M. F., Yang, L. *et al.* (2009). The microRNA-regulated SBP-box transcription factor SPL3 is a direct upstream activator of *LEAFY, FRUITFULL*, and *APETALA1*. *Developmental Cell*, 17: 268–278.

Yamaguchi, T., Lee, D. Y., Miyao, A. *et al.* (2006). Functional diversification of the two C-class MADS box genes *OSMADS3* and *OSMADS58* in *Oryza sativa*. *Plant Cell*, 18: 15–28.

Yamaguchi, T., Yano, S. & Tsukaya, H. (2010). Genetic framework for flattened leaf blade formation in unifacial leaves of *Juncus prismatocarpus*. *Plant Cell*, 22: 2141–2155.

Yamaki, S., Nagato, Y., Kurata, N. & Nonomura, K.-I. (2011). Ovule is a lateral organ finally differentiated from the terminating floral meristem in rice. *Developmental Biology*, 351: 208–216.

Yant, L., Mathieu, J. & Schmid, M. (2009). Just say no: floral repressors help *Arabidopsis* bide the time. *Current Opinion in Plant Biology*, 12: 580–586.

Yephremov, A., Wisman, E., Huijser, P. *et al.* (1999). Characterization of the *FIDDLEHEAD* gene of *Arabidopsis* reveals a link between adhesion response and cell differentiation in the epidermis. *Plant Cell*, 11: 2187–2201.

Yockteng, R. B., Almeida, A. M. R., Morioka, K., Alvarez-Buylla, E. R. & Specht, C. D. (2013). Molecular evolution and patterns of duplications in the *SEP/AGL6*-like lineage of the Zingiberales: a proposed mechanism for floral diversification. *Molecular Biology and Evolution*, 30: 2401–2422.

Yogeeswaran, K., Frary, A., York, T. L. *et al.* (2005). Comparative genome analyses of *Arabidopsis* spp.: inferring chromosomal rearrangement events in the evolutionary history of *A. thaliana*. *Genome Research*, 15: 505–515.

Yoo, M.-J., Bell, C. D., Soltis, P. S. & Soltis, D. E. (2005). Divergence times and historical biogeography of Nymphaeales. *Systematic Botany*, 30: 693–704.

Yoo, M.-J., Soltis, P. S. & Soltis, D. E. (2010). Expression of floral MADS-box genes in two divergent water lilies: Nymphaeales and *Nelumbo*. *International Journal of Plant Sciences*, 171: 121–146.

Yoshida, S., Barbier de Reuille, P., Lane, B. *et al.* (2014). Genetic control of plant development by overriding a geometric division rule. *Developmental Cell*, 29: 75–87.

Yoshida, S., Cui, S., Ichihashi, Y. & Shirasu, K. (2016). The haustorium, a specialized invasive organ in parasitic plants. *Annual Review of Plant Biology*, 67: 643–667.

Yu, J., Hu, S., Wang, J. *et al.* (2002). A draft sequence of the rice genome (*Oryza sativa* L. ssp. *indica*). *Science*, 296: 79–92.

Yuan, Z., Gao, S., Xue, D. W. *et al.* (2009). *RETARDED PALEA1* controls palea development and floral zygomorphy in rice. *Plant Physiology*, 149: 235–244.

Žádníková, P. & Simon, R. (2014). How boundaries control plant development. *Current Opinion in Plant Biology*, 17: 116–125.

Zagotta, M. T., Shannon, S. & Jacobs, C. (1992). Early-flowering mutants of *Arabidopsis thaliana*. *Australian Journal of Plant Physiology*, 19: 411–418.

Zahn, L. M., Kong, H., Leebens-Mack, J. H. *et al.* (2005). The evolution of the *SEPALLATA* subfamily of MADS-box genes: a preangiosperm origin with multiple duplications throughout angiosperm history. *Genetics*, 169: 2209–2223.

Zanis, M. J., Soltis, P. S., Qiu, Y.-L., Zimmer, E. & Soltis, D. E. (2003). Phylogenetic analyses and perianth evolution in basal angiosperms. *Annals of the Missouri Botanical Garden*, 90: 129–150.

Zhang, J. Z., Li, Z. M., Mei, L., Yao, J. L. & Hu, C. G. (2009). *PtFLC* homolog from trifoliate orange (*Poncirus trifoliata*) is regulated by alternative splicing and experiences seasonal fluctuation in expression level. *Planta*, 229: 847–859.

Zhang, N., Wen, J. & Zimmer, E. A. (2015). Expression patterns of *AP1*, *FUL*, *FT* and *LEAFY* orthologs in Vitaceae support the homology of tendrils and inflorescences throughout the grape family. *Journal of Systematics and Evolution*, 53: 469–476.

Zhang, R., Guo, C. C., Zhang, W. G. *et al.* (2013a). Disruption of the petal identity gene *APETALA3-3* is highly correlated with loss of petals within the buttercup family (Ranunculaceae). *Proceedings of the National Academy of Sciences of the United States of America*, 110: 5074–5079.

Zhang, W., Kramer, E. M. & Davis, C. C. (2010). Floral symmetry genes and the origin and maintenance of zygomorphy in a plant pollinator mutualism. *Proceedings of the National Academy of Sciences of the United States of America*, 107: 6388–6393.

Zhang, W., Kramer, E. M. & Davis, C. C. (2016). Differential expression of *CYC2* genes and the elaboration of floral morphologies in *Hiptage*, an Old World genus of Malpighiaceae. *International Journal of Plant Sciences*, 177: 551–558.

Zhang, W., Steinmann, V. W., Nikolov, L., Kramer, E. M. & Davies, C. C. (2013b). Divergent genetic mechanisms underlie reversals to radial floral symmetry from diverse zygomorphic flowered ancestors. *Frontiers in Plant Science*, 4: 302.

Zhao, D., Yu, Q., Chen, C. & Ma, H. (2001). Genetic control of reproductive meristems. In *Meristematic Tissues in Plant Growth and Development*, eds. M. T. McManus & B. Veit. Sheffield: Sheffield Academic Press, pp. 89–142.

Zhong, J. & Kellogg, E. A. (2015). Stepwise evolution of corolla symmetry in *CYCLOIDEA2*-like and *RADIALIS*-like gene expression patterns in Lamiales. *American Journal of Botany*, 102: 1260–1267.

Zhong, J., Powell, S. & Preston, J. C. (2016). Organ boundary NAC-domain transcription factors are implicated in the evolution of petal fusion. *Plant Biology*, 18: 893–902

Zhong, R. & Ye, Z. H. (2004). Molecular and biochemical characterization of three WD-repeat-domain-containing inositol polyphosphate 5-phosphatases in *Arabidopsis thaliana*. *Plant Cell Physiology*, 45: 1720–1728.

Zhou, Q., Wang, Y. & Xiaobai, J. (2002). Ontogeny of floral organs and morphology of floral apex in *Phellodendron amurense* (Rutaceae). *Australian Journal of Botany*, 50: 633–644.

Zhou, X.-R., Wang, Y.-Z., Smith, J. F. & Chen, R. (2008). Altered expression patterns of TCP and MYB genes relating to the floral developmental transition from initial zygomorphy to actinomorphy in *Bournea* (Gesneriaceae). *New Phytologist*, 178: 532–543.

Zimmerman, R. H., Hackett, W. P. & Pharis, R. P. (1985). Hormonal aspects of phase change and precocious flowering. *Encyclopaedia of Plant Physiology*, 11: 79–115.

Zluvova, J., Nicolas, M., Berger, A. *et al.* (2006). Premature arrest of the male flower meristem precedes sexual dimorphism in the dioecious plant *Silene latifolia*. *Proceedings of the National Academy of Sciences of the United States of America*, 103: 18854–18859.

Zobell, O., Faigl, W., Saedler, H. & Münster, T. (2010). MIKC MADS-box proteins: conserved regulators of the gametophytic generation of land plants. *Molecular Biology and Evolution*, 27: 1201–1211.

Taxonomic Index

Note: Extinct taxa are marked with a dagger (†)

436

Subject Index

Printed in the United States
by Baker & Taylor Publisher Services